SOCIETY FOR GENERAL MICROBIOLOGY. Microbes and biological productivity; 21st symposium, London, 1971, ed. by D. E. Hughes and A. H. Rose. Cambridge, 1971. 378p il tab. 16.00 ISBN 0-521-08112-2
Includes papers in the fields of microbial technology and ecology ranging from research on the use of biochemicals formed as a result of microorganism activity to use of industrial by-products and wastes as sources of food for microorganisms, to use of algae and bacteria or their biochemical products as food or food additives, to microbial diseases and plant or animal productivity.   Articles on microbial ecology are from the areas of aquatic environment, oceans, polar regions, soil, association of microbes and roots (nitrogen fixation and other phenomenon), intestinal flora (ruminant and non-ruminant), antimicrobial food preservation and control of pest insects.   This symposium would be useful to researchers in the various fields discussed or as additional reading for a graduate course in microbial technology or microbial ecology. Most of the papers are well written but the interest of many microbiology students in this book will be limited.   Papers presented at the *Conference on global impacts of applied microbiology* (1964) edited by M. P. Starr comprise an older but comparable volume. Index.

MICROBES AND BIOLOGICAL
PRODUCTIVITY

Other Publications of the
*Society for General Microbiology**

THE JOURNAL OF GENERAL MICROBIOLOGY
THE JOURNAL OF GENERAL VIROLOGY

SYMPOSIA

I   THE NATURE OF THE BACTERIAL SURFACE

2   THE NATURE OF VIRUS MULTIPLICATION

3   ADAPTATION IN MICRO-ORGANISMS

4   AUTOTROPHIC MICRO-ORGANISMS

5   MECHANISMS OF MICROBIAL PATHOGENICITY

6   BACTERIAL ANATOMY

7   MICROBIAL ECOLOGY

8   THE STRATEGY OF CHEMOTHERAPY

9   VIRUS GROWTH AND VARIATION

10 MICROBIAL GENETICS

11 MICROBIAL REACTION TO ENVIRONMENT

12 MICROBIAL CLASSIFICATION

13 SYMBIOTIC ASSOCIATIONS

14 MICROBIAL BEHAVIOUR, 'IN VIVO' AND 'IN VITRO'

15 FUNCTION AND STRUCTURE IN MICRO-ORGANISMS

16 BIOCHEMICAL STUDIES OF ANTIMICROBIAL DRUGS

17 AIRBORNE MICROBES

18 THE MOLECULAR BIOLOGY OF VIRUSES

19 MICROBIAL GROWTH

20 ORGANIZATION AND CONTROL IN PROKARYOTIC
    AND EUKARYOTIC CELLS

* Published by the Cambridge University Press, except for the first Symposium, which was published by Blackwell's Scientific Publications Limited.

MICROBES AND BIOLOGICAL PRODUCTIVITY

TWENTY-FIRST SYMPOSIUM OF THE
SOCIETY FOR GENERAL MICROBIOLOGY
HELD AT
UNIVERSITY COLLEGE LONDON
APRIL 1971

CAMBRIDGE
Published for the Society for General Microbiology
AT THE UNIVERSITY PRESS
1971

Published by the Syndics of the Cambridge University Press
Bentley House, 200 Euston Road, London, N.W.1
American Branch: 32 East 57th Street, New York, N.Y.10022

© The Society for General Microbiology 1971

ISBN: 0 521 08112 2

Printed in Great Britain
at the University Printing House, Cambridge
(Brooke Crutchley, University Printer)

CONTRIBUTORS

BUNT, J., Institute of Marine Sciences, University of Miami, Miami, Florida.

CROWDY, S. H., Department of Botany, University of Southampton.

DEMAIN, ARNOLD L., Department of Nutrition and Food Science, M.I.T., Cambridge, Massachusetts.

DERBYSHIRE, J. B., Institute for Research on Animal Diseases, Compton, Newbury, Berkshire.

GRAY, T. R. G., Hartley Botanical Laboratories, University of Liverpool.

HARLEY, J. L., Department of Forestry, University of Oxford.

HEDÉN, C.-G., Karolinska Institutet, Stockholm.

LEWIS, D., Department of Applied Biochemistry and Nutrition, School of Agriculture, University of Nottingham, Sutton Bonington, Loughborough.

MANNERS, J. G., Department of Botany, University of Southampton.

MOLIN, N., Karolinska Institutet, Stockholm.

MOSSEL, D. A. A., Central Institute for Nutrition and Food Research TNO, Zeist, The Netherlands and the Catholic University, Louvain, Belgium.

NORRIS, J. R., Shell Research Ltd., Sittingbourne, Kent.

POSTGATE, J. R., University of Sussex, Falmer, Brighton.

STRICKLAND, J. D. H., Institute of Marine Resources, Scripps Institution of Oceanography, La Jolla, California.

SWAN, H., University of Nottingham, Department of Applied Biochemistry and Nutrition, School of Agriculture, Sutton Bonington, Loughborough.

VINCENT, W. A., Sherborne, Dorset.

VISHNIAC, W., Department of Biology, University of Rochester, Rochester, New York.

WILKINSON, J. F., Department of Microbiology, University of Edinburgh.

WILLIAMS, S. T., Hartley Botanical Laboratories, University of Liverpool.

CONTENTS

Editors' Preface   *page* ix

C.-G. HEDÉN AND N. MOLIN:
The productivity of microorganisms – a catalytic factor for
research and transdisciplinary cooperation   1

J. F. WILKINSON:
Hydrocarbons as a source of single-cell protein   15

W. A. VINCENT:
Algae and lithotrophic bacteria as food sources   47

ARNOLD L. DEMAIN:
Microbial production of food additives   77

S. H. CROWDY AND J. G. MANNERS:
Microbial disease and plant productivity   103

J. B. DERBYSHIRE:
Microbial diseases and animal productivity   125

D. LEWIS AND H. SWAN:
The role of intestinal flora in animal nutrition   149

D. A. A. MOSSEL:
Ecological essentials of antimicrobial food preservation   177

J. R. NORRIS:
Microbes as biological control agents   197

J. D. H. STRICKLAND:
Microbial activity in aquatic environments   231

T. R. G. GRAY AND S. T. WILLIAMS:
Microbial productivity in soil   255

J. R. POSTGATE:
Relevant aspects of the physiological chemistry of nitrogen
fixation   287

J. L. HARLEY:
Associations of microbes and roots   309

J. BUNT:
Microbial productivity in polar regions 333

W. VISHNIAC
Limits of microbial productivity in the ocean 355

INDEX 367

EDITORS' PREFACE

The choice of topic for the 1971 Society Symposium and the production of this book have taken what is now the accepted route; that is, the subject was selected from a number submitted to the Council of the Society. Subsequently, contributions were chosen by a small committee with Council's final approval. It is difficult now to pin-point precisely the reasons for the present choice of subject. But there is no doubt in our minds as Editors that, once presented with a choice of this title, we were influenced to a greater or lesser degree by its relevance to the widespread discussion of the role of science and technology in the human predicament. That is, the situation brought about by the advice, or perhaps order, to our primogenitor to 'be fruitful and multiply and replenish the earth and subdue it and have dominion...over every living thing...' (Gen. i. 28). Maybe the process now accelerates too rapidly and amidst cries of doom and call for Luddism in Science and Technology, our symposium reinforces the idea that man's wits used wisely will ensure his continued survival.

Two themes emerged in our discussions. First there is a group conveniently labelled 'Ecological'. Some of the topics have been partly covered by previous symposia of the Society (Microbial Associations and Symbiotic Associations), and are now brought up to date.

Secondly, there is what might be called 'Microbial Technology', whereby the outstanding advances in our ability to control microbial growth, reproduction and chemistry are discussed in terms of large-scale production of food or food additives.

One main theme is lacking for completion, namely the direct interaction of microbes in the overall productivity of the human crop. This aspect, however, has indirectly been the subject of more recent symposia of the Society, e.g. those on Antimicrobial Drugs and Airborne Microbes; in any case this aspect could not be dealt with adequately here. Even so, there is a decidedly anthroprocentric theme to our discussion here since it can be assumed that, although we are discussing the effect of microbes on the growth of other living systems, the ultimate effects on human productivity are not far from the forefront of our minds.

It is noteworthy that in the ecological contributions to this volume there is an attempt to quantify complex associations. This is essential to any ecological model in order that it should not be merely descriptive but become predictive in terms of overall yields and oscillations. The difficulties in this transformation can be seen especially in the soil systems described by Drs Gray and Williams, and the root associations described by Professor Harley. The same attempt is seen in the description of aquatic systems by the late Dr Strickland, and the restricted but important polar regions by Dr Bunt. In all of these articles it is encouraging to see that the tools to tackle these complex systems of interacting microbes are now developing rapidly.

The deleterious intervention of single or mixed microbial systems on growing plants discussed by Professor Crowdy and Dr Manners, and in animals by Dr Derbyshire, and the preservation of stored food from microbial attack by Professor Mossel completes a trio of contributions which extend the previous quantifying into economic terms. It is clear that this translation into profit and loss is not easy, as is well brought out by Crowdy and Manners, who show that the partial destruction of a crop by microbes may under certain conditions be economically advantageous to the producer.

The balance of intestinal microbes in farm animals is a delicate one. The article by Professor Lewis and Dr Swan may be seen as a connecting link between our two themes. This is especially so when the cow is seen as a continuous fermentor for the interconversion of relatively recalcitrant material into high-quality protein at the optimal rate.

The common linking theme seen in the contributions of Héden and Molin, Wilkinson, Norris, Demain, and Vincent, is the application of the rapid growth in fundamental knowledge and hence control over microbial growth, biomass production as single-cell protein and chemical productivity. Yet, as is implied in Wilkinson's invocation of Elliott, the conception is comparatively old, and foreseen, for instance, by Kluvyer amongst others. Yet the practical reality in large-scale technology is only now generally realizable. The important role of the pilot plant, not only in productivity but also in training, as discussed by Héden, is relatively new to Microbiology but well established in Chemical Engineering.

As pointed out by Professor Postgate, Biological Productivity may be viewed in overall chemical terms as the conversion of inorganic substrates into the more highly organized polymers of living creatures. Nitrogen fixation is uniquely a microbial process replaced only in part by the Haber process. The movement between Biology, Chemistry and Economics, which is developed in this contibution, is also mirrored throughout this volume and has never been wholly lacking throughout the History of Microbiology.

We would like to thank those at the Cambridge University Press who guided us throughout the production of this volume and our office staff who accepted its considerable and additional burden. Members of the Society Committee and Council who selected the topic and contributors must receive mention as must also the contributors who survived our urgent and often hectoring demands for manuscript.

D. E. HUGHES
A. H. ROSE

University College Cardiff
Bath University

THE PRODUCTIVITY OF MICROORGANISMS–A CATALYTIC FACTOR FOR RESEARCH AND TRANSDISCIPLINARY COOPERATION

CARL-GÖRAN HEDÉN AND NILS MOLIN

Department of Applied Microbiology,
Karolinska Institutet, Stockholm, Sweden

INTRODUCTION

The natural environments in which microbial cells divide or produce extracellular materials very rarely permit those processes to run at their maximum speeds, and when they do, it is only for brief periods of time. In artificial environments created by man, on the other hand, the pH and the P_{O_2} can be controlled, the concentration of nutrients can be kept constant, various sorts of feedback inhibition can be eliminated, etc. Since the control can also be maintained over long periods of time this relation between microbe and man can often be regarded as a very successful symbiosis. The microbe can run its life processes at full throttle, and in the bargain man obtains many valuable substances. Since the life processes follow similar patterns in all living systems man can also harvest a wide range of substances, including many which were earlier associated only with higher animals or with plants. The high metabolic rates possible for cells with a large surface-to-volume ratio and the versatility inherent in a genetic apparatus that is easy to manipulate, have forged microorganisms into essential tools not only for bacteriologists and mycologists but also for biochemists, bio-engineers and a whole range of other specialists. A brief discussion of how their professional productivity and transdisciplinary activities have been influenced by the utilization of the potential of microorganisms will be the topic of this paper. It will discuss some trends in industrial fermentations, but particular attention will be given to the increasing number of pilot plants designed to supply the large quantities of material which are now often needed for research purposes. The activities of a few units will be reviewed in order to provide some guidance for departments planning new facilities.

SOME TRENDS IN THE INDUSTRIAL
USE OF MICROORGANISMS

Biomass production

Yeasts grown on carbohydrates have been used as animal feed and for human consumption for a long time. At present the production of yeasts may be in the order of 0·4 million tons a year. It is generally agreed that yeast and bacterial protein represents an important potential protein resource, which may help to fill an existing shortage and growing needs in the world. Both cheap carbohydrates and hydrocarbons will certainly be used as carbon sources for the production of biomass. Thus it has been forecast that the quantity of yeast produced from mineral oil alone will reach the level of 1 million tons a year as early as 1980. This will represent a significant source of protein during the subsequent decade.

In developing new processes for the production of microbial proteins, it is essential that pilot plants are available for testing new technical approaches. In many instances integrated processes (Hedén, 1969a, b) and new types of reactors, such as tower fermentors, will no doubt improve the economy (Falch & Gaden, 1970; Kitai, Tone & Ozaki, 1969; Wang, 1970). Sophisticated equipment for control of growth parameters and for the recirculation of media will also be employed. It should be a challenge for food technologists and nutritionists to find the most efficient and economic ways to use this new raw material. Encouraging results have been published. Thus, Cremer (1951) reported successful results in the addition of yeasts to bread. A supplement of a few per cent increased the biological value from about 40·5 to 46·5. Yeasts and yeast extracts are of course also being widely used in foods such as soups, dressings, sausages and drinks (Klapka, Duby & Pavcek, 1958; von Loesecke, 1946; Lyall, 1963; Paul, Frueh & Ohlsson, 1948; Reiff, Lindemann & Holle, 1962).

There is, however, an obstacle to the use of microbial protein for human consumption, namely their comparatively high content of nucleic acids (Waslien, Calloway & Margen, 1968). The intake of nucleic acids should not exceed 2 g. per day, which corresponds to about 10–15 g. of yeasts. The relation between physiological factors and the nucleic acid content, however, indicates how this problem might be solved and the development of economic methods for large-scale extraction of nucleic acids and nucleotides from microorganisms will also help to increase the use of yeasts and bacterial biomass for human consumption.

Fermented foods

Another area where biotechnological techniques and basic knowledge in microbiological physiology could open up new industrial applications is the conversion of foodstuffs by microorganisms. The main advantage of microbiologically produced foods is a high nutritive value which makes them suitable as food supplements. Microorganisms may also increase the shelf-life of the product by means of metabolites formed during the fermentation process, improve the digestibility and sometimes also change the taste, flavour and texture in a desirable direction (Hesseltine & Wang, 1969). Production of genetically improved starter organisms for use both in the food industry and for distribution in developing countries will probably become a new task for the development work in some pilot plants (Hedén, 1969b). The call for more rapid and reliable processes for the production of fermented foods has been reinforced by the development of continuous processes for beverages (Hough, 1969), cheese (Tchieret, 1969; Wränghede, 1969) and bread (Fortman, 1969).

Microbial cells as raw material for the extraction of biochemicals

The production of biochemicals of different kinds from microbial biomass is still in its infancy, but it will certainly become a substantial commercial activity in the future. One of the main reasons is that the development of new techniques for the separation of cell constituents have substantially changed the efficiency and economy of such processes (Lilly & Dunnill, 1969; Edebo, 1969; Edwards, 1969). In particular, the developments in enzyme technology are rapid and may expand into new applications as a consequence of a growing experience of hydrocarbon fermentations and the appearance of other cheap carbon sources like methane and methanol (Hedén, 1969b).

The rapid development of what might be called the biotechnology of nucleic acids and derivatives is another striking feature of the progress in this area. The use of nucleotides for enhancing the natural flavours of foods is now an established procedure, which has a considerable industrial interest (Chambers, 1964; Demain, 1968). The synergistic effect of monosodiumglutamate and 5'-purinribonucleotides has also received attention (Demain, 1968). The antioxidative effect of certain nucleotides in food and the use of adenosine and inosine for pharmaceutical purposes has finally been reported (Boettge, Jaeger & Mittenzwei, 1957).

In principle, all the metabolites which participate in the various physiological activities of microbial cells can be isolated, but the quantities are often small. This may make it necessary to use mutants with defective regulatory genes or cultivation techniques which counteract feedback inhibition or optimize the cultivation parameters which favour the accumulation of a desired metabolite. A logical tool in this connexion is provided by continuous cultivation. However, there are still doubts about the advantages of this technique for some products (Royston, 1968). Experimental pilot plant establishments where new continuous processes can be developed to a semi-commercial stage will certainly play an essential part in a rapid expansion of this field.

Antibiotics

If powerful new antibiotics active against common virus infections can be found, a great increase in antibiotic production for human needs would probably come about. Substances that are toxic to certain viruses but harmless to animals exist, but their rarity and the difficulties involved in screening for antiviral antibiotics makes progress slow. On the other hand the development of special antibiotics set aside for husbandry and for the use against bacterial and rickettsial diseases in animals may be rapid. Already antibiotics like tylosine, neftine and variomycine are used in balanced animal feed. Monensin, which is the major factor in a group of closely related biologically active compounds produced by fermentation, has been reported to be effective against coccidiosis in chicken (Stark, 1969). The present world market for therapeutic agents for coccidiosis is in fact estimated to be as much as 35 million U.S. dollars.

A breakthrough for the use of novel antibiotics in animal therapeutics and growth could be very profitable and will undoubtedly challenge the fermentation technologists.

Vaccines

The production of vaccines is a routine of long standing in most countries; a fact which may help to explain a certain technical rigidity in the field. In 1965 Möller, for instance, reported the results of a survey indicating that only 2 out of 24 manufacturers making 10 different vaccines used continuous-cultivation techniques. Their potential has been discussed by Holme (1968). Vaccine production is also normally based on laboratory-scale vessels which have taken little advantage of the advances in fermentation technology. In this field there obviously exist a considerable lag-period between the developments on the research level and their application for routine production. However, commercial devices and procedures for the handling of pathogens on a large scale now exist, and the appearance of specialized pilot plant facilities in different parts of the world may initiate a rapid change (see Plate 1).

The growing awareness of the limitations and possibilities of various construction material may also be helpful (Brookes, 1969), for instance, in the tissue-culture field. The potential of human diploid cells for the production of viral vaccines is very great, and as soon as the new approach has been accepted by the licencing authorities (Perkins, 1969) a rapid development is likely, since several methods are available for scaling-up the cultivation technique (van Hemert, 1964; Molin & Hedén, 1968). Those methods may of course also become important for the production of the raw materials needed for the isolation of enzymes and hormones applicable in medicine. In particular, the use of carriers and microencapsulation may increase the demand considerably (Chang, 1969).

RESEARCH AS A LARGE-SCALE USER
OF MICROBIOLOGICAL MATERIALS

Up until the last couple of decades the possibilities to handle micro-organisms on a large scale was restricted to the research laboratories of the fermentation industry. There was, however, a growing awareness of the need for similar facilities at many university departments and some large institutes, notably the Istituto Superiore di Sanità in Rome, built pilot plants to satisfy the needs of their scientists.

Many biologically active substances like antibiotics, toxins and enzymes are present in so small concentrations that efforts at isolation must start with sizeable quantities in order to yield reasonable amounts of material. It is true that remarkable achievements were early recorded, where batteries of Erlenmeyer flasks or large metal trays provided the necessary starting material, but the advantages of stirred fermentors became more and more obvious. This was particularly true in cases where careful control and homogeneous batches were desired, as in the transfer of an isotope from the medium to specific microbial product. However, as long as deficient fermentor constructions involved a risk of losing expensive batches due to contamination, many scientists felt that it was prudent not to put all one's eggs in one basket. During the last ten years this policy is no longer tenable, since many commercial designs have become available that make the sterility problem insignificant. Since they also permit the use of continuous cultivation and dialysis techniques (Schultz & Gerhardt, 1969) which greatly increases the yield per equipment volume, the traditional techniques are rapidly losing ground. In fact there is now a very rapid increase in the number of general-purpose pilot plants all over the world, and the demand on existing facilities has been increasing steadily. At the pilot plant of the Microbiological Research Establishment in England, for instance, an average of 16 projects a year were undertaken prior to 1968, but 28 projects in 1968 and 33 in 1969.

In order to give an idea about the scope of their work, six representative installations will be mentioned: The New England Enzyme Center in Boston (S. E. Charm), the NIH Pilot plant at Bethesda (H. A. Sober), the Fermentation Pilot Plant at Imperial College of Science and Technology in London (E. B. Chain and G. T. Banks, see Plate 1), the Microbial Products Section of the Microbiological Research Establishment at Porton (C. E. Gordon-Smith), the Department of Technical Microbiology at the Institute of Microbiology, the Czechoskovakian Academy of Sciences (I. Málek, A. Prokop and F. Machek), and the Fermentation Pilot Plant at Karolinska Institutet in Stockholm (C.-G. Hedén, see Plate 1). The background information was kindly provided by the individuals mentioned in parentheses.

The pilot plants mentioned are all fairly flexible, but at the same time they represent different lines of specialization and emphasis, ranging from single-cell protein (Prague) and enzymes (Boston) to the large-scale handling of pathogens or semi-pathogens (Porton and Stockholm). Also the desired balance between the six impact areas of a fermentation pilot plant can of course vary (Table 1).

Table 1. *Interdisciplinary aspects of studies on large-scale fermentors*

	Research	Education
Extramural stimulation	Biochemistry, industrial collaboration	Transdisciplinary training
Intramural stimulation	Microbial chemistry and physiology	Post-graduate training
Autostimulation	Fundamental fermentation research, biotechnology and bioengineering	Training of fermentation technologists

Obviously direct numerical comparisons in terms of medium throughput, etc., are impossible. Also batch and continuous cultivations are difficult to compare, and the techniques involved in handling *Escherichia coli* and fastidious pathogens are of course quite different. Keeping this reservation in mind it might, however, be of interest to list a few figures (Table 2).

Some 0·4 % of the total medium may be converted into cells, but the yields of course vary greatly, depending on the microorganisms used. At Bethesda the 59,000 l. represented 20 different microorganisms yielding about 235 kg. of cells, whereas in Stockholm the 16,000 l. represented 12 agents weighing about 130 kg. In both cases the numerous cultivations in 10 l. fermentors were excluded from the comparison.

Table 2. *Fermentor size and throughput at various centres*

Fermentor size (litres)*		Culture throughput per year (l.)
Bethesda	1×10 2×380 $1 \times 1,135$	59,000
Boston	1×29 $1 \times 755†$	53,500
London	$18 \times 3·5$ 8×60 7×400 $1 (+2) \times 2,700$	150,000
Porton	$13 \times 0·5$ to 100 for continuous and 9×10 to 400 for batch cultivation	In 1969 7,500 l. including many continuous cultivations
Prague	15×3 and 3×20 for continuous and 2×200 (and $1 \times 1,000$) for batch cultivation	In 1970 regular pilot plant started up
Stockholm	2×10 1×100 1×500 $2 \times 1,000$ $1 \times 3,000$	16,000

* In addition there are normally a number of agitated tanks, some of which can often be used for anaerobic cultivations.

† Estimated on the basis that the normal culture volumes (20 and 530 l.) represent about 70 % of the total vessel volume.

The number of runs involved in reaching the throughput volumes listed of course depends on the sizes of the fermentors used. In the case of the Boston laboratory the figure 53,500 l. was for instance made up of 96 runs of 530 l. and 35 runs of 20 l. This reflects a very efficient use of the equipment and an emphasis on standard runs. In fact 21 out of the 51 fermentations (38 large and 13 small) carried out during the period October 1969 to March 1970 involved mutant strains of *E. coli*, 16 were aimed at phage infected cells and two at the production of coli-phages. Only the remaining 11 were for a wider variety of bacteria. It is of interest to note that there was a substantial increase in phage-related products in recent years, no doubt a reflexion of the dynamic character of molecular biology and lowered technical barriers against the search for minor microbial components, needed for structural, biosynthetic and biological studies. To obtain 4 g. of enzyme the Boston laboratory for instance last year extracted no less than 200 kg. of *E. coli*.

Special equipment for disintegration of cells is occasionally used – for

Table 3. *Examples of microorganisms cultivated in recent years*

Bethesda

Acanthamoeba castellanii
Aerobacter aerogenes
Bacillus amyloliquefaciens
B. megatherium
B. subtilis
Clostridium sticklandii
Cl. thermosaccharolyticum
Escherichia coli
Lactobacillus acidophilus
Methanomonas vannielli
Pseudomonas ATCC II 2996
Pseudomonas sp.
Saccharomyces cerevisiae
Salmonella typhimurium
S. typhosa
Serratia marcescens
Tetrahymena pyriformis
Ustilago sphaerogena
Various phages and phage-infected cells

Stockholm

Azotobacter sp.
B. subtilis
B. pumilus
Clostridium botulinum
Cl. tetani
Cytophaga johnsoni
E. coli
Lactobacillus sp.
Listeria monocytogenes
Methanomonas sp.
Mucor Rouxii
Salmonella typhimurium
Staphylococcus sp.
Streptococcus sp.

instance, a large-scale freeze press with a capacity of 100 g. frozen paste per min. at the Stockholm pilot plant (Edebo, 1969; Hedén, 1969 b) but quite often the focus has been on extracellular enzymes; for instance, from the *E. coli* and *Bacillus amyloliquefaciens* strains cultivated at Bethesda. Quite often preparations are delivered after various degrees of purification; for instance, the synthetases, dehydrogenases, asparaginase and catalase prepared at Porton. All the pilot plants split their activities on a relatively large number of projects last year. At Porton some 33 projects involving Amoebae, Bacilli, *E. coli*, Micrococci, moulds, Proteus, Rhodopseudomonas, trypanosomes and vibrios, as well as viruses and phages, were for instance undertaken for the Medical Research Council, the Ministry of Health, universities, etc. Also, in this case an increase in the sophistication of the demands has been noted in the form of requests for enzymes and for products of pathogenic microorganisms.

The increase in the sophistication of requests mentioned is reflected in the complexity of the auxiliary equipment now needed in any pilot plant where one wants to make optimal use of the cultivation facilities. This need often starts a difficult exercise in balancing costs. On the one hand there is the wish to provide the fermentors with instrumentation for measuring, recording and controlling pH, dissolved oxygen and redoxpotential as well as the composition of the exhaust gas. On the other hand there is the cost of the centrifugal separators and extractors, filters, evaporators, freeze-driers, disintegrators and separation columns needed for the utilization of the full capacity of the cultivation equipment. No hard and fast rules can be given, but considering the trends in current pilot plant requests, many planners would probably now trade their biggest fermentor for an efficient disintegrator or a zonal centrifuge of the type being installed at Bethesda (Model K). However, the equipment needs vary enormously due to differences in specialization.

Almost all of the auxiliary facilities mentioned must be available in a pilot plant as large as that in London, where the staff at any time might be engaged in the following projects.

(a) Large-scale production and isolation of fungal viruses (and RNA derived therefrom) using various *Penicillium* and *Aspergillus* species.

(b) Production and extraction of various antibiotics produced by bacteria.

(c) Production and extraction of various ergot alkaloids produced by *Claviceps* species.

(d) Production, extraction and purification of the enzyme asparaginase, an antitumour agent, from various bacteria.

(e) Production of bacterial insect toxins.

(f) Production, extraction and purification of the phytotoxin Fusicoccin A, produced by the fungus *Fusicoccum amygdali*.

(g) Production of intermediates in the biosynthesis of bacterial cell walls on a large scale.

(h) Various miscellaneous small projects in connexion with the departmental research programme.

(i) Various miscellaneous fermentations, etc., as a service for institutes lacking adequate fermentation pilot-plant facilities.

THE INTERNATIONAL SIGNIFICANCE OF THE PAIRING OF MICROBIOLOGY AND BIOENGINEERING

As has been amply documented at the Conferences on Global Impacts of Applied Microbiology (GIAM) (1964, 1969 a, b, see also Hedén & Starr, 1964), moulds, bacteria and viruses are of great significance for the developing countries, and the future of those areas in the world is certainly closely linked to the fate of the whole of mankind (Hedén, 1967, 1969 c, 1969 d). The fight against communicable diseases which traditionally have been regarded as the main microbiological problem in the developing parts of the world is only one factor in a complex equation (Hedén, 1961) where the productivity of microorganisms influences most of the terms. The effects of vaccines and antibiotics on man must of course be considered, but also their significance for the meat production via domestic animals. The reduction of waste and the biological control of insects and pests which attack our food crops is equally important. Finally, the likely impact of single cell proteins as fodder and food additives must be taken into account.

Most disciplines should exercise some soul-searching along the lines proposed by Platt (1969), who takes a dim view of the many 'over-studied' areas of science. In this vein, one of the authors of this paper has made a case for more integration of the basic and the applied research in microbiology (GIAM, 1969b), pointing at the needs of the developing countries: they 'had to put up with colonialism; how much colonialism will they still have to suffer from the microbiologists in the rich countries?' (GIAM, 1969b). The phrasing of this question did not intend to detract from the critically important advances of microbial genetics made possible by the use of special coli-strains. It rather wanted to make the point that the time has now come for similar concerted efforts on microorganisms with a range of practically important attributes.

The knowledge now available about regulation mechanisms and enzyme technology also opens up new areas for applications in technically advanced countries. But the manufacturing industry as well as the environmental 'recirculation' technologists require solid bioengineering support. This is rapidly emerging and this gives strength to a new structure aimed at stimulating the transfer of relevant knowledge to the developing countries. This has taken the form of a network of cooperating laboratories recognized by UNESCO and operating under the name of International Organization of Biotechnology and Bioengineering (IOBB). Considering the ease with which a case can be made for biotechnology as a factor in advancing the technically underdeveloped countries (Hedén, 1964) it is safe to predict that this body, which already links a dozen or so of fermentation pilot plants, will assume great importance.

CONCLUSIONS

The productivity of microorganisms is normally considered in quantitative terms: material turnover per unit of time. However, one might also emphasize a qualitative aspect related to the complexity of the molecules synthesized or to the flexibility of the metabolic response of the cells. This aspect brings out a unique catalytic effect on transdisciplinary research and development. The improvement of established products like yeast, vaccines and antibiotics stimulated the biochemical industry and triggered the solution of many cultivation problems. Consequently the potential of microorganisms is now available to biochemists and molecular biologists in universities and research institutes all over the world. Well aware of the rapid developments in microbial physiology and genetics, they now find themselves forced to build pilot plants which meet a growing need. Those plants will certainly act as potent stimuli to transdisciplinary contacts and international activities. The productivity of microorganisms is obviously a fundamentally important element in the catalysis of research.

REFERENCES

BOETTGE, K., JAEGER, K. H. & MITTENZWEI, H. (1957). The adenylic acid system; new results and problems. *Arzneimittel-Forsch.* **7**, 24.

BROOKES, R. (1969). Properties of materials suitable for the cultivation and handling of micro-organisms. In *Methods in Microbiology* **1**, 21. Ed. J. R. Norris & D. W. Ribbons. London: Academic Press.

CHAMBERS, W. S. (1964). Ribonucleotides – new flavouring compounds. *Food Technol. Austr.* **16**, 250.

CHANG, T. M. S. (1969). Clinical potential of enzyme technology. *Science Tools* **16**, 33.

CREMER, H. D., LANG, K., HUBBE, I. & KULIK, V. (1951). Versuche zur Aufbesserung der biologischen Wertigkeit von Weizeneiweiss durch Lysin oder Hefe und ein Vergleich mit Hafereiweiss. *Biochem. Z.* **322**, 58.

DEMAIN, A. L. (1968). Production of purine nucleotides by fermentation. *Progr. Industr. Microbiol.* **8**, 35.

EDEBO, L. (1969). Disintegration of cells. In *Fermentation Advances*, p. 249. Ed. D. Perlman. New York: Academic Press.

EDWARDS, V. H. (1969). Separation techniques for the recovery of materials from aqueous solutions. In *Fermentation Advances*, p. 273. Ed. D. Perlman. New York: Academic Press.

FALCH, E. & GADEN, E. L. JR. (1970). A continuous, multistage tower fermentor. II. Analyses of reactor performance. *Biotechnol. Bioeng.* **12**, 465.

FORTMANN, K. L. (1969). Innovations in continuous processing methods. *Bakers' Dig.* **43**, 64.

GLOBAL IMPACTS OF APPLIED MICROBIOLOGY (1964). Proceedings of the first conference in Stockholm 1963. Ed. M. P. Starr. Stockholm: Almqvist and Wiksell.

GLOBAL IMPACTS OF APPLIED MICROBIOLOGY (1969a). Proceedings of the second conference in Addis Ababa, 1967. In *Biotechnol. Bioeng. Symp.* **1**. Ed. E. L. Gaden. New York: Interscience Publishers.

GLOBAL IMPACTS OF APPLIED MICROBIOLOGY (1969b). Proceedings of the third conference in Bombay, 1969. (In the Press.)

HEDÉN, C.-G. (1961). Paper read at symposium. IUBS General Assembly, 14 July 1961. *TVF* **32**, 297.

HEDÉN, C.-G. (1964). Research in biotechnology: a factor in advancing the technically underdeveloped countries. In *The Population Crisis and the Use of World Resources*, p. 478. Ed. S. Mudd. Den Haag: W. Junk.

HEDÉN, C.-G. (1967). Microbiology in world affairs. *Impact* **17**, 187.

HEDÉN, C.-G. (1969a). Ferment or perish: Future role of applied microbiology in world affairs. In *Fermentation Advances*, p. 861. Ed. D. Perlman. New York: Academic Press.

HEDÉN, C.-G. (1969*b*). The potential of microbial food resources. In *All-Congress Symposium on World Food Supply*, 28 August 1969. *XI Int. Bot. Cong., Seattle*. Washington: Allis-Chalmers.

HEDÉN, C.-G. (1969*c*). The social responsibility of microbiologists. *Biotechnol. Bioeng. Symp.* **1**, 7.

HEDÉN, C.-G. (1969*d*). The conferences on Global Impacts of Applied Microbiology and the developing nations. In *Third Int. Conf. on Global Impacts of Applied Microbiology, Souvenir*, 7–12 December 1969. Bombay.

HEDÉN, C.-G. & STARR, M. P. (1964). Global Impacts of Applied Microbiology; an appraisal. *Adv. Appl. Microbiol.* **6**, 1.

HEMERT, P. VAN (1964). The Bilthoven Unit for submerged cultivation of microorganisms. *Biotechnol. Bioeng.* **6**, 381.

HESSELTINE, C. W. & WANG, H. L. (1969). Fermented soybean foods. *Proc. Third Int. Conf. Global Impacts of Applied Microbiology*, 7–14 December. Bombay. (In the Press.)

HOLME, T. (1968). Application of continuous culture in the production of vaccines. In *Continuous Cultivation of Micro-organisms*, p. 385. Ed. I. Málek, K. Beran, Z. Fencl, V. Munk, J. Řičica & H. Smrčková. Prague: Academia.

HOUGH, J. S. (1969). Continuous fermentation of beverages. In *Dechema-Monographien*, **63**, 255. Berlin: Verlag Chemie.

KITAI, A., TONE, H. & OZAKI, A. (1969). Performance of a perforated plate column as a multistage continuous fermentor. *Biotechnol. Bioeng* **11**, 911.

KLAPKA, M. R., DUBY, G. A. & PAVCEK, P. L. (1958). Torula yeast as a dietary supplement. *J. Am. Dietet. Ass.* **34**, 1317.

LILLY, M. D. & DUNNILL, P. (1969). Isolation of intercellular enzymes from microorganisms. The development of a continuous process. In *Fermentation Advances*, p. 225. Ed. D. Perlman. New York: Academic Press.

LOESECKE, H. W. VON (1946). Controversial aspects: Yeast in human nutrition. *J. Am. Dietet. Ass.* **22**, 485.

LYALL, N. (1963). Yeast products in the food industry. II. The increasing demand for yeasts extracts. *Fd Trade Rev.* **33** (12), 57.

MOLIN, O. & HEDÉN, C.-G. (1968). Large scale cultivation of human diploid cells on titanium discs in a special apparatus. *Progr. Immunobiol, Standard.* **3**, 106.

MÖLLER, Å. (1965). The present state and the future of human bacterial vaccine production. In *Symp. Series Immunobiol. Standard.* **3**, 11.

PAUL, P., FRUEH, M. & OHLSSON, M. A. (1948). Corn meal and macaroni products containing dry primary yeast. I. Palatability and acceptability. *J. Am. Dietet. Ass.* **24**, 673.

PERKINS, F. T. (1969). Dispute grows over vaccine substrate. *Sci. J.* **5**A (2), 5.

PLATT, J. (1969). What we must do. *Science, N.Y.* **166**, 1115.

REIFF, F., LINDEMANN, M. & HOLLE, K. (1962). Hefe in Nahrungsmitteln. In *Die Hefen*, vol. 2, p. 755. Ed. F. Reiff *et al.* Nürnberg: Hans Carl.

ROYSTON, M. G. (1968). Continuous food processes. *Process. Biochem.* **3** (9), 58.

SCHULTZ, J. S. & GERHARDT, P. (1969). Dialysis culture of microorganisms: Design, theory and results. *Bact. Rev.* **33**, 1.

STARK, W. M. (1969). Monensin, a new biologically active compound produced by a fermentation process. In *Fermentation Advances*, p. 517. Ed. D. Perlman. New York: Academic Press.

TSCHIERET, F. X. (1969). Continuous manufacture of curd for cheesemaking. In *Dechema-Monogr.* **63**, 269. Berlin: Verlag-Chemie.

WANG, D. I. C. (1970). Engineering aspects of SCP production from hydrocarbons. Abstracts *Xth Int. Congr. Microbiol.*, Mexico City, 10 August.

WASLIEN, C. I., CALLOWAY, D. H. & MARGEN, S. (1968). Uric acid production of men fed graded amounts of egg protein and yeast nucleic acid. *Am. J. Clin. Nutr.* **21**, 892.

WRÄNGHEDE, K. (1969). Kontinuerlig framställning av kvarg. *Svenska Mejeritidn.* **61**, 108.

EXPLANATION OF PLATE

Fig. 1. Pilot plant for handling pathogens.
Fig. 2. Fermentation pilot plant at Imperial College of Science and Technology, London.
Fig. 3. Fermentation pilot plant at Karolinska Institutet, Stockholm.

PLATE 1

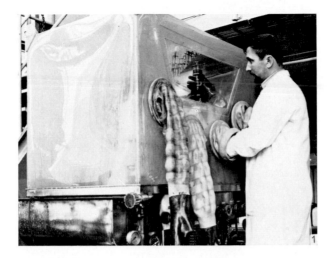

HYDROCARBONS AS A SOURCE
OF SINGLE-CELL PROTEIN

J. F. WILKINSON

Department of Microbiology, University of Edinburgh

Between the idea
And the reality
Between the motion
And the act
Falls the shadow
T. S. ELIOT (*The Hollow Men*)

INTRODUCTION

If heterotrophic microorganisms are to be used as a source of single-cell protein, the only substrates likely to be economically feasible at the moment are industrial by-products and wastes (mainly carbohydrate), some fractions of crude oil and natural gas. Although it has been known since the beginning of this century that microorganisms can grow on petroleum hydrocarbons, the potentialities of this discovery have only been realized recently. These potentialities are underlined by the calculations of Johnson (1967) that it is technologically possible to use between 15% and 20% of the world's petroleum production per annum to produce 100% of the protein required by the world's population. Another way of expressing this is that about 10% of crude oil is readily utilizable by microorganisms and can be converted to protein at about 50% carbon-to-carbon efficiency. This means a potential for producing about 50 million tons of protein per annum.

Why, then, has this source not been fully utilized? As pointed out by Humphrey (1970), although there is now intense investigation by most oil companies and many research laboratories throughout the world, there is still an imperfect knowledge of the nutritional and toxicological properties of protein foods derived from hydrocarbons and of the economics of their production. However, since this Symposium is organized by the Society for General Microbiology, this review will deal primarily with microbiological aspects of the subject and only in passing with engineering and commercial problems. The author has found it very difficult to obtain any reliable statements on the economics of present or proposed industrial projects. Although this difficulty is partly related to the local and sometimes artificial nature of many of the parameters used, it is nevertheless a world rife with rumours and counter-rumours; hot and cold water are poured on projects with a grim abandon. The reviewer can only hope to give some idea of the properties and potentialities of the microorganisms involved from the ivory-towered and campus-encompassed Valhalla of the academic world in the realization that he is not involved in the responsibility of making key decisions concerning the worth or worthlessness of a new process. However, the impression is gained that often too little attention is paid to microbiology in the fermentation industry and for this state of affairs both industries and academic institutions are to blame.

No attempt will be made in this article to review the wide range of different hydrocarbons that can be attacked nor the many microorganisms that have this ability. This has been very adequately done elsewhere (Beerstecher, 1954; Fuhs, 1961; Foster, 1962; Davis, 1967). In practice, the substrates most favoured for industrial processes on the basis of their value relative to other oil functions and their susceptibility to microbial attack, are the aliphatic n-alkanes. Of these, C2 to C10 are not so susceptible to microbial attack which also tends to be too specific; alkanes of C20 upwards present difficulties since they are solid at usual fermentation temperatures. The two types that have been most studied from an industrial point of view and which are the subject of this review are: (a) methane in the form of natural gas which is attacked by a variety of bacteria incapable of utilizing alkanes from C2 upwards. (b) liquid n-alkanes from C11 to C19. Foster (1962) has generalized that this group is attacked with the greatest frequency by bacteria, filamentous fungi and yeasts.

Although by far the greatest amount of information is available on the second group of alkanes, this review will be concerned with the methane fermentation in at least as great a length since it has not been nearly so well documented and reviewed and much of the information is recent. The author has had more personal experience in this field and believes that in the long run it may have greater industrial potential.

n-ALKANE FERMENTATION

A wide variety of microorganisms, ubiquitous in nature and particularly common in soil, can grow on liquid n-alkanes as the sole carbon and energy source. Bacteria are predominantly species of *Mycobacteria*, *Pseudomonas*, *Nocardia*, *Corynebacterium* and *Micrococcus*; the number of them recorded far exceeds that of other groups, probably reflecting the relative amount of work done on them. Yeasts are mainly *Candida*, *Torulopsis*, *Rhodotomula*, *Pichia* and *Debaromyces* and a wide variety of filamentous fungi have also been described which have attracted little attention. It must be stressed that only some species of the genera listed above are capable of alkane-utilization.

Mechanism of alkane utilization

The mode of attack on n-alkanes has been adequately reviewed (van der Linden & Thijsse, 1965; McKenna & Kallio, 1965; Klug & Markovetz, 1970). To sum up, there are differences between various microorganisms, but the initial attack on the hydrocarbon is usually a hydroxylation reaction generally at C_1 and catalysed by a mixed-function oxidase (monooxygenase)

$$CH_3(CH_2)_nCH_3 + NADH_2 + O_2 \rightarrow CH_3(CH_2)_nCH_2OH + NAD + H_2O.$$

(Throughout this review $NADH_2$ and NAD will represent the reduced and oxidized forms of either nicotinamide adenine dinucleotide or its phosphate.)

The enzyme system has been studied in cell-free extracts and probably involves a series of electron carriers; in *Pseudomonas oleovorans*, the following system occurs (Peterson, Kusunose, Kusunose & Coon, 1967).

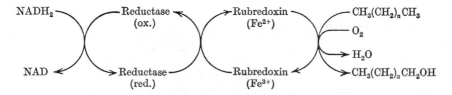

In other organisms, a cytochrome P-450 functions as an intermediate electron carrier (Cardini & Jurtshuk, 1968).

The initial oxygenase reaction may involve a hydroperoxidation stage:

$$CH_3(CH_2)_nCH_3 + O_2 \rightarrow CH_3(CH_2)_nCH_2OOH.$$

This is followed by transfer of an oxygen atom on to the electron donor.

Another possibility which has great attractions to those seeking an economic industrial process for alkane-utilization, is dehydrogenation to the corresponding alk-1-ene:

$$CH_3(CH_2)_nCH_3 + NAD \rightarrow CH_3(CH_2)_{n-1}CH = CH_2 + NADH_2.$$

This reaction could be either followed by the addition of water to the double bond or by oxygenation to give a 1,2 epoxide which could be reduced to the primary alcohol. However, there is some doubt about the thermodynamic feasibility of this type of reaction and Lebeault, Roche,

Duvnjak & Azoolay (1970) have shown that in mitochondrial preparations of *Candida tropicalis*, ATP is necessary, possibly by a 'reverse
electron flow' mechanism.

However, whatever the actual reaction involved, there is little doubt
that the initial oxidation is usually accompanied by the incorporation of
molecular oxygen as has been shown by $^{18}O_2$ experiments (Leadbetter
& Foster, 1959a). If a mixed function oxidase is involved, the highly
exergonic reaction may account for some of the oxygen requirement and
heat production in alkane fermentations (see p. 21). Even if it were
possible to obtain an anaerobic fermentation of alkanes, and there is
some evidence that this can happen (Traxler & Bernard, 1969), it is
doubtful whether the process would be commercially feasible due to
lower yield values presumably obtained.

The initial hydroxylation to the corresponding alcohol is normally
followed by two NAD-linked dehydrogenation steps to the corresponding acid which is attacked by β-oxidation in the normal manner:

$$CH_3(CH_2)_n CH_2OH \xrightarrow{\text{NAD \quad NADH}_2} CH_3(CH_2)_n CHO \xrightarrow{\text{NAD \quad NADH}_2} CH_3(CH_2)_n COOH$$

An important question is what happens to alkanes with an odd number
of carbon atoms? Is the corresponding acid incorporated into cell lipids
possibly producing a toxic component? The evidence suggests that there
may be considerable variation among microorganisms in this respect.
For example, Wagner, Kleeman & Zahn (1969) showed that the fatty
acids of *Nocardia* species tended to reflect the chain length of the parent
alkanes whereas those of *Candida lipolytica* did not and presumably
arose by *de novo* synthesis.

Although oxidation on C1 is probably the usual mechanism for
alkane utilization, diterminal α,ω-oxidation has also been reported to
give the corresponding dicarboxylic acid. Another possibility is oxidation in the 2 position to give the corresponding secondary alcohol and
ketone:

$$CH_3(CH_2)_n CH_3 \begin{cases} CH_3(CH_2)_n COOH \longrightarrow COOH(CH_2)_n COOH \\ CH_3(CH_2)_{n-1} CHOH.CH_3 \longrightarrow CH_3(CH_2)_{n-1} CO.CH_3 \end{cases}$$

An important and interesting question concerning the oxidation of
liquid *n*-alkanes is how can compounds of such low solubility be
attacked? Certainly it is impossible to explain the observed growth rates

by diffusion of the alkanes through the aqueous phase. McAuliffe (1969) and Johnson (1964) seem to have been the first to suggest an attachment of the microbial cell to the surface of the alkane followed by solubilization in the lipid of the cytoplasmic membrane. In support of this theory, Bos & de Boer (1968) showed that yeasts cling to droplets of emulsified n-hexadecane and unpublished experiments from this Department by S. Davies and A. G. McLee have shown that in slide cultures only those microorganisms actually touching a hydrocarbon droplet proliferated continuously.

How, then, is the barrier of the cell wall overcome? There seem to be no separate analyses of cell-wall lipid, although Bos & de Boer (1968) found some evidence from electron micrographs of hydrocarbon-grown yeast of a 'lipophilic' wall absent in glucose-grown cells. Ludvik, Munk & Dostalek (1968) noticed structural changes in the cytoplasmic membrane of alkane-utilizing yeasts leading to a considerable increase in the surface area. However, Davies & Whittenbury (1970) found that bacteria growing on higher alkanes had normal membranes and walls in contrast to the massive membranous systems of methane-utilizers (p. 31). In view of the solubility of alkanes in the cytoplasmic membrane lipid it might be expected that no permeases would be required; there is nevertheless evidence for specific active transport mechanisms from a study of the uptake of labelled hydrocarbons under anaerobic conditions (Munk, Volfova, Kotyk & Dostalek, 1969).

The relevance of these results to possible industrial processes is evident. The amount of growth may be proportional to the surface area of the hydrocarbon droplet and, if this is limiting, linear growth will occur. Consequently some dispersion of the alkane is necessary but high shearing rate must be avoided; in some cases the microorganisms themselves may produce an emulsifying agent.

Requirements for industrial microorganisms

Too often in the past a microorganism used in an industrial process has been chosen more by accident (sometimes happy!) than by design. What are the main requirements for an organism to be used for single-cell protein production from n-alkanes?

Cheap nutrients for growth

Given a particular carbon source, the lower the cost of other nutrients the better. In practice a microorganism requiring no preformed amino acids, nucleotides, growth factors or other additives is sought, usually by isolation from nature by enrichment in presence of the carbon source together with a suitable inorganic salt mixture. The nitrogen source is usually ammonia since there seems to be little industrial potential in a nitrogen-fixing hydrocarbon-utilizer.

In general bacteria are more suitable in their nutrient requirements to yeasts since the latter usually require the addition of a source of growth factors. (However, the yield of cells from hydrocarbons may be increased by the addition of growth factors (see p. 21).) It should be noted that to produce the high densities growing at near the minimum generation time required for efficient continuous culture, additional inorganic growth requirements may come to light.

High rate of growth at high density

There is no doubt that continuous culture is the most suitable method for the efficient and economic growth of hydrocarbon-utilizing micro-organisms; it requires a much smaller plant and gives a more reproducible product. The chemostat is the method generally adopted, and it is necessary to obtain the highest output (density × dilution rate) provided no loss of efficiency is entailed. In practice this means seeking a dilution rate of about 0·1 (generation time about 7 h.) at a density of about 10 g. dry weight/l. Both bacteria and yeast are potentially capable of this performance on n-alkanes.

Granted the desired output, is it preferable to grow cells at a high density and a low dilution rate or vice versa? Various points should be considered.

(a) The higher the dilution rate, the higher the ribosome content and thus the level of ribosomal protein and RNA, both of which may be undesirable in diets, the former because of the change in amino acid ratios and the latter because of the possibility of a high blood uric acid level if used for human consumption.

(b) The oxygen consumption per unit time may differ in the two extremes.

(c) In a high-density situation, there may be an increased concentration of extracellular by-products which, apart from making the process less efficient, may alter the cellular compositions and hence give increased (or lower) concentrations of potentially toxic compounds.

(d) In a complex nutritional system such as the gas-oil process in which there is a preferential utilization of certain alkanes (Dostalek, Munk, Volfova & Pecka, 1969; Munk et al. 1969), then growth at different rates may alter the utilization pattern and hence the cell composition.

(e) A low-density situation increases the cost of harvesting.

These are some of the considerations involved and in practice some sort of compromise must be reached.

Oxygen requirement and cell yield

A critical factor in the economics of an aerobic fermentation is the requirement of oxygen and hence of aeration costs which may account for up to 20% of the cost of the product. In a well-designed process, almost all the oxygen that can be transferred from the gas to the liquid phase is used, provided the steady-state level of dissolved oxygen does not become the growth-rate limiting factor. In practice, oxygen has a solubility of about 10 p.p.m., while microorganisms fortunately often have a K_m for oxygen of as low as 0·1 p.p.m. Clearly it is advantageous to choose an organism with as low a K_m as possible.

Table 1. *The effect of substrate and cell yield on the oxygen-transfer and heat-removal requirements*

From Humphrey (1970).

Microorganisms	Substrate	Cell yield (g./g.)	g. O_2 required/ 100 g. cells	kcal. heat produced/ 100 g. cells
Yeast	Carbohydrate	0·5	67	383
Yeast	*n*-Alkane	1·00	197	799
Bacteria	*n*-Alkane	1·00	172	780

Unfortunately, the oxygen requirement and the related heat production per unit weight of cells produced is considerably higher when hydrocarbons are the substrate compared to carbohydrates (Table 1). These high figures for hydrocarbons relate partly to the inefficiency of the first oxygenation stage of alkane utilization where a considerable amount of energy is given out as heat; this also results in a greater cell yield per unit amount of oxygen used as the chain length of the hydrocarbon goes up (Miller, Lie & Johnson, 1964). These figures for oxygen requirement and heat production are related to the efficiency of conversion of alkane to cells and values between 30% and 120% can be obtained on a weight for weight basis corresponding to at least a fivefold difference in oxygen consumption and heat production. Although the results vary from organism to organism, the range of values obtained reflects primarily differences in the cultural conditions used. Very high yields can be obtained by incorporating some additional organic growth factors to the medium. Thus Wagner, Kleeman & Zahn (1969) obtained conversion figures for bacteria growing on alkanes of as high as 142% (weight for weight) and 97% (carbon to carbon) by the addition of urea, yeast extract and cornsteep liquor. There was practically no CO_2 production and it is inferred that the oxygen demand and heat production was correspondingly lower.

Suitable optimum temperature

The cost of cooling a large fermentor may become prohibitively high in areas where water of a temperature little different from that of the fermentor is available and refrigeration is necessary. Ideally the optimum temperature of the microorganisms should be chosen to suit the particular fermentation, the fermentor and the ambient external temperature so as to minimize the cooling. In practice, this means obtaining thermophilic alkane-utilizers with optimum growth temperatures between 40° and 60°, a comparatively easy task with bacteria. For example, Matales, Baruah & Tannenbaum (1967) obtained a thermophilic alkane-utilizing strain of *Bacillus stearo-thermophilus* which grew in the range of 60–70° and had a protein content (based on total N) of over 70 %. No figures were given for the yield which may be lower than in mesophiles. It is more of a problem to obtain yeasts growing above about 40°.

Ease of harvesting

Most harvesting methods ultimately depend upon centrifugation and clearly it is more costly to use bacteria with a typical diameter of $1\,\mu$ than yeast at $5\,\mu$, resulting in a 25-fold difference in ease of sedimentation. (Wang (1967) has provided an assessment of the costs and difficulties involved.) However, concentration by other methods such as flotation and foam concentration or flocculation can also be considered and a microorganism suitable for these purposes should be sought.

Suitable composition

Clearly it is desirable to obtain as high a content of protein as possible containing a high level of the essential amino acids. The average range of protein contents are 50–75 % for bacteria, 20–45 % for fungi, 40–60 % for yeasts and 30–60 % for algae; bacteria are clearly most suitable in this respect. However, most protein estimations on alkane-utilizers in the literature are based on total-nitrogen content which can vary between about 5 % and 13 % of the dry weight, and takes no account of nucleic acid, or of other nitrogenous polymers such as amino-polysaccharides or, in prokaryotes, of mucopeptide. It is important to note that mucopeptides, although adding to amino acid composition, are probably of no nutritional value and the D-amino acids produced by hydrolysis may even be inhibitory. There is a considerable difference in the protein content of a single microorganism grown under a variety of different environmental conditions. Evans (1969) quotes figures for an alkane-

Table 2. *The contents of some essential amino acids in microorganisms*

Microorganisms grown on liquid *n*-alkanes and compared with other sources and with the FAO standards.

Organism	Substrate	Amino acid content (g./100 g.)							Reference
		Lysine	Valine	Leucine	Iso-leucine	Theonine	Methionine	Phenyl-alanine	
Yeast	Liquid alkane (a)	4·6	3·6	4·7	3·1	3·2	1·2	3·1	Evans (1969)
	(b)	5·3	3·9	5·6	3·7	3·6	1·2	2·9	Evans (1969)
Bacteria	Liquid alkane (c)	4·0	4·7	6·7	3·7	4·7	3·2	2·9	Wagner et al. (1969)
	(d)	2·4	3·7	5·6	1·8	3·8	3·1	3·1	Wagner et al. (1969)
Bacteria	Methane	3·1	.	9·0	.	.	0·9	.	Wolnak et al. (1967)
Bacteria	Methane	5·3	.	.	.	4·5	3·4	6·2	Vary & Johnson (1967)
Bacteria	Methane	5·9	6·6	8·2	4·6	4·6	1·5	4·8	Klass et al. (1969)
Fish meal	—	4·6	3·4	4·7	3·0	2·7	1·7	2·6	—
Soybean meal	—	2·9	2·3	3·5	2·5	1·8	0·6	2·3	—
FAO standard	—	4·2	4·2	4·8	4·2	2·8	2·2	2·8	—

(a) Purified liquid alkane process; (b) gas oil process; (c) alkane + inorganic medium; (d) urea, yeast extract and cornsteep liquor added.

utilizing yeast which vary between 24 % and 60 % for protein, 8 % and 31 % for lipid and 10 % and 22 % for carbohydrates; under these conditions, the yield of cells varied between 30 % and 110 %. In particular, cells grown in conditions of carbon and energy source excess may have considerable amounts of storage polymers such as polysaccharide, poly-β-hydroxybutyrate and lipid accumulating (Wilkinson & Munro, 1967). For this reason growth in the chemostat is usually limited by carbon and energy availability, thereby also minimizing difficulties in removing excess of the carbon source remaining.

Average figures for the content of some important amino acids are given in Table 2. Compared with the FAO standard, yeasts have the disadvantage over bacteria of a deficiency in methionine with which a diet may need to be supplemented (p. 28). There can be appreciable changes in the proportions of individual amino acids under different cultural conditions. Wagner *et al.* (1969) analysed the total nitrogen content, nitrogenous cell-wall components (glucosamine, muramic acid and diaminopimelic acid) and a range of amino acids in *Nocardia* species grown on different media and found considerable variations (Table 2). However, the organisms were grown in batch culture and no evidence is given of the phase of growth or the nature of the limiting nutrient. Yamada *et al.* (1967) also provide some data for the amino acid composition of microorganisms grown on alkanes under different environmental conditions.

Nutritional availability and lack of toxicity

So far we have considered the paper specification of a product and this is fairly easy to determine. However, there are two further stages to consider, the testing of which is costly and time-consuming.

(*a*) Nutritional testing for the availability of the amino acids, vitamins and other components for utilization by the animal. Normally testing is carried out by adding to an animal feed the amount of protein which would be used in commercial practice (10–20 % of the feed) and comparing the results with known standards. In general, availability of amino acids is related to the ability of intestinal enzymes to hydrolyse the protein and this may vary according to the physical state of the feed. Thus Tannenbaum & Matales (1968) have pointed out that rupturing cells of *Bacillus megaterium* increases their digestibility. The use of enzymic or physical methods for cell rupture may also become more important if pure protein is required.

(*b*) Toxicological tests which are normally done at a higher level in the feed (about 40 %) and require at least a two-year period for a study of possible chronic effects, carcinogenic activity and action over many generations.

So far most of our knowledge has been gained with yeasts (see p. 27). Some of the experiments reported for bacteria are not promising (Waslien, Galloway & Margen, 1969), at least for direct feeding to man. Trials on poultry by Ko & Yu (1967) with bacteria grown on *n*-alkanes have been satisfactory.

Susceptibility to microorganisms to attack by predators

The ideal microorganism should be completely resistant to attack by viruses or any other predators, particularly if it is to be used under non-aseptic conditions. This consideration has tended to rule out the use of bacteria except under aseptic conditions, because of the danger of phage attack as well as by other lytic predators such as *Bdellovibrio*. It is possible that suitable genetic manipulation could produce an organism *generally* resistant to phage attack although such an organism might have a lower growth rate compared to the wild type. In the absence of such a variant, it may be useful to have on hand a variety of *specific* phage-resistant strains of the same organism and carry out a planned rotation in their use.

Yeasts present little problem in this respect, a fortunate state of affairs for those of us enjoying alcoholic drinks. Indeed, it is possible to design a non-aseptic continuous alkane fermentation by yeasts by choosing environmental conditions unfavourable for most microorganisms except that in use.

Suitability for genetic manipulation

An important consideration, particularly with regard to future industrial potential, is the selection of microorganisms suitable for genetic recombination or those in which stable extrachromosomal DNA can be incorporated, thus allowing addition of new or of duplicated genes. Such considerations are likely to become more essential since it should be possible to obtain variants of suitable organisms which over-produce a particular protein, resulting in an alteration in the amino acid composition and possibly the total protein content in the desired direction. The potentialities for the production of specific enzymes or of enzyme products are even greater but are outside the scope of this review.

Bacteria are clearly more suited to genetic manipulation, at least at our present state of knowledge.

Commercial processes for alkane fermentation

The first concerted effort to grow microorganisms on hydrocarbons with a view to commercial protein production was initiated in the late 1950s by Champagnat and his associates at the Société Français des Petroles B.P. (the French associate of the British Petroleum Company). The work originated in the development of a method to dewax oil (remove *n*-alkanes) with the production of microorganisms as a secondary consideration. Since that time, there has been a vast increase of work in the field and many semi-commercial plants have been developed (for examples, see Humphrey, 1970). Two main general methods are being used as illustrated in the work of the British Petroleum Company, who appear, from published information available to the author, to have the furthest developed processes (Bennet, Hondermarck & Todd, 1969; Evans, 1967, 1969). These processes both involve yeasts although other groups have tested bacteria under similar conditions. The main phases are given in Fig. 1.

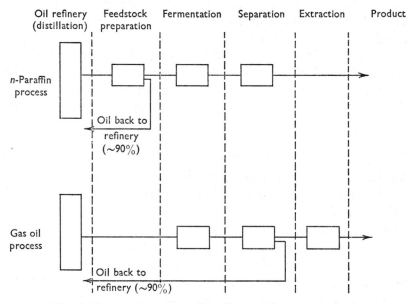

Fig. 1. Comparation process flow diagrams for yeast production from *n*-alkanes (Evans, 1969).

The gas-oil process

This process was developed from the original dewaxing work and uses as the growth substrate gas oil, a petroleum fraction distilling in the diesel fuel and lubricating oil range, and containing about 10 %

n-alkanes. The other components of the gas oil are not utilized by the microorganisms, at least under the conditions used for the fermentation. Growth on a complex substrate of this type may be polyauxic (Dostalek *et al.* 1969) and the yield in a continuous system depends upon the dilution rate and the concentration of gas oil (Munk *et al.* 1969). The B.P. process uses a fermentor under non-aseptic conditions, thus reducing the growth costs appreciably; their yeast remains the dominant organism over a long period of time in the environments used and contaminants (particularly bacteria) are discouraged partly by an appropriate temperature and pH in the fermentor. However, there is clearly the possibility of a different dominant microorganism in another area of the world and also of a toxigenic contaminant which could cause difficulties even if present at a low level.

The yeast is removed from the gas-oil remaining after growth (*c.* 90 %) and this is done by washing and centrifuging. However, some traces of oil will remain in the cells, particularly in the lipid, and this can only be removed by an efficient and costly solvent extraction with a combination of polar and non-polar solvents such as propanol and hexane. It is desirable, therefore, that the microorganism has a low lipid content.

Crude oil with a high wax (alkane) content can also be used as a substrate when the primary object is to upgrade the crude oil.

n-*Alkane process*

In this method a purified liquid *n*-alkane fraction of up to C18 is obtained by a molecular sieve process in the refinery. Under conditions of carbon and energy deficiency in the chemostat, nearly all the alkane is removed in the fermentor. Thus the expensive solvent extraction is avoided. Furthermore, since the substrate is much more closely defined, tighter quality control can be maintained. On the other hand, these gains are offset by the requirement of an aseptic fermentation and by the increased cost of the purified *n*-alkanes. The cost of gas oil is about £10 per ton compared with a figure of more than double this for the purified *n*-alkane fraction (these values tend to be artificial and to depend upon the type of oil used, the refinery facilities available and the potential market at any one time for the various oil fractions). These costs must be compared to the price of about £100 per ton at which a crude single-cell protein product containing about 60 % protein would need to sell to compete with oil-seed or fish meals.

Extensive feeding trials using mainly pigs and poultry have been carried out on the yeast produced by these two processes and the products have been shown to be both safe and useful components of animal feeds (Shacklady, 1969). The true digestibility of the protein was 94–96% and the biological value was 91–95%, the last figures being obtained in presence of 0·3% DL-methionine. No toxic factors were found in long-term tests; the tendency of the products of the early stage of development work to cause liver enlargement were eradicated in later studies.

An example of the scale of plants at present being developed or in action, BP have built a demonstration unit at Grangemouth, Scotland, producing 4,000 tons per year. A further increase in size of individual fermentors is obviously desirable, although larger plants would undoubtedly contain multiple fermentors. The gas-oil unit built at Lavera, France, has a capacity of 16,000 tons per year and aseptic precautions are not necessary.

In spite of pessimistic sounds in some quarters, it is probable that these processes and others similar to them will provide a valuable and economic source of protein in the world, at least for animals. Direct feeding to human beings remains for the future; the products themselves are tasteless but incorporation into a form unlikely to encounter consumer resistance remains a problem.

METHANE FERMENTATION

Methane is potentially a more suitable substrate for single-cell protein production than higher n-alkanes. It is cheaper (from £0–8 per ton, depending on site and circumstances), plentiful and has the great advantages of presenting no direct toxicity problems. It is the most reduced form of carbon available and theoretically should give the highest yield of cells per unit weight of substrate. A production unit can be set up without requiring any additional facilities such as a refinery. Indeed, methane produced by the anaerobic decomposition of waste products such as sewage could be used although it is doubtful if the output would be large enough to make a protein plant an economic proposition.

Consequently much interest has been shown recently in methane-utilizing microorganisms. At the same time it has been realized that while the isolation and growth of organisms utilizing higher alkanes presents a fairly easy microbiological problem, methane-utilizers are difficult, especially in an industry accustomed to the use of fungi and, in particular, of yeasts. In this section an attempt will be made to summarize what we know about the basic properties of these microorganisms and their potentiality for industrial exploitation.

The isolation of methane-utilizing microorganisms

Although clear accounts of the occurrence of methane-utilizing bacteria were given in the early 1900s, their isolation has proved a difficult task. This is surprising since in theory an enrichment in an inorganic medium with air and methane (the only carbon and energy source) in the gas phase should yield only methane-utilizing microorganisms or organisms growing on their extracellular or autolysis products. Yet until very recently and in spite of their potential industrial importance, only three species were in pure culture: *Pseudomonas* (*Methanomonas*) *methanica* (Dworkin & Foster, 1956), *Methanomonas methanooxidans* (Brown, Strawinski & McClesky, 1964), *Methylococcus capsulatus* (Foster and Davis, 1966).

All these are Gram-negative rods with an obligate requirement for methane or methanol as the carbon and energy source. Lack of success in obtaining other pure cultures can be in part ascribed to peculiarities of the organisms themselves and partly to using methane contaminated with higher hydrocarbons (e.g. natural gas), to using agar contaminated with organic nutrients and to lack of an adequate understanding of microbiological enrichment techniques. There have been many other reports of methane-utilizing microorganisms, but none of them have been sufficiently well described to allow definite characterization and in many cases they have probably not been in pure culture. Indeed, it has sometimes been claimed or implied that methane-utilizers can only grow adequately in mixed culture, the contaminants presumably producing either an essential nutrient or removing a toxic inhibitor produced during growth of the methane-utilizer. Claims that myco-bacteria and related bacteria which can utilize higher hydrocarbons also utilize methane must be treated with some scepticism until it can be shown that the culture is pure and that it will grow in a medium containing methane as the *sole* carbon and energy source. If the culture can be grown in the absence of methane, then the ability to utilize methane should not be lost on subculture; if there is such a loss, it suggests growth of a contaminant.

As a start to a general study of methane-utilizing microorganisms, we have isolated over 100 strains in pure culture (Whittenbury, Phillips & Wilkinson, 1970*b*). All of these isolates had the following character-istics in common.

(*a*) They are Gram-negative, strictly aerobic bacteria.

(*b*) They have an obligate requirement for methane, methanol or dimethylether (see later) as the carbon and energy source for growth.

(c) They have no absolute requirements for any organic growth factors, amino acids, etc. However, since they were all isolated after enrichment in a medium containing no organic additions with the exception of methane, this is perhaps not surprising.

(d) The majority form a resting stage.

(e) They have a complex internal membranous structure.

So far there appears to have been no substantiated case of a methane-utilizing eukaryote.

Distribution

The distribution of methane-utilizing bacteria is ubiquitous. We have isolated them from mud, water (from ponds, streams, ditches, estuaries and marine sources) and soil samples in a variety of countries (Whittenbury *et al.* 1970*b*). Indeed, after enrichment in bottles containing inorganic salts and a gaseous environment of air and methane, four or five different morphological types were often isolated from each enrichment and only rarely did a single type occur. This distribution is not surprising in view of the widespread occurrence of methane in nature produced as a result of anaerobic respiration. Methane-utilizers would be expected to occur in aerobic environments above or near areas in which anaerobic microbial decomposition of organic matter is occurring. This wide distribution makes it impossible to use them for prospecting.

Nomenclature

The nomenclature of methane-utilizing bacteria has been confused by the use of the prefix *Methan-* to describe both methane-producing and methane-utilizing genuses. Because of this, Dworkin & Foster (1956) renamed *Methanomonas methanica* as *Pseudonomas methanica*, saying that morphological considerations should have preference over methane-utilizing ability. However, it was later argued (Foster & Davies, 1966) that in view of the obligate nature of methane-utilization separate genera should be given to methane-utilizers; in order to avoid confusion with methane-producers for which the prefix *Methano-* was commonly employed, methane-utilizers should have the prefix *Methylo-*. We have adopted this convention and although attempts at a formal classification are obviously premature at this stage due to lack of differential information from conventional tests, we have drawn up the provisional classification outlined in Table 3, based on variations in morphology, fine structure and type of resting-stage formed. It is also proposed for the rest of this review that all these organisms capable of utilizing methane as their sole source of carbon and energy be called the methylobacteria.

Table 3. *A simplified basis for the classification of methylobacteria*

Based on Whittenbury, Phillips & Wilkinson (1970*b*).

Group	Morphology	Membrane* Motility type	Pathway of C uptake	Resting stage
'Methylosinus'	Rod- or pear-shaped	+ I	Serine pathway	Exposure
'Methylocystis'	Rod or vibrioid	− I	Serine pathway	Lipid cyst
'Methylomonas' 'Methylobacter' } Rod		± II	Allulose pathway	{ Azotobacter-type or immature cyst
'Methylococcus'	Coccus	− II	Allulose pathway	Immature cyst

* See below.

Morphology

The shape of the methylobacteria varies widely from vibrioid and pear-shaped cells to rods and cocci; they also show a range in size from about $5 \times 2\mu$ in some '*Methylobacter*' varieties to about $1 \times 1\mu$ in '*Methylococcus minimus*' (Whittenbury *et al.* 1970*b*). Thus although small compared with eukaryotic cells, some are large for prokaryotes and would not present too great problems for industrial harvesting. Further, some methylobacteria agglutinate to form rosettes. A few species are motile (see Table 3), producing polar tufts of flagella ('*Methylosinus*') or polar flagella ('*Methylomonas*' and '*Methylobacter*'). They may produce characteristic resting forms (see Table 3 and p. 32).

One of the most characteristic properties of methylobacteria are the complex membranous systems seen in their sections (Whittenbury, 1969; Proctor, Norris & Ribbons, 1969; Davies & Whittenbury, 1970). They consist of pairs of unit membranes, probably all derived by invagination of the cytoplasmic membrane. Two patterns of organization are found:

(*a*) Type II membrane (Plate 1*a*). These are in the form of parallel closely packed bundles of membranes consisting of a number of disc-shaped vesicles arranged in a highly orientated manner.

(*b*) Type I membrane (Plate 1*b*). The internal arrangement of these membranes is more variable and less ordered than those of type I. The paired membranes are not so close-packed and bound a lumen or tube of varying dimensions. They are usually concentrated round the periphery of the cell.

Since these membranes are not found in methanol-oxidizing hetero-trophs, their universal occurrence in methylobacteria may be related to the ability to oxidize methane specifically. It is true that methylobacteria grown on methanol still possess membranous systems but the value in nature of their derepression or induction by methane is doubtful. Since similar membranous systems are found in most lithotrophic bacteria it is possible that their presence is related to a strict lithotrophic existence (see p. 38). It is very interesting that the type of membrane is correlated with the method employed for carbon uptake (see p. 36).

Other less characteristic properties of the morphology of the methylobacteria are the complex wall structures found in some (Plate 2*a*,*b*), poly-β-hydroxybutyrate granules (Plate 2*c*) and gas vacuoles.

Resting stages

As mentioned previously, methylobacteria are peculiar in that the majority of those studied form some type of resting stage (Whittenbury, Davies & Davy, 1970*a*) which may be in form of spores or cysts.

Exospores are formed by '*Methylosinus*' varieties. As cultures enter the stationary phase of growth in batch culture, a proportion of organisms elongate or taper to a pear shape ('*Methylosinus trichosporium*') or to a comma shape ('*Methylosinus sporium*'). This change is followed by the production of an exospore as a rounded body budded off from the tapered end of the cell. In '*M. trichosporium*' the spore is surrounded by a large capsule of a fine fibrous nature. In fine structure, these exospores are similar in many respects to bacterial endospores. They are surrounded by an outer coat (derived from the host cell), a laminated inner coat, an area similar to the cortex of the endospore and a membrane; they are resistant to heat and desiccation but do not contain dipicolinic acid.

The majority of non-sporing methylobacteria produce one of three types of cyst which have varying degrees of desiccation resistance and cell specialization in their formation.

(*a*) '*Lipid*' *cyst found in* '*Methylocystis parvus*'. As the organisms go into the stationary phase of growth in batch culture, they form large lipid inclusions of poly-β-hydroxybutyrate, become rounded and increase in size. This change is associated with an increase in complexity of the cell wall and a loss of the internal membranous system.

(*b*) '*Azotobacter-type*' *cyst*. These cysts, produced by '*Methylobacter*' strains, are indistinguishable in microscopic appearance from those formed by *Azotobacter* species. The formation of the cyst is associated with rounding, an increase in cell size and sometimes the production of pigment.

(*c*) '*Immature cysts*'. The cysts, produced in '*Methylomonas*' and '*Methylococcus*' strains are also spherical in shape, but have a wall structure intermediate between that of a vegetative cell and a mature cyst.

The significance of this variety of resting stages in nature may be related to reliance on a single and unique carbon and energy source for growth. In practice in the laboratory, difficulty may be encountered in continuous culture studies, since a change in the environment caused by an alteration in the medium or in the dilution rate may induce encystment or sporulation which is associated with a calamitous drop in the rate of growth; cultures may take a long time to recover to normal vegetative growth.

The pathway of methane oxidation

The pathway of methane oxidation is usually assumed to involve the series of reactions given below; Ribbons, Harrison & Wadzinski (1970) have calculated the standard free-energy changes for individual two electron steps in kcal. mole^{-1} at pH 7:

$$CH_4 \xrightarrow[-26\cdot12]{} CH_3OH \xrightarrow[-44\cdot81]{} HCHO \xrightarrow[-57\cdot15]{} HCOOH \xrightarrow[-58\cdot25]{} CO_2$$

The evidence for this pathway, which has been recently reviewed by Ribbons et al. (1970), can be summarized as follows.

(a) The proposed intermediates can be oxidized by washed suspensions of methylobacteria and can accumulate during methane oxidation, particularly in presence of inhibitors and trapping agents (Leadbetter & Foster, 1958, 1959b; Brown et al. 1964; Higgins & Quayle, 1970).

(b) Methanol can act as an alternative carbon source in all the methylobacteria tested, although it may be toxic at relatively low concentrations in some strains (Whittenbury et al. 1970b).

(c) Formaldehyde is a key compound for C1 fixation into cell compounds (see p. 36).

(d) Methanol dehydrogenase occurs in methylobacteria and is similar to that of other C1-utilizers in its inability to transfer electrons to NAD, the non-participation of H_2O_2 and a specific requirement for ammonium ions. NAD-linked dehydrogenases for formaldehyde and formic acid have also been demonstrated in methylobacteria (Harrington & Kallio, 1960; Johnson & Quayle, 1964). The evidence, although largely indirect, suggests that the proposed intermediates can act in methane oxidation to CO_2. Attention has been particularly focused on the first oxidation stage of methane to explain some of the puzzling differences between methane-utilizers and methanol-utilizers and because of the key it might provide to some of the difficulties involved in the growth of methylobacteria.

Is free oxygen involved in the initial hydroxylation of methane? Experiments by Leadbetter & Foster (1959a) showed that there was greater uptake of $^{18}O_2$ into cells when methylobacteria were grown on methane than on methanol although it was less than would be expected from comparison with ethane and propane in organisms utilizing higher alkanes. Recently Higgins & Quayle (1970) have unequivocally confirmed the participation of free oxygen by showing that methylobacteria grown on methane and $^{18}O_2$ are able to accumulate appreciable quantities of methanol by suitable changes in the pH and ionic strength during growth. The oxygen of methanol was derived solely from O_2, suggesting a mixed function oxidase similar to those involved in higher alkane utilization. This mechanism implies that two out of the four oxidation stages in the conversion of methane to CO_2 would be unproductive of ATP (one to provide $NADH_2$) and that growth on methanol would be more efficient than on methane. On the contrary, the growth yields on methane are higher than those on methanol (p. 40) and would be unobtainable by the above mechanism. Is there any way we can explain the incorporation of O_2 into CH_3OH and at the same time allow for the production of ATP in the initial oxidation stage?

Recent work in this department (Bryan-Jones & Wilkinson, 1971) has indicated the possibility that dimethylether may be an intermediate. The evidence is as follows.

(a) Dimethylether is oxidized and can act as the sole carbon and energy source for all methylobacteria tested. It is inert to methanol-utilizers.

(b) Diethylether and ethane can be oxidized but cannot act as the sole carbon and energy sources for methylobacteria. Ethane-utilizers cannot attack diethylether. This suggests that the mechanism of the oxidation of ethane is different in methane- and ethane-utilizers although in both cases it can lead to the eventual production of acetaldehyde and possibly acetate (p. 42). The inference is that methane is oxidized by a unique mechanism different from that in organisms growing on higher alkanes.

(c) Under special conditions, a small amount of dimethylether can accumulate from cells growing on methane.

It is true that this is some way from proving that dimethylether *is* an intermediate, but the initial oxidation of methane may occur as follows:

$$2CH_4 + O_2 \longrightarrow CH_3OCH_3 + H_2O.$$

If this is so, how is dimethylether attacked? $^{18}O_2$-incorporation experiments into methylobacteria grown on either methane, dimethylether or methanol as the sole carbon and energy sources have shown that, like methanol but unlike methane, there is practically no $^{18}O_2$ incorporation with dimethylether (Bryan-Jones & Wilkinson, 1971). This result together with those of Higgins & Quayle (1970) suggests that if dimethylether is an intermediate, it is not oxidized by a reaction involving molecular oxygen (unless a product exchanges its oxygen with that in water), nor is it converted into two molecules of methanol. An attack by a mixed function oxidase to methanol and formaldehyde similar to that of the O-dimethylase found by Ribbons (1970) in *Pseudomonas aeruginosa* seems unlikely because of the inefficiency that would result in the overall energy transfer. We must assume the following type of mechanism:

$$CH_3OCH_3 + XH \longrightarrow CH_3OH + CH_3X.$$

The molar growth yields on methane can be explained if additional ATP is either formed during the production of dimethylether or we must assume that the conversion of CH_3X to formaldehyde or its equivalent produces more ATP than the oxidation of methanol to formaldehyde.

Much work needs to be done on this pathway.

Mechanism of the biosynthesis of cell components

In an elegant series of papers, Quayle and his co-workers (Kemp & Quayle, 1967; Heptinstall & Quayle, 1970; Lawrence, Kemp & Quayle, 1970) have shown that there are two alternative pathways concerned in carbon utilization during the growth of microorganisms on reduced C1 carbon and energy sources. These pathways have been elucidated by the use of kinetic studies on ^{14}C uptake, the determination of isotope distribution in individual C atoms of isolated metabolites and the demonstration of certain key enzyme systems.

(*a*) The allulose pathway. As shown in Fig. 2, the key enzyme in this scheme catalyses a condensation of formaldehyde and ribose-5-phosphate to give allulose-6-phosphate which is metabolized by a series of transaldolase–transketolase reactions analogous to those concerned in the uptake of CO_2 in lithotrophs.

(*b*) The serine pathway. As shown in Fig. 2, the key reaction is the hydroxymethylation of glycine by methylene-tetrahydrofolic acid derived from an intermediate in the pathway of methane oxidation (? formaldehyde) to give serine. This is followed by the carboxylation of phosphoenolpyruvate derived from serine to give oxalacetate. Glycine is resynthesized by a mechanism not yet fully understood and involving the condensation of an intermediate in methane oxidation with CO_2, either by a *de novo* synthesis or by a cyclic regeneration.

Recent experiments by Lawrence & Quayle (1971) have shown a clear distinction amongst methylobacteria between those with a type I-membrane system which use an allulose pathway which is unique to them and those with a type II membrane system having a serine pathway which is shared by heterotrophs capable of utilizing reduced C1 compounds (methanol, formaldehyde, formate, methylamine). These

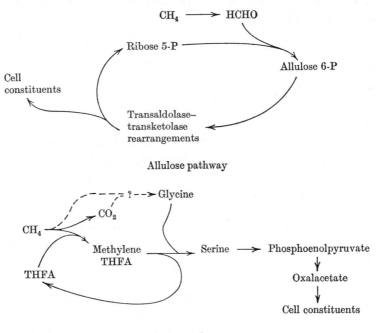

Fig. 2. The allulose and the serine pathway for the uptake of methane carbon into cell carbon.

pathways appear to be analogous in respect to efficiency of energy conversion since unpublished experiments in this laboratory have shown similar growth yields in methylobacteria using either pathway.

The possible evolutionary significance of the two pathways will be discussed later.

Growth requirements

Methylobacteria are able to grow in presence of a suitable carbon and energy source (methane, methanol or dimethylether), a mixture of inorganic salts and oxygen. No organic growth factors are required. A variety of salt mixtures have been used by different workers usually based on that published by Leadbetter & Foster (1958). However, there has been very little systematic work on the levels of the individual inorganic ions required with the exception of the work of Wolnak, Andreen, Chisholm & Saaden (1967) using a mixed culture growing very slowly. This is surprising in view of the notorious difficulty experienced in growing methylobacteria and may to some extent reflect the difficulty of obtaining reproducible results with these organisms.

Ammonium salts can be used as the nitrogen source for all the methylobacteria tested, nitrite and nitrate by the majority and casamino acids by some (Whittenbury *et al.* 1970*b*). No direct assay of nitrogen-fixing ability was made, but all strains were tested by the acetylene-reduction method for nitrogenase activity. Although most organisms had a low level of nitrogenase, only one showed active reduction and that was a strain of '*M. trichosporium*' originally isolated as a nitrogen-fixing methane-utilizer (Coty, 1967). Some interesting results can be seen in comparing the effect of ammonia and nitrate as nitrogen sources (Whittenbury *et al.* 1970*b*).

(*a*) The yields in adequately buffered media are higher on ammonia, possibly due to the energy expended in nitrate reduction.

(*b*) If media are inadequately buffered, growth is better on nitrate. This is largely a pH effect.

(*c*) Methylobacteria have the ability to oxidize ammonia to nitrite.

(*d*) Ammonia has an inhibitory effect on the *rate* of growth and of methane oxidation. This effect cannot be ascribed to a pH change nor to nitrite accumulation, but rather to apparent competition between NH_4^+ and CH_4 for methane oxidase, an idea supported by the finding that neither growth on methanol nor oxidation of methanol is affected by ammonia. Indeed, as pointed out previously (p. 33), methanol dehydrogenase requires ammonium ions for full activity. Apparently we have the peculiar position of the first stage of methane oxidation being inhibited by ammonium ions and the second stage being activated. Perhaps the growth difficulties of methylobacteria are caused by psychiatric disturbances brought on by this state of affairs?

(*e*) Although there is considerable variation in this ammonia inhibition, work in this department has shown that if an ammonium salt is used as the nitrogen source for continuous culture, there must be some means to control its level to near the minimum steady state level in the environment required to avoid nitrogen deficiency.

There is very little published information concerning the factor or factors limiting the total amount of growth attainable in batch culture systems although in many cases authors describe a maximum growth of less than 1 g. dry wt./l. Sometimes limitation of methane or oxygen may be responsible and Mueller (1969*a*) has shown with a mixed culture system that the growth rate was directly proportional to the working volume of his cyclone fermentor; this in turn was inversely proportional to the oxygen transfer rate. It is evident that the linear rate of growth he describes is a result of gas-transfer limitation but whether the stationary phase is brought on by oxygen limitation is unclear. In many other instances, it is likely that there is a limitation in the medium of an essential inorganic ion which may lead to autolysis and the production of a resting form rather than simple cessation of growth. Indeed, methylobacteria appear to be very prone to autolysis, particularly in the absence of methane, and this may also cause difficulty in the maintenance of stock cultures because the products of autolysis may themselves prove inhibitory. Intermediates in methane oxidation have not been shown to be produced in sufficient quantities to inhibit growth nor is there direct evidence for the accumulation of toxic end-products of metabolism. There is little published information available about the growth of methylobacteria in continuous culture, although work of this type is being undertaken by various laboratories.

Most of the cultures isolated by Whittenbury *et al.* (1970*b*) grew at 30°, some failed to grow at 37° whereas others grew at 45° and 55°. It is probable that a wide range of different temperature requirements can be obtained by using enrichments incubated at appropriate temperatures from appropriate environments. Certainly there are thermophilic varieties in some groups of methylobacteria and these should be of particular value in industry.

The effect of organic compounds and the relationship of methylobacteria to lithotrophic bacteria

Methylobacteria show a similarity to some lithotrophic bacteria in their membrane systems (see p. 31) and in their response to organic compounds.

(*a*) Organic compounds above C1 cannot be utilized as the sole carbon and energy sources. However, they may be co-oxidized (Leadbetter & Foster, 1958, 1960) and D. S. Hoare (personal communication) has shown that *P. methanica* can assimilate ^{14}C-labelled acetate into a variety of cellular compounds with a labelling pattern similar to that of chemolithotrophic bacteria and dissimilar to that of chemoorganotrophic bacteria.

(*b*) Although some organic compounds can stimulate growth at low concentrations, an inhibitory effect is more common (Erohsin, Harwood & Pirt, 1968). We have found a considerable variation in this inhibition among the methylobacteria (Whittenbury *et al.* 1970*b*).

(c) The allulose pathway used by some methylobacteria for C assimilation is a modified form of the ribulose diphosphate pathway found in lithotrophs.

(d) One reason for strict chemolithotrophy has been assumed to lie in metabolic defects causing the inability to utilize organic compounds. Lack of oxoglutaric acid dehydrogenase and lack of an intact tricarboxylic acid cycle has been implicated (Patel, Hoare & Taylor, 1969) and unpublished work by J. Davey in this department has shown that this enzyme is indeed lacking in numerous methylobacteria. Smith, Loudon & Stanier (1967) has also implicated a lack of NADH oxidase in strict chemolithotrophs. As Ribbons et al. (1970) point out, NADH oxidase estimations in the literature must be interpreted with caution since the assays are not unequivocal and are often poorly documented, but both chemolithotrophs and methylobacteria seem to show a low activity.

(e) Phosphatidyl choline occurs in the membranes of several methylobacteria, a component occurring in some lithotrophs but not most heterotrophs (Ribbons et al. 1970).

The significance of this similarity could be due to an evolutionary origin of the methylobacteria from chemolithotrophic bacteria or to the inevitable imposition of some aspects of lithotrophy caused by the ability to grow on methane. The question is obviously complicated by the two distinct membrane and C assimilation pathways suggesting two quite distinct evolutionary mechanisms.

Consider the first possibility. As pointed out previously (p. 37), the oxidation of methane by methylobacteria is inhibited by ammonia in an apparently competitive manner and that the organisms can also oxidize ammonia. Could methylobacteria, or at least those using the allulose pathway, have been evolved by a mutation in the ammonia oxidase allowing methane oxidation to occur?

If this is so, what about the methylobacteria utilizing a serine pathway for carbon assimilation. If they have been derived from organisms using methanol or other reduced C1 compounds as carbon and energy sources by the same pathway, why should the gain in the ability to oxidize methane be associated with the acquisition of so many of the properties of lithotrophic bacteria? The study of methane oxidase has so far presented no clues to resolve the question and an answer may have to wait until we fully understand the nature of strict lithotrophy.

Clearly the evolutionary origin of the methylobacteria presents some intriguing problems and their lithotrophic properties may provide a clue to some of the difficulties in growing them.

Cell yield

A range of figures have been given in the literature for the efficiency of conversion of methane into microbial cells (Table 4); they should be compared with a theoretical maximum of about 160 % if all the methane carbon is converted to cell carbon (assuming 47 % dry wt. as C). Clearly values of over 100 % can be obtained and it is probable that the low figures are due to extensive extracellular polysaccharide production. These results are encouraging for the industrial use of the methane fermentation and compare favourably with the figures obtained for higher alkanes. Indeed, work in this department has produced yields of about 110 % in a stable continuous culture of methylobacteria growing at a density of about 3 g. dry wt./l. and at a dilution rate of 0·1 (K. C. Phillips, R. Whittenbury & J. F. Wilkinson, unpublished results).

Table 4. *The yield of methylobacteria from methane*

Organism	Yield (g. cells/g. methane)	Reference
Pseudomonas methanica	56%	Dworkin & Foster (1956)
Methanomonas methanooxidans	110%	Brown et al. (1964)
Methylococcus capsulatus	110%	Foster & Davis (1966)
Mixed culture	65%	Vary & Johnson (1967)
1GT-10 (? '*Methylosinus*')	144%	Klass et al. (1969)
Various methylobacteria	110%	Whittenbury et al. (1970b)

Experiments to measure the stoichiometry of methane and methanol utilization by growing methylobacteria (Whittenbury et al. 1970b) can be expressed in the following equations:

$$1CH_4 + 1·0 \text{ to } 1·1 \, O_2 \longrightarrow 0·2 \text{ to } 0·3 \, CO_2 + \text{cells}$$

or

$$1CH_3OH + 1·0 \text{ to } 1·1 \, O_2 \longrightarrow 0·5 \text{ to } 0·6 \, CO_2 + \text{cells}.$$

Under these conditions the sole products from methane carbon are CO_2 and cells. The differences between methane and methanol were borne out by the yield of cells which were about 20 % lower with methanol than with methane when calculated on a molar basis. Since the yield on methanol is similar to that obtained with bacteria other than the methylobacteria, it is doubtful whether a commercially feasible process could be based on a chemical conversion of methane to methanol which is then used as a substrate for microbial growth, although methanol-utilizers are easy to obtain, present a wide spectrum of different microbes and are easy to work with.

Composition and nutritional value

It is difficult to find in the literature reliable estimations of the composition of methylobacteria and the figures given are often obtained with mixed cultures and dubious or undocumented analytical methods.

Values for protein (many based on total N values) vary widely between 30 % and 65 % (e.g. Coty, 1967; Wolnak *et al.* 1967; Vary & Johnson, 1967; Klass, Iandolo & Knabel, 1969; Whittenbury, 1969). Some analyses of individual amino acids have been carried out and are given in Table 2, and these compare favourably with other cheap protein sources. Figures are also available for the vitamin content of an unidentified 'Bacillus' species (Wolnak *et al.* 1967). Figures for lipid and polysaccharide are generally valueless due to the analytical methods used. Poly-β-hydroxybutyrate can accumulate in some methylobacteria, particularly under nitrogen limitation (Coty, 1967; Whittenbury *et al.* 1970*b*).

In the absence of any good published data, the only conclusion that can be made is that the methylobacteria probably have a similar compositional range to other Gram-negative bacteria, the actual amounts depending upon the organism and cultural conditions employed. The only mention of feeding trials is by Klass *et al.* (1969), who say that their organism (? '*Methylosinus*' variety) contained no toxic or undesirable components. However, it is perhaps unfortunate that the methylobacteria are probably all Gram-negative organisms and may well have the characteristic chemical components of a Gram-negative cell wall. The effect of this in long-term feeding trials has yet to be assessed.

Genetics and phage sensitivity

To my knowledge there is no published evidence for genetic recombination in the methylobacteria, for the extrachromosomal genes or for attack by bacteriophages. This lack of knowledge presumably reflects lack of work carried out rather than some special peculiarity of the organisms. Indeed, the wide range of bacteria capable of methane oxidation suggests transfer of this ability between unrelated organisms during their evolution.

The utilization of natural gas

Natural gas is a mixture of low molecular weight hydrocarbons, the proportions varying with the origin and with the processing methods. Although it may contain over 90 % methane, it is usually contaminated with ethane, propane, butane and other impurities which may amount to as high as 40 % of the gas (Davis, 1967). It is important, therefore, to consider what effect these higher hydrocarbons will have on the methane fermentation, especially since nearly all the results described previously have been carried out with pure methane. Mueller (1969b) has shown preferential utilization of methane from natural gas, but with a mixed population probably containing ethane-utilizers.

Leadbetter & Foster (1958, 1960) showed that although C_2—C_4 hydrocarbons cannot be utilized as the sole source of carbon and energy for growth by methylobacteria, they can apparently be cooxidized through the aldehyde to the corresponding acid; ^{14}C-labelling experiments also showed assimilation into cell carbon in presence of methane. How does this cooxidation of the C_2—C_4 hydrocarbons in natural gas affect the yield and, more important, the oxygen requirement and heat production figures? This is an important question, particularly if natural gas with a high level of higher hydrocarbons is to be used, but the answer is not yet available. Not all methylobacteria are able to oxidize ethane, propane and butane through to the corresponding acid; Patel et al. (1969) have found that Methylococcus capsulatus, and as shown here, most methylobacteria are unable to oxidize beyond the corresponding aldehydes. Methylobacteria of this type are likely to present a problem if natural gas is used since the aldehydes may well prove toxic if allowed to accumulate quite apart from the effect of this oxidation on heat production and oxygen demand. Possibly the answer lies in mixed cultures of a methylobacterium with a C_2—C_{14} alkane-utilizer.

Conclusions

Can anything useful be said about the feasibility of a commercial process for single-cell protein production from natural gas based on our present knowledge? Klass et al. (1969) have attempted an analysis of the factors involved and have concluded that a process is required giving about 20 g./l. dry weight at a dilution rate of 0·1. Oxygen-demand problems might prohibit such a high output unless a pressurized fermentor is used, but some of the assumptions made are pessimistic. Certainly, if a process is to be developed, two previously unmentioned problems must be considered.

(a) An explosive mixture must be avoided. This problem is discussed by Hamer, Heden & Carenberg (1967) and the answer depends upon using a mixture of methane and oxygen outside the explosive range (i.e. outside 5–15 % by volume of methane in air).

(b) As methane passes in the gas stream through the fermentor, probably no more than half will be removed by the culture. Although a recirculation system can be devised, it would involve topping up with natural gas and with pure oxygen rather than air. Problems could arise by the accumulation of toxic qualities of the higher gaseous alkanes in natural gas if they cannot be removed at a sufficiently high rate.

In spite of the many difficulties outlined above, the author remains optimistic about the future development of the methane fermentation as a means of single-cell protein production. Should this optimism prove unfounded, there may still be a considerable potential for the production of metabolites or enzymes.

REFERENCES

BEERSTECHER, E. (1954). *Petroleum Microbiology*. Amsterdam: Elsevier.

BENNET, I. C., HONDERMARCK, J. C. & TODD, J. R. (1969). How B.P. makes protein from hydrocarbons. In *Hydrocarbon Processing*. Gulf Publishing Co.

BOS, P. & DE BOER, W. E. (1968). Some aspects of the utilization of hydrocarbons by yeast. *Antonie van Leeuwenhoek* **34**, 241.

BROWN, L. R., STRAWINSKI, R. J. & McCLESKY, C. S. (1964). The isolation and characterization of *Methanomonas methanooxidans* Brown and Strawinski. *Can. J. Microbiol.* **10**, 791.

BRYAN-JONES, G. & WILKINSON, J. F. (1971). The utilization of dimethylether by methylobacteria. *J. gen. Microbiol.* (In the Press.)

CARDINI, G. & JURTSHUK, P. (1968). Cytochrome P-450 involvement in the oxidation of *n*-octane by cell-free extracts of *Corynebacterium* sp. strain 7E1C. *J. biol. Chem.* **243**, 607 D.

COTY, V. F. (1967). Atmospheric nitrogen fixation by hydrocarbon-oxidizing bacteria. *Biotechnol. Bioengng* **9**, 25.

DAVIES, S. L. & WHITTENBURY, R. (1970). Fine structure of methane and other hydrocarbon-utilizing bacteria. *J. gen. Microbiol.* **64**, 227.

DAVIS, J. B. (1967). *Petroleum Microbiology*. Amsterdam: Elsevier.

DOSTALEK, M., MUNK, V., VOLFOVA, O. & PECKA, K. (1969). Cultivation of the yeast *Candida lipolytica* on hydrocarbons. I. Degradation of *n*-alkanes in batch fermentation of gas oil. *Biotechnol. Bioengng* **10**, 33.

DWORKIN, M. & FOSTER, J. W. (1956). Studies on *Pseudomonas methanica* (Söhngen) *nov.comb. J. Bact.* **72**, 646.

EROSHIN, V. K., HARWOOD, J. H. & PIRT, S. J. (1968). Influence of amino acids, carboxylic acids and sugars on the growth of *Methylococcus capsulatus* on methane. *J. appl. Bacteriol.* **31**, 560.

EVANS, G. H. (1967). Industrial experience-hydrocarbon fermentation. Massachusetts Institute of Technology Symposium on *Single Cell Protein*, p. 243. Ed. R. I. Matales and S. R. Tannenbaum. Cambridge, Massachusetts: M.I.T. Press.

EVANS, G. H. (1969). *Protein form Petroleum*. Fourth Petroleum Symposium, U.N. Economic Commission for Asia and the Far East.

FOSTER, J. W. (1962). Hydrocarbons as substrates for microorganisms. *Antonie van Leeuwenhoek* **28**, 241.

FOSTER, J. W. & DAVIS, R. H. (1966). A methane-dependant coccus, with notes on classification and nomenclature of obligate, methane-utilizing bacteria. *J. Bact.* **91**, 1924.

FUHS, G. W. (1961). The microbial degradation of hydrocarbons. *Arch. Mikrobiol.* **39**, 374.

HAMER, G., HEDEN, C. G. & CARENBERG. (1967). Methane as a carbon substrate for the production of microbial cells. *Biotechnol. Bioengng* **9**, 499.

HARRINGTON, A. A. & KALLIO, R. E. (1960). Oxidation of methanol and formaldehyde by *Pseudomonas methanica. Can. J. Microbiol.* **6**, 1.

HEPTINSTALL, J. & QUAYLE, J. R. (1970). Pathways leading to and from serine during growth of *Pseudomonas* AM1 on C_1 compounds or succinate. *Biochem. J.* **117**, 563.

HIGGINS, I. J. & QUAYLE, J. R. (1970). Oxygenation of methane by methane-grown *Pseudomonas methanica* and *Methanomonas methanooxidans. Biochem. J.*

HUMPHREY, A. E. (1970). Microbial protein from petroleum. *Process Biochem.* **5** (6), 19.

JOHNSON, M. J. (1964). Utilization of hydrocarbons by microorganisms. *Chemy Ind.* **36**, 1532.

JOHNSON, M. J. (1967). Growth of microbial cells on hydrocarbons. *Science, N.Y.* **155**, 1515.

JOHNSON, P. A. & QUAYLE, J. R. (1964). Microbial growth on C_1 compounds of oxidation of methanol, formaldehyde and formate by methanol-grown *Pseudomonas* AM 1. *Biochem. J.* **93**, 281.

KEMP, M. B. & QUAYLE, J. R. (1967). Uptake of [^{14}C]formaldehyde and [^{14}C]formate by methane-grown *Pseudomonas methanica* and determination of the hexose labelling pattern after brief incubation with [^{14}C]methanol. *Biochem. J.* **102**, 94.

KLASS, D. L., IANDOLO, J. J. & KNABEL, S. J. (1969). Key process factors in the conversion of methane to protein. *Chem. Engng Prog. Symp. Ser.* **93**, 72.

KLUG, M. J. & MARKOVETZ, A. J. (1970). Utilization of aliphatic hydrocarbons by microorganisms. *Adv. Microb. Phys.* **5**, (in the Press).

KO, P. C. & YU, Y. (1967). Production of SCP from hydrocarbons: Taiwan. From *Single Cell Protein*, p. 255. Ed. R. I. Matales & S. R. Tannenbaum. Cambridge, Massachusetts: M.I.T. Press.

LAWRENCE, A. J., KEMP, M. B. & QUAYLE, J. R. (1970). Synthesis of cell constituents of methane-grown *Methylococcus capsulatus* and *Methanomonas methanoxidans. Biochem. J.* **116**, 631.

LAWRENCE, A. J. & QUAYLE, J. R. (1971). Alternative carbon assimilation pathways in methane-utilizing bacteria. *J. gen. Microbiol.* (In the Press.)

LEADBETTER, E. R. & FOSTER, J. W. (1958). Studies on some methane utilizing bacteria. *Arch. Mikrobiol.* **30**, 91.

LEADBETTER, E. R. & FOSTER, J. W. (1959*a*). Incorporation of molecular oxygen in bacterial cells utilizing hydrocarbons for growth. *Nature, Lond.* **184**, 1428.

LEADBETTER, E. R. & FOSTER, J. W. (1959*b*). Oxidation products formed from gaseous alkanes by the bacterium *Pseudomonas methanica. Archs Biochem. Biophys.* **82**, 491.

LEADBETTER, E. R. & FOSTER, J. W. (1960). Bacterial oxidation of gaseous alkanes. *Arch. Mikrobiol.* **35**, 92.

LEBEAULT, J. M., ROCHE, B., DUVNJAK, Z. & AZOOLAY, E. (1970). Isolation and study of the enzymes involved in the metabolism of hydrocarbons by *Candida tropicalis. Arch. Mikrobiol.* **72**, 140.

LUDVIK, J., MUNK, V. & DOSTALEK, M. (1968). Ultrastructural changes in the yeast *Candida lipolytica* caused by penetration of hydrocarbons into the cell. *Experientia* **24**, 1066.

MCAULIFFE, C. (1969). Solubility in water of normal C_9 and C_{10} alkane hydrocarbons. *Science, N.Y.* **163**, 478.

MATALES, R. I., BARUAH, J. N. & TANNENBAUM, S. R. (1967). Growth of a thermophilic bacterium on hydrocarbons; a new source of single-cell protein. *Science*, *N. Y.*, **157**, 1322.

MCKENNA, E. J. & KALLIO, R. E. (1965). The biology of hydrocarbons. *Ann. Rev. Microbiol.* **19**, 183.

MILLER, T. L., LIE, S. & JOHNSON, M. J. (1964). Growth of a yeast of normal alkanes. *Biotechnol. Bioengng* **6**, 299.

MUELLER, J. C. (1969a). Fermentation of natural gas with a cyclone column fermenter. *Can. J. Microbiol.* **15**, 1047.

MUELLER, J. C. (1969b). Preferential utilization of the methane component of natural gas by a mixed culture of bacteria. *Can. J. Microbiol.* **15**, 1114.

MUNK, V., VOLFOVA, O., KOTYK, A. & DOSTALEK, M. (1969). Cultivation of yeast on gas oil. *Biotechnol. Bioengng* **11**, 383.

PATEL, R., HOARE, D. S. & TAYLOR, B. F. (1969). Biochemical basis for the obligate C-1 dependance of *Methylococcus capsulatus*. *Bacteriol. Proc.* **128**.

PETERSON, J. A., KUSUNOSE, M., KUSUNOSE, E. & COON, M. J. (1967). Enzymatic *w*-oxidation. II. Function of rubredoxin as the electron carrier in *w*-hydroxylation. *J. biol. Chem.* **242**, 4334.

PROCTOR, H. M., NORRIS, J. R. & RIBBONS, D. W. (1969). Fine structure of methane-utilizing bacteria. *J. appl. Bact.* **32**, 118.

RIBBONS, D. W. (1970). Stoichemistry of *O*-demethylase activity in *Pseudomonas aeriginosa*. *Fed. Euro. Biochem. Soc. Letters* **8**, 101.

RIBBONS, D. W., HARRISON, J. E. & WADZINSKI, A. M. (1970). Metabolism of single carbon compounds. *A. Rev. Microbiol.*

SHACKLADY, C. A. (1969). Further observations on the use of hydrocarbon grown yeast in pig and poultry feeds. *3rd Int. Congr. Global Impacts appl. Microbiol.*, Bombay.

SMITH, A. J., LOUDON, J. & STANIER, R. Y. (1967). Biochemical basis of obligate autotrophy in blue-green algae and thiobacilli. *J. Bact.* **94**, 972.

TANNENBAUM, S. R. & MATALES, R. I. (1968). Single cell protein. *Sci. J.* **4**, 87.

TRAXLER, R. W. & BERNARD, J. M. (1969). The utilization of *n*-alkanes by *Pseudomonas aeriginosa* under conditions of anaerobiosis. Part I. Preliminary observation. *Int. Biodeterioration Bull.* **5**, 21.

VAN DER LINDEN, A. C. & THIJSSE, G. J. E. (1965). The mechanisms of microbial oxidations of petroleum hydrocarbons. *Adv. Enzym.* **27**, 469.

VARY, P. S. & JOHNSON, M. J. (1967). Bacterial cell yields from methane. Abstract, 154th Meeting of American Chemical Society, Chicago.

WAGNER, F., KLEEMAN, T. & ZAHN, W. (1969). Microbial transformation of hydrocarbons. II. Growth constants and cell composition of microbial cells derived from *n*-alkanes. *Biotechnol. Bioengng* **11**, 393.

WANG, D. I. C. (1967). Cell recovery. In *Single Cell Protein*, p. 217. Ed. R. I. Matales and S. R. Tannenbaum. Cambridge, Massachusetts: M.I.T. Press.

WASLIEN, C. L., GALLOWAY, D. H. & MARGEN, S. (1969). Human intolerance to bacteria as food. *Nature, Lond.* **221**, 84.

WHITTENBURY, R. (1969). Microbial utilization of methane. *Proc. Biochem.* **4** (1), 51.

WHITTENBURY, R., DAVIES, S. & DAVY, J. F. (1970a). Exospores and cysts formed by methane-utilizing bacteria. *J. gen. Microbiol.* **61**, 219.

WHITTENBURY, R., PHILLIPS, K. C. & WILKINSON, J. F. (1970b). Enrichment, isolation and some properties of methane-utilizing bacteria. *J. gen. Microbiol.* **61**, 205.

WILKINSON, J. F. & MUNRO, A. L. S. (1967). In *Microbial Physiology and Continuous Culture*. London: H.M.S.O.

WOLNAK, B., ANDREEN, B. H., CHISHOLM, J. A. & SAADEN, M. (1967). Fermentation of methane. *Biotechnol. Bioengng* **9**, 57.

YAMADA, K., TAKAHASHI, J. KAWABATA, Y., OKADA, T. & ONIHARA, T. (1967). SCP from yeast and bacteria grown on hydrocarbons. In *Single Cell Protein*, p. 192. Ed. R. I. Matales and S. R. Tannenbaum. Cambridge, Massachusetts: M.I.T. Press.

EXPLANATION OF PLATES

These electron micrographs were kindly taken by
Dr S. W. Watson using cultures isolated in my department.

PLATE 1

(a) The type II Membrane system in '*Methylobacter vinelandii*'. The membranes are in the form of stacked discs, some cut to show the surface of the discs ($\times 41,500$).

(b) The type I Membrane system in '*Methylosinus trichosporium*'. The membrane pairs are much less regular and occur circumferentially in the cell. Poly-β-hydroxybutyrate granules also occur. The cell is probably in process of forming an exospore ($\times 50,000$).

PLATE 2

(a) Periodic structures in the cell wall of '*Methylomonas agile*' ($\times 232,550$).

(b) Periodic structures in the cell wall of '*Methylomonas albus*' ($\times 243,000$).

(c) Poly-β-hydroxybutyrate storage in '*Methylococcus* capsulatus' ($\times 55,000$).

PLATE 1

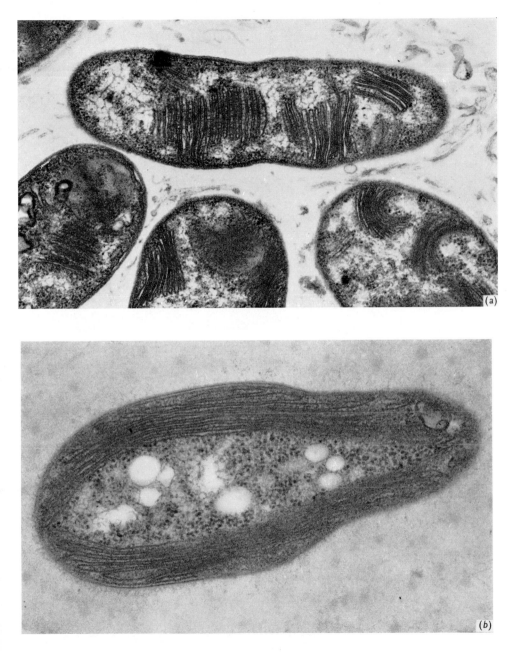

PLATE 2

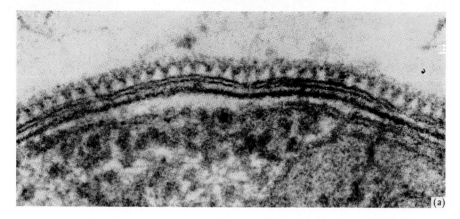

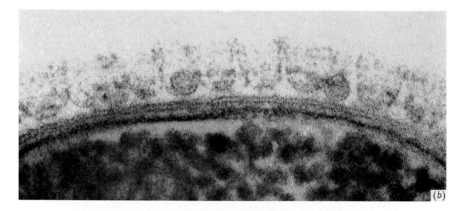

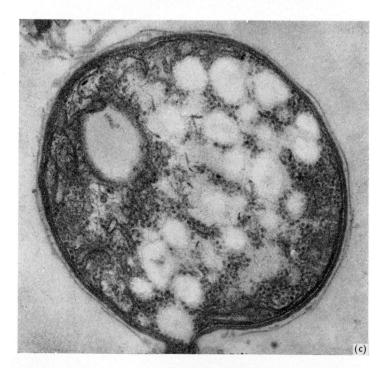

ALGAE AND LITHOTROPHIC BACTERIA AS FOOD SOURCES

W. A. VINCENT

Sherborne, Dorset

MICROBIAL FOODS IN GENERAL

Introduction: the limitations of single cell protein

Microorganisms in general offer an appealing contribution to the World supply of food because of their rapid growth rate and because of their ability to 'upgrade' the materials of their growth medium, with the net conversion of inorganic nitrogen to protein. If direct use of microbial cells as human food is to be envisaged, their digestibility and biological value must be high. This is not the case with present microbial foods in their whole-cell form. In addition, microbial cultivation is capital-intensive and highly sophisticated, needing a lot of capital and a high level of technology or at least well-trained personnel. It is of no use developing sophisticated techniques for eventual application in areas where there is no foreign-exchange capital to buy equipment and not enough trained personnel to run the plant; neither is it intelligent to urge the use of techniques based on those fossil-fuel substrates which have a limited distribution or a limited life. The North Sea gas deposits, for instance, are at the moment thought to have only enough reserves for 25 years of use at the predicted rates.

A more general problem which affects all microbial foods is that, to be eaten in large quantities (thinking of the amount which would furnish about 80–90 g. of protein daily), the cell walls must be removed because the human alimentary canal has not the enzymes to digest them. This is true whether we are dealing with walls in which the mechanical elements are composed of cellulose or of murein. The intact walls prevent digestion of most of the potentially valuable contents. In addition, the unabsorbed walls or contents are fermented in the large intestine producing not only flatus or diarrhoea but pressor amines deleterious to health.

If the walls can be broken down we can separate from them the cell contents including the desired protein. It may be necessary to separate protein from nucleic acid in view of the reputed effects of purine overload, but I can find little concrete evidence on this last point.

Fractionation of microorganisms

The above considerations point to the necessity of fractionation for direct use as human food of any of the microorganisms at present being considered. Fractionation is probably as difficult as the actual cultivation, and suffers from the same problems of investment required and sophistication. The problems of storage and distribution are increased because of the generally poorer keeping qualities of a fractionated product. However, in spite of these difficulties, it may be possible to produce a nutritious food which is still reasonably cheap after passing through both cultivation and fractionation stages. It has been claimed that normal economics do not apply in a World food shortage but, in fact, people die of starvation even in the developed countries because they have no money to buy food. The real truth is that any large-scale provision of food for areas which are chronically famine-ridden can only be made by establishing this supply on an economic footing. Every additional stage or additional sophistication in the manufacture or cultivation of a new food product makes less likely its eventual adoption as an important mass food simply because of the increased cost.

Algae contrasted with other microorganisms

Advocates of algae as a source of cheap protein have held up photosynthesis as a prime advantage of algae. In fact, although the carbon is provided 'free', a high price will have to be paid to perform the cultivation in such a way as to let in light and carbon dioxide. A high price may have to be paid for supplementary carbon dioxide and for agitation machinery if the highest yields are thought to be necessary. To some extent, a vicious circle may be created in which the high growth rates are necessary in order to justify the expensive plant which makes them possible. In terms of the cost of carbon used by the culture, it may be cheaper to grow yeast with its higher specific growth rate and to pay for the substrate used than to invest the money in special equipment for algal cultivation. This is certainly indicated as a harsh commercial reality in Japan where the major producer of algae (as a source of vitamins, growth factors for lactobacilli, and green colouring matter for a fermented drink) uses heterotrophic cultivation on a glucose-containing medium rather than photosynthetic growth (Enebo, 1969). From my personal experience it is relatively simple to grow cultures containing 30–40 g. dry weight/l. in heterotrophic growth of *Chlorella pyrenoidosa* strain 7–11–05, and the problems are the normal ones such as maintenance of sterility and gas transfer. To provide protein, however, yeast would be a more attractive proposition under such conditions because of its higher growth rate. Thus in general it may be concluded that the ability to fix carbon dioxide may not qualify algae as the thriftiest source of single-cell protein. To claim carbon dioxide fixation as an advantage only makes economic sense if there is in fact a shortage of fixed carbon. On a World scale we have molasses, starch, cellulose and hemicellulose, sulphite liquor, whey and various vegetable wastes which are not even completely used at the moment, not to mention fossil fuels.

Is there then a need for more fixed carbon? The answer lies in the distribution, evenness of supply and ready fermentability of these sources of carbon, coupled again with the technical sophistication and investment necessary to exploit them. The geographical distribution of fresh and fossil fermentable materials is not ideal. The World is divided by political barriers and, in any case, transport is prohibitively expensive. It may be that to use local mineral resources to the full we have to construct new food-yielding ecosystems with the most efficient carbon fixer available plus a nitrogen-fixation step integrated into the system (although the nitrogen-fixation step might be chemical). Locally, the use of fossil fuels plus inorganic nitrogen to produce yeasts or bacteria might be the best solution. Paradoxically this seems to be happening where the food supplies are already good, simply because of the relative abundance of investment capital and the advanced technology available. Elsewhere, with a surplus of fermentable carbohydrate material, a microbial upgrading such as suggested for cassava might be ideal. Evidently local conditions define the optimum solution. Where there is a deficiency of both fossil fuels and fermentable materials – and many areas of the World suffer from a calorie as well as from a protein deficiency – the carbon-fixation step is the limiting one in local food supplies. Given the fixed carbon, many solutions are possible. We must thus look to the most efficient carbon fixer to give a better all-round food supply. This naturally leads us to consider not only algae but higher plants, the basis of conventional farming, which in fact have a huge development potential in terms of possible increases in yield. The alternatives offered are summarized in Fig. 1.

Algae contrasted with higher plants

The highest net photosynthetic efficiency of higher plants recorded in experimental field plots is about 4 % conversion of the incident light energy. The average in normal farming practice is about 1 % (Duckham, 1968). Photosynthetic algal growth is, broadly speaking, subject to the same limiting factors as is that of higher plants. It is no surprise, therefore, that the peak yield-rates of dry matter from algal lagoons when properly managed (Enebo, 1969) are about the same as the yields obtained for a short time at the height of the growth season from higher plants (N. W. Pirie, personal communication; Vincent, 1969). For both these systems, the highest actual weight harvested may be calculated by postulating a kind of continuous culture in which the removal of

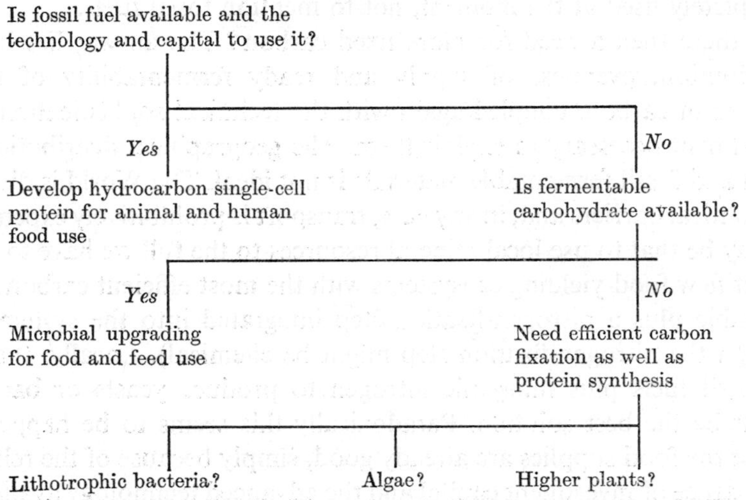

Is fossil fuel available and the
technology and capital to use it?

Yes | *No*

Develop hydrocarbon single-cell
protein for animal and human
food use

                         Is fermentable
carbohydrate available?

Yes | *No*

Microbial upgrading
for food and feed use

                         Need efficient carbon
fixation as well as
protein synthesis

Lithotrophic bacteria?   Algae?   Higher plants?

Fig. 1. Alternative protein sources. The protein source to be obtained at minimum cost in any particular region is indicated by the choice network.

freshly grown material is so adjusted as to keep the optimum mass of photosynthesizing tissue per unit area. Myers (1969) discusses results showing sharp optima in productivity of continuous-culture photosynthetic algal systems when considered as a function of dilution rate. At suboptimal dilution rates, the culture is dense and light penetration is poor, so that losses due to respiration are high. This is true even for continuously illuminated cultures because of the limited volume of the culture which will receive light at intensities above the compensating intensity at which photosynthesis balances respiration. At supraoptimal dilution rates, the population falls as some factor other than light becomes the limiting one, and the light utilization becomes less efficient so that overall output rates are less even with much less net respiratory loss.

The essential point in considering lagoon culture of unicellular microalgae is that the optimum culture-density for photosynthetic yield is much lower than the optimum density for economic harvesting: an intermediate density will represent the overall cost minimum for algal mass freed of culture medium. This is so mainly because the cost of centrifuging depends on the volume of liquid handled rather than on the mass of heavy phase removed (Golueke & Oswald, 1965). Ragonese & Williams (1968) found that the logarithmic (i.e. low population, not light limited) and linear (higher population with light limitation) phases of photosynthetic algal growth in a batch culture, grown from a small inoculum, could both be described by a single equation based on the Einstein law of photochemical equivalence. This is only another way of saying that the algal growth is proportional to the light absorbed. One tacitly ignores the respiratory loss in this system but, in spite of this simplification, the model they derived fitted actual growth curves fairly closely, even at high cell densities. Thus, at light intensities between the saturation and compensation points, the production rate depends, by definition, on the light intensity. There is a region above the light saturation intensity for algae where the higher plant has a definite advantage because of the optical properties of the plant itself, of its leaves and of their orientation and partial mutual shading. The overall effect at high light intensities is to attenuate the light and diffuse it over a larger area of leaf than the superficial area of the ground occupied by the plant. This is especially true of plants with vertical leaves (Pirie, 1969). Static algal cultures, at the same intensity of incident light, tend to absorb it in the superficial layers, even suffering damage in the process. Turbulent algal cultures make better use of bright light, but the energy required to agitate the culture adequately more than outweighs the carbon-fixation advantage. It may be remarked in passing that those plants which use the C_4 dicarboxylic acid pathway of Hatch & Slack (see Slack, Hatch & Goodchild, 1969) have an additional advantage in their high efficiency of light utilization and their high values for light saturation intensity.

The main points in favour of algal systems depend on their growth kinetics and their protein content. The higher specific growth rate than that of higher plants is presumably a function of the transport problems inside the structures of higher plants and also the amount of the photosynthetic product which is used to form mechanical tissues rather than photosynthetic ones.

The algal population reaches the 'ground-cover' state in a well-managed lagoon more quickly than does a field or greenhouse crop of higher plants from a comparable inoculum. The optimum population-density may be established or re-established more quickly for algae than for higher plants in any 'seasonal' situation where the culture has to be closed down periodically, or where the culture cycle is disrupted by a drought or a flood. Both algal culture growing photosynthetically and field crops of higher plants exhibit growth curves which flatten out as the population rises towards one which is using all the available light or carbon dioxide. At low population density, there is little interaction between organisms and the kinetics are those of the logarithmic phase, with the doubling time determined by the specific growth rate for isolated individuals. The instantaneous growth rate thus depends only on the mass already present (ignoring the effects of non-photosynthetic parts of the higher plants). A simple exponential relation models the growth in this phase:

$$P_t = P_0 e^{\mu t}$$

where P_0 = the size of the original population, P_t = that at time t, μ = specific growth rate $(\mathrm{d}P/\mathrm{d}t)(1/P)$.

We are ignoring a lot of modifying factors such as the effect of advancing differentiation on the higher plant and the increasing respiration loss as mutual shading increases. Both of these factors introduce negative terms that are dependent on values of P on the right-hand side of the equation.

As soon as the population density is such that mutual shading occurs, the growth-rate constant is affected and exponential kinetics cease to apply, the culture entering the phase where the behaviour can be described by the Einstein equation of photochemical equivalence (see Calvert & Pitts, 1966) in which no term appears for population density:

$$\mathrm{d}P/\mathrm{d}t = kIa,$$

where Ia is light absorbed, and k is a constant. This equation can only relate to thin layers of algae and to higher plants in which much of the tissue is photosynthesizing. The growth rate then relates to the surface area *of the tissue*. It follows that this equation must hold equally well for light and dense populations, and can be used experimentally if we can only measure the light stopped by the organisms themselves. The maximum yield-rate from optimum populations is probably therefore similar for both algae and higher plants given similar absorption co-efficients for their photosynthetic apparatus and similar photosynthetic efficiency. A maximum yield-rate for higher plants is quoted as 50 g./m.²/day (measuring total dry matter; Verduin, 1953) and the highest dry-matter yield for filamentous green algae is 50·4 g./m.²/day given for *Ulothrix* spp. by Hindak & Pribil (1968).

Photosynthetic efficiency is nearly similar in algae and in higher plants that use the 'Calvin' photosynthetic pathway. Plants using the

Hatch and Slack pathway are a different matter when we are considering high light intensities because they have a measurably higher efficiency at high light intensity as well as a higher light saturation intensity. The 'Hatch and Slack' plants will be ignored for the moment, and we will concentrate on the more common and more generally useful Calvin-cycle plants.

We may accept that the maximum cropping rate per unit area is determined by the light intensity and the carbon dioxide concentration. The dry-weight gains from similar areas of maximum populations of algae and of higher plants are generally very close in size. The point of confusion in reconciling this with the higher specific growth rate of algae is that the population of higher plants would have a higher mass

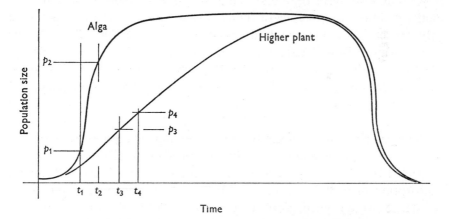

Fig. 2. Effect of growth kinetics on seasonal yields of algae and
higher plants. For explanation see text.

per unit area. Harvesting a similar proportion of each population, or perhaps an equal weight of protein from each population, would allow regrowth. The growth curves shown in Fig. 2 indicate what happens in such a case. It is evident that the population with the higher specific growth rate will replace the material removed more quickly. In the algal curve in Fig. 2, the amount removed $(p_2 - p_1)$ will take a time $t_2 - t_1$ to be replaced. In the case of the population of higher plants, regrowth of a much smaller amount of material $(p_4 - p_3)$ would take the same length of time $(t_4 - t_3 = t_2 - t_1)$. It is evident that excursions up and down the population scale will be made using repeated harvesting and regrowth, so that the difference in yield shown here will be multiplied several fold during the whole growing season.

The other important difference between algae and higher plants is the protein content. The average protein content of the strains of filamentous algae examined by Hindak & Pribil (1968) approached 50 %, some strains exceeding this value. This gives an appreciable advantage over leguminous plants (a favourable example from the point of view of protein content) even at equal dry-matter yield per unit area. Algae have, in general, at least a threefold advantage in terms of protein content. In terms of dry-matter yield, algae will give 2–10 times more depending on the length of the season and the individual plants considered. In terms of overall protein yield, the algal culture will therefore give between 6 and 30 times as much protein (see also Verduin & Schmid, 1966).

Fractionation of algal cells from the economic standpoint

Up till now we have been discussing weight gains from photosynthetic activity almost irrespective of the chemical nature of the product. The dry matter of the alga contains more protein than that of the higher plant. Ways exist to fractionate the higher plant (Pirie, 1961; Woodham, 1969) to produce a protein-rich fraction which may be used directly. In terms of the energy required for harvesting, transport, cell breakage and separation of the products, the algal source involves the handling of less total material for a unit weight of protein. There are two additional differences since the weaker tissues of the algae require less shearing energy for cell breakage, and higher recoveries of protein will be expected with less fibrous and other insoluble tissues for adsorption and absorption of proteins from solution. Fractionation procedures are in existence which yield a protein-enriched product from algal cells (Enebo, 1969; Mitsuda, Yasumoto & Nakamura, 1969).

It can be seen, however, that any method of fractionation will enormously increase the costs on a protein basis. The problem of extracting a 'clean' protein from a photosynthetic organism is considerably more difficult than that of extracting the relatively accessible storage protein from an oilseed meal. This is presumably an inherent difficulty because of the close spatial relationship between the mainly enzymic protein and the metabolically active lipid component. Since soya *protein* is increased in cost by a factor of about seven (a calculation based on the prices of soya processing companies; U.S.A., 1970) we should expect algal or higher-plant protein to be increased in cost by at least such a factor.

We may therefore conclude that, on economic grounds, algal and other microbial cells have a better chance of adoption if fractionation is not necessary. We will therefore consider the possibility of their direct use as human foods.

Use of algae and bacteria as human food without any fractionation
Acceptance

Tamiya (1959) discounted the problems of acceptance of algae by humans, but he was considering marine algae such as *Porphyra tenera* which are traditional foods in Japan. This alga is cultivated in a labour-intensive way and yields about 75 lb. dry weight/acre/year, so that it does not seem to be an economic proposition for more widespread use. Freshwater algae have much higher potential yields but they are much less acceptable, and even when fresh they tend to have the strong smell found in bodies of water undergoing eutrophication and thought to be due to partially oxidized algal lipids. This smell will be present in any real situation involving masses of algae and since, on a large scale, immediate metabolic arrest is by 'blanching' is not practicable, some further deterioration is inevitable. The natural texture of algal masses is also slimy and uninviting. These facts and the inherently low digestibility militate against the direct use of algae for human foods. I shall only mention briefly the use of whole bacterial cells. Since the comprehensive review by Lachance (1968), little of note has been published to change the opinion that whole bacterial cells provoke grave alimentary canal disturbances in any quantity. To use them to provide most or all of the dietary protein requirement for the adult human is not conceivable without fractionation to break and remove the indigestible cell walls (see also Anon, 1968*a*, *b*; Shapira & Mandel, 1968).

Digestibility and nutritive value

Fresh, dried or solvent-extracted *Chlorella* is digested hardly at all by humans (Mitsuda *et al.* 1969) or by other non-ruminant animals (Meffert & Pabst, 1963) although the isolated algal protein is easily digested by non-ruminants (Mitsuda *et al.* 1969). Its cellulose cell wall makes it completely resistant to animal enzymes except perhaps those of the snail. Theriault (1965) found that the β-carotene of *Chlorella* cells was not available to chickens (absorption was detected by egg coloration) unless the cells were ball-milled before use. Milder methods of handling dried cells were ineffective in making them permeable to β-carotene. Although this lipophilic compound might behave differently from proteins, Theriault's results seem generally applicable.

Apart from the digestive disturbances produced by undigested and therefore unabsorbed food, the net result of low digestibility, whether caused by the impermeable cell walls or by the nature of the protein itself, is a low net-protein utilization. It is doubtful whether algae could be induced to autolyse to any useful extent since, in normal cultures of *Chlorella*, there are persistent empty cell walls left in the medium which must represent a considerable loss of fixed carbon to the organism.

At first sight, *Spirulina* spp. seem to offer an exception to our generalization of indigestibility for algae, since they are reputed to have been used for generations as a staple protein source in periods of food shortage, in the Lake Tchad region from which the first isolates were recovered (Institut Français de Petrole, 1967). Some special factor in the harvesting, drying or cooking in this area might affect their digestibility in a favourable manner. However, in a study with rats using cells grown in France, it was found that even broken cells were nutritionally poor; protein digestibility was 76 %; biological value was 62 %; test animals fed *ad libitum* gained no weight whilst casein-fed controls gained 30 g. (Clement, Giddey & Menzi, 1967). This result throws open to doubt the digestibility of whole cells as well as of the isolated protein of *Spirulina*. Unicellular green algae in general are thought to be tolerated in the diet up to a generally agreed level of about 30 g. dry cells daily for a human adult, although 100 g. daily has been quoted. Even at the higher figure, the diet should still be in negative nitrogen balance if no other source of amino acids was used. Powell, Nevels & McDowell (1961) noted that gastro-intestinal upsets were produced by *Chlorella* in the diet for the first few days, but that these tended to subside if the diet was continued. No evidence of digestibility in humans has ever been produced, nor have we any real idea of the biological value of the isolated protein. Indications are given by the work with rats and by some experiments on pigs (Hintz & Heitman, 1967). In trials with ruminants and pigs (Hintz *et al*. 1966) a mixture of *Scenedesmus* and *Chlorella* was digested to an extent of 73 % by cattle but only 54 % by pigs. They concluded that algae in general had potential as a livestock feed. Various data exist indicating that methionine supplementation improves the value of *Chlorella* for rats, but no such work has been attempted for ruminants, nor would one expect supplementation to be markedly beneficial in the presence of adequate minerals in the diet.

To summarize, there seems to be no strong evidence that *Chlorella* and several other algal species are harmful in the diet of Man or animal, but there is a fair weight of evidence that for non-ruminants the algal cell, fresh or dried, is too resistant to allow the protein to be used to any useful extent. We are thus left with the alternatives of: (*a*) fractionation of the cells for use by humans and other non-ruminants; (*b*) direct use by ruminants; (*c*) use as a fermentation substrate. Of these possibilities we will consider only the first two.

Pretreatment for direct use as human food

Reasons for fractionation

The shortcomings of whole algal cells as food have been touched upon in the preceding section, but the explicit reasons for the desirability of fractionation if use as human food is mandatory are as follows:

(*a*) The indigestible cell wall must be removed or at least breached to allow the digestive enzymes access to the protein inside. As a secondary aim the indigestible part should also be decreased in quantity so as to limit intestinal trouble and also to make the product more concentrated, since it is almost axiomatic that as soon as we consider fractionation we must also consider transport to and from a central plant.

(*b*) The lipid fraction includes the chlorophylls and carotenoids, which give the strong characteristic colour, as well as the unsaturated triglycerides which produce the 'off' flavours when they are oxidized.

(*c*) The nucleic acid content is not very high since we are considering a relatively slow-growing microorganism when compared with most bacteria for instance (see Maynard Smith, 1969) but nevertheless some decrease in the nucleic acid content might be shown to be important for long-term use.

The mechanical and chemical problems of fractionation

The rupture of algal cells presents little problem on the laboratory scale. Most of the standard methods will work and the cell contents are readily released. The problems of preparing purified protein in bulk from such a source are twofold. First, the microbial contamination fostered by any process not operated under sterile conditions is such that rapid spoilage will occur under most climatic conditions. Sterile operation is out of the question from cost considerations alone. Any process to be acceptable must have either an environment not allowing bacterial and fungal growth or else a very short treatment time. The product must then be stabilized quickly by some treatment such as sterilization, drying, salting, or acidification. Secondly, as with most photosynthetic organisms, the physicochemical problem in fractionation really consists of taking apart the photosynthetic apparatus. Most of the enzymically active protein must be associated with the pigment in a functional and intimate liaison. The chlorophylls and other lipids tend to be released from the cell as natural or artefactual lipoproteins which are difficult to dissociate. If cells are broken down without an immediate and complete separation of the proteins and the lipids, enzymes present and still active will cause lipolysis and fat oxidation, producing the characteristic 'off' flavours.

It is possible to conceive of treatment methods which avoid these difficulties, such as solvent treatment to remove the lipids or at least to inactivate the enzymes; isopropyl alcohol is very active in this latter respect and would also act as a drying stage. The difficulty with all solvent treatments is that the protein becomes denatured and therefore difficult to extract, although the use of concentrated urea solution and other hydrogen-bond solvents will dissolve even denatured protein, e.g. from fish protein concentrate. Such an extraction however involves the cost of the extractant a swell as that of its eventual removal.

Current fractionation methods

Mechanical methods. Hedenskog *et al.* (1969) compared mechanical, enzymic and chemical methods of increasing the availability of the protein of *Scenedesmus obliquus* by degrading the cell-wall structure. Mechanical disintegration in a ball-mill (after Novotny, 1964) containing glass beads gave an *in vitro* digestibility by pepsin which was raised appreciably and in proportion to the number of cells seen by microscopic observation to be broken. Breakage was very efficient, and the bacterial count also fell by 95 % during the treatment, a fact beneficial for the keeping qualities of the product. This continuous-flow method is thus applicable to any bacterial and probably yeast protein source. It must inevitably suffer from contamination of the product by powdered glass. The flow rates used were of a useful size (100 l./h. = 2·5 kg. dry weight of algae) at least for pilot scale work.

As mentioned earlier, Theriault (1965) increased considerably the availability of the xanthophylls of heterotrophically grown *Chlorella pyrenoidosa* by ball-milling the dried cells. Evidently the proteins in such a preparation are denatured and not easily separable from the other constituents. Vacuum drying, boiling or roller drying have all been found to increase the availability of algal protein for rats (Meffert & Pabst, 1963). It is evident that although mechanical methods will increase the protein availability they do not permit any direct separation of the purified protein.

Chemical extraction methods. Hydrogen peroxide treatment (Reynolds, 1966) was tried (Hedenskog *et al.* 1969) without any appreciable release of protein. Sodium hydroxide extracts *Chlorella* cells effectively (Mitsuda, 1965), but protein degradation occurs. Mitsuda (1965) also tried autolysis induced by butanol but this has never been tried on a large scale. An untried possibility for chemical attack of the cellulose is the use of hydrogen peroxide and ascorbic acid in combination (Mac-Cabe & King, 1952) which is very effective in degrading some complex polysaccharides, for instance gastric mucin (W. A. Vincent, personal observation). A cuprammonium treatment of cell walls would not be acceptable for a food process, apart from cost considerations.

Various hydrogen bond-breaking agents have been used for protein extraction, including phenol (Westphal, Luderitz & Bister, 1952), formic acid (Lovern, 1966) and urea (Mitsuda *et al.* 1969). Urea seems to be the least impractical for food use. Recovery of protein from *Chlorella ellipsoidea* cells, using urea extraction, after a preliminary wall breakage step was up to 40 % expressed as protein in urea solution/total protein but further losses of protein would be introduced by the expensively complicated process of urea removal. Dialysis was used to remove the urea but this was only done on a small scale without any consideration of the additional engineering necessary for the recovery of urea. Recovery would have been necessary for economic large-scale use. Protein precipitation with *meta*phosphoric and trichloroacetic acids was tried as a way of separating protein and urea, and was fairly effective, but it is obviously not very applicable for an eventual production process. One of the interesting possibilities offered by urea or similar extraction of protein, is the direct texturization of the protein as it is precipitated by dilution, pH change or by heating. The change in pH could be induced by the use of urease, provided perhaps by crude soya meal. However the coagulation is induced, the textured product can subsequently be washed free of urea.

Enzymic lysis of algae. A cellulase preparation has been used on *Scenedesmus obliquus* cells but no appreciable improvement was seen in protein extractability (Hedenskog *et al.* 1969) when compared with spray-dried cells. This result seems to confirm the impression that the cell walls of at least some members of the Chlorophyceae contain not only fibrils of α-cellulose but a continuous matrix of lower molecular-weight polysaccharides, perhaps similar to hemicellulose (Northcote, Goulding & Horne, 1958). Evidently a preliminary physical or chemical treatment would be necessary to expose the cellulose to enzymic degradation. The gastric juice from snail (*Helix pomatia*) will attack isolated walls of *Chlorella pyrenoidosa* (Northcote *et al.* 1958), but it was found to be ineffective against whole cells (King & Shefner, 1962). In his review of techniques for enzymic breakdown of cells, Wiseman (1969) found no enzymic method active for whole algae. Undoubtedly a microorganism could be found and cultivated which would produce extracellular enzymes capable of lysing algal cells in an *in vitro* system.

Large-scale production of such enzymes could then be carried out in a similar way to that for detergent enzymes or antibiotics, in an established fermentation industry. Such a method has already been developed for bacterial and yeast cells (Kyowa Hakko Kogyo, 1968).

An alternative to the use of isolated enzymes is to ferment the algal mass with a microorganism. This has the possibility of improving the flavour as is already done in the fermentation of milk, soymilk and soybeans and, more recently, cassava. Various organisms have already been used to ferment algal masses (Becker & Shefner, 1963) including *Rhizopus nigrificans* (Hedenskog *et al.* 1969) although no nutritional follow-up seems to have been done.

Fat removal and decolorization. These two factors are considered together as a complementary pair. Removal of lipids, including pigments, generally gives a bland acceptable product which is also much more stable to both bacterial attack and to oxidation. Tamiya (1959) cites the use of ethanol to remove the bulk of the lipids and to stabilize the *Chlorella* produced by the Japan Nutrition Association on their 3,000 m.² lagoon. From Tamiya's figures one can calculate that the ethanol extraction about doubles the cost of the product on a protein basis, giving a final price of U.S.A \$2·2 per kg. This of course is the 1959 cost figure. Solvent costs comprise about 16 % only of the total process costs; use of isopropanol would therefore have little effect on the overall cost, although the process would probably be more efficient.

A solvent-extracted product tends to be more easily digested by non-ruminants, probably because the walls are made permeable to enzymes in the alimentary canal, just as isolated walls are more susceptible to snail-gut enzymes than are whole *Chlorella* cells (Northcote *et al.* 1958; King & Shefner, 1962). No reference has been found to work on combinations of solvent extraction followed by enzymic treatment. This may offer a promising line of approach to purified proteins from algae, although the cost would be high. A decolorizing method not involving solvent extraction is claimed by Chapman *et al.* (1965). According to this, a concentrated suspension of *C. pyrenoidosa* is subjected to treatment by artificial white light at 5,000 foot-candles (about 50,000 Lux) in the absence of carbon dioxide and with a high partial pressure of oxygen. The process involves the consumption of oxygen and takes 8–16 h., so it appears quite uneconomic other than for astronautical applications. The white-to-tan coloured powder produced by this process is bland and odourless, so this method points to cheaper variants such as hydrogen peroxide treatment which might give an acceptable product.

Total proteolysis and use of isolated amino acids. One possible method of avoiding many of the difficulties of acceptance of unconventional proteins is by effecting total proteolysis with subsequent recovery of free amino acids and their formulation into foods directly. Most of the problems of acceptance of amino acid mixtures have now been overcome (Winitz *et al.* 1965; Lachance & Berry, 1967). Digestion and absorption would present no problem for such a food. The technique would be equally applicable to any protein source, microbial or otherwise, and the mass production of proteolytic enzymes already renders them cheap enough to consider for use on such a scale.

Texturization and presentation

Most microbial and vegetable protein materials tend to be unacceptable as food in their crude form. Even for those with a completely bland taste and absence of offensive colour, the consistency, being that of an impalpable powder, can only be made satisfactory by mixing the supplement into a familiar food at a relatively low concentration. Texture can be imposed on such products by spinning or extrusion techniques, more or less as practised in the textile industry, using coagulation by heat, chemical precipitation using a binding agent, or other method of preserving the form imposed on the material. Boyer (1954) described the essential features of the spinning process which can be used to produce finely fibrous textures of varying degrees of internal orientation and hence of chewiness. He also described a means of binding the fibres together in meat-like structures and the application of various simple flavouring and colouring procedures and ways of building up fat-containing macrostructures to simulate different types of meat. This has since been done with casein, soya protein, wheat gluten, peanut protein, microbial protein and combinations of these materials.

In general, spinning of fibres can only be done with native proteins which have been previously separated at a minimum of about 80 % purity. The purification and spinning steps increase the cost. Soya protein, for example, costs about 15 times as much in the spun fibre form as in the form of flour still containing polysaccharide.* This increase in price would be still larger for proteins which are difficult to isolate or texturize.

A texturizing process for less refined materials, containing less protein and not necessarily completely soluble, depends on the extrusion

---

* Soy grits cost 3c./lb., therefore at 44 % protein, the protein is 6·8c./lb. Spun protein is 100c./lb. with 80 % protein, so protein is 125c./lb. Price increase on a protein basis is therefore × 18·4. These are 1969 season prices in U.S.A., average of four sources.

through a die of a slurry or paste of material which is at some stage of its passage heated so as to denature that part of the protein which is in solution. A preliminary 'plasticizing' treatment, such as intense mixing at high pH values, provides a consistency suitable for extrusion and liberates some protein into the suspending medium. This type of treatment seems to be equally valid for crude protein sources such as soy flour (Mustakas, Griffin & Sohns, 1966) as for pasta foods such as spaghetti. Other methods which work with the minimum of refining of the material beforehand are rolling, tearing or slicing of doughy mixtures to produce flakes or sheets, shreds or slices of product which can afterwards be cooked in oil, by steam or microwave techniques, to produce variously shaped, more-or-less puffed forms. The simple extruded form is familiar as that used for many pet foods, while the puffed forms are beloved of the snack food makers.

Table 1. *Texturizing methods possible with algal preparations using solvent-extracted or bleached cells*

Dispersion	Coagulation	After Treatment
Water and gelifiant (e.g. alginate, carrageenate)	Addition of Ca^{2+} and/or change in pH value	Drying. Stable to heat but not to high pH values
Agar solution (hot)	On cooling	Not stable to heat
Urea solution	On heating	Needs washing
In soy protein solution	On heating	Stable and probably completely digestible

If inadequate amounts of protein or starch are released into solution during the preparation of dough or slurry before the texturizing step, it is obvious that some additional binding agent will be needed which either gels on heating, by combination with another component of the mixture, or reversibly on cooling, although this last would be of limited application. The application of such techniques to microbial proteins is beset with various problems, apart from the initial extraction of the protein. The material obtained in solution may not be very easily coagulable. Evidently such methods have been tried with yeasts, bacteria and other organisms, and various products obtained from them. The particular problem with algae is that no amount of texturizing will render acceptable an algal food unless the colour is first removed. The summary of texturizing methods in Table 1 therefore only refers to preparations which have been so treated. For solvent-extraction procedures, this means that the protein is denatured and therefore insoluble except by hydrolysis or dispersion in a hydrogen-bonding solution such as concentrated aqueous urea. Bleached cells (Chapman *et al.* 1965) would probably behave in the same way. Protein extracted by the method of Mitsuda *et al.* (1969) in a concentrated urea solution can generally be textured by cooking extrusion in the normal way, but evidently the urea must be washed out of the solidified product afterwards. Alternatively the urea-extracted protein could be freed of urea beforehand and used in its amorphous insoluble form, with a binding agent to impose a structure on it during extrusion.

Nutritional value of algal proteins when used directly by non-ruminants

Extraction of the protein or treatment of the cells so as to break their walls or increase their permeability will, as expected, increase the digestibility of the protein. Roller drying raises the digestibility of *Scenedesmus obliquus* from 11·1–33·4 % up to 70–75·3 % in rats (Meffert & Pabst, 1963) whilst the extracted protein is apparently 85·6–87·4 % digestible (Tamiya, 1955). For *Chlorella* spp., fresh cells have an *in vitro* digestibility by trypsin of 46·2 %, dried cells have 62·5–65·5 %, decolorized cells have 75·1 %, and the extracted protein (not guaranteed to be *all* the protein) is apparently 85·6–87·4 % digestible (all from Mitsuda, 1965). Given the preliminary requirement of cell breakage, the protein itself seems to be easily broken down. Its biological value seems to be mediocre in the absence of amino acid supplementation. Table 2 compares the protein efficiency ratios of several common proteins with isolated proteins from *C. vulgaris* and *C. pyrenoidosa* (Johnson, 1967; Fink, 1955).

Table 2. *Protein efficiency ratios (weight gain/wt. of available protein used)*

Whole egg	3·5	*Chlorella vulgaris*	1·84
Fish flour	3·0	Dried brewer's yeast	1.69
Dried skim milk	2·83	Peanut meal	1·34
Whole milk	2·7	Wholewheat bread	1·1
Chorella TX 7-11-05	2·17	Cereal flour	0·03
Soy flour	2·04	Gelatin	Negative

Algae, in general, have like other microorganisms a rather low methionine content. Reliance on protein connective tissue, skin and muscles gives man a large demand for dietary sulphur-containing amino acids which are not provided by microorganisms which have other structural arrangements. Supplementation of microbial food sources with fish flour, sesame meal or synthetic amino acids is possible but is not logistically a very elegant solution to the problem. If we consider feeding ruminant animals rather than Man, the problems largely disappear because ruminants have less stringent requirements for particular amino acids, rumen microorganisms doing most of the necessary synthesis.

Use as animal feed

One may assume from the nutritional findings for non-ruminants that ruminant animals would find algae a valuable food source without the need for preliminary protein extraction. Compression of an algal paste into cakes or other palpable form would probably be beneficial, as would the inclusion therein of supplementary minerals and perhaps of an additional nitrogen source such as urea. The use of partly depolymerized cellulose as a support matrix for unicellular algae has been considered, forming a sort of artificial grass (Toyama, 1969). Yields of protein obtained from various large-scale algal and conventional farm management methods are given, for comparison, in Table 3.

Table 3. *Productivity: actual yield figures, protein/unit area/season*

From Institut Français de Petrole (1967); Tamiya (1959); N. W. Pirie 1969.

Protein source	Yield (dry weight of protein kg./ha./yr)
Spirulina platensis	24,300
Chlorella pyrenoidosa (Emerson)	15,700
Clover leaf	1,680
Grass	670
Peanuts	470
Peas	395
Wheat	300
Milk from cattle on grassland	100
Meat from cattle on grassland	60

The figures given for meat and milk yields are those cited by Tamiya (1959) and are evidently a little low by today's standards for Europe. However, the difference in protein yield between grassland and algal cultivation indicates that much higher yields of animal protein could be obtained using algae instead of grass as the primary food source. The data in Table 4 have been calculated in order to compare the known meat and milk yields with those we would expect on substituting algae for grass. The weight yields of algal protein were obtained from large lagoon cultures in the cases of *Spirulina platensis* (Institut Français de Petrole, 1967) and of *Chlorella pyrenoidosa* (Tamiya, 1959). The figures for filamentous algae are extrapolations to large scale of the results of Hindak & Pribil (1968), using an average yield figure amongst the strains of algae they used. The algal protein yields are in each case converted to hypothetical yields of beef protein by multiplying by 0·09 which is the factor we find linking the grass and associated beef protein yields. Other factors would come into play in determining the actual yield, such as the digestibility of the algae and their content of other nutrients, but the figures given are probably a reasonable approximation.

Table 4. *Extrapolated animal protein yields, based on the use of various feeds*

Primary protein source	Yield of protein (kg. dry wt/ha./yr)	Beef protein obtained using such a primary protein source (kg./ha./yr)
Grassland	670	60
Filamentous algae* (average of results of Hindak & Pribil, 1968)	20,000	1,800
Spirulina platensis	24,300	2,180
Chlorella pyrenoidosa strain TX 7–11–05	44,700	4,030

* The filamentous algae referred to are: *Hormidium, Ulothrix, Uronema* and *Stigeoclonium* spp.

The production of animal protein using almost any of the known forms of intensive algal cultivation would thus increase the productivity of a farming system yielding a more nutritious and more acceptable food than is algal protein itself. Several other algal species offer even higher potential yields, but under conditions difficult or expensive to realize in practice and also with the harvesting problems inherent in the use of unicellular algae. These other species are listed in Table 5, but no suggestion is made that they are practically realizable food sources.

Table 5. *Extrapolated yields of some algae, based on laboratory growth*

Species	Optimum temperature (°C)	Kg. dry protein produced ha./yr
Chlorella pyrenoidosa (Emerson)	25	15,700
Filamentous algae (Hindak & Pribil, 1968)	20–36	20,000
Spirulina maxima	27	24,300
Chlamydomonas mundana	33	41,500
Chlorella pyrenoidosa TX 7–11–05	30	47,000
Synechococcus lividus	70	51,500
Anacystis nidulans	41	59,200

It appears at first sight that the use of some of the higher-yielding species would increase still further the possible animal protein yields. The range of temperature optima is also very useful in that a series of different algae could be used during a season with a changing mean temperature. These 'extrapolated' yields, however, are not proven by any large-scale work and would probably not be achieved in full. The figures given are derived from the results of Hoogenhout & Amesz (1965) for laboratory growth rates. The laboratory figure was then multiplied for each algal species by a factor derived by considering *Chlorella pyrenoidosa* (Emerson strain) which has a well-documented *in vitro* and lagoon performance. Probably limitation of light penetration and carbon dioxide transfer would reduce the yields to something much nearer to the figures found for the less exacting algae. Use for animal feeding would still give high protein yields however, but the increase would probably not be justified when the additional costs of culturing unicellular algae, as contrasted to filamentous algae, were considered. We must therefore consider in detail the costs of algal production by different methods.

Cost in relation to culture method
Existing cost estimates

The relative costs of different methods of algal production and harvesting have been examined elsewhere (Fisher, 1955; Verduin & Schmid, 1966; Golueke & Oswald, 1965; Vincent, 1969). Fodder yeast grown on sulphite liquor costs about 22 U.S. cents per kilogram of dry cells. This figure should be borne in mind when discussing other sources of protein. One could estimate roughly that the cost of dry protein would be about 80 U.S. cents per pound from this source, assuming that 50 % of the dry weight turns up as extracted protein and that the cost is doubled by the extraction procedure. Quotations for algal cells vary between 17 cents and U.S. $1 per pound, i.e. 38 cents and $2·2 per kilogram (Tamiya, 1959; Fisher, 1955). One could take a very rough average estimate of 50 cents per pound of dry algae or $4·4 per kilogram of dry extracted protein, assuming an average composition of the cells and a doubling of cost by the extraction process. For unfractionated cells one would assume a cost of $2·2 per kilogram of protein, but such a figure would only be truly valid for ruminant animals which render fractionation unnecessary, and the figure is a very high one for an animal feed concentrate. Table 6 compares this cost with that of other protein sources.

Table 6. *Comparative costs of different protein sources*

Material	Price/kg. (U.S. cents)	Protein on dry matter (%)	Cost in U.S. cents per kilogram protein
Soy grits	6·6	44	15
Fish protein concentrate	28	85	33
Petroleum yeast	25 (projected)	60	41
Textured soy flour (extruded)	17·6	44	44
Feed yeast cells	22	50	88
Untreated *Chlorella* *pyrenoidosa* (Fisher, 1955)	44	50	88
Spun soya protein	220	80	275
Decolorized *Chorella* *ellipsoidea* (Tamiya, 1955)	247	50	494
Lean beef (U.S.A.)	110	17	650

Basis for cost comparisons

The first comment on the data in Table 6 must be the very high price of algal protein, even in unfractionated cells. In simple cost terms, the production of *Chlorella* cells would not be competitive with that of soya protein. The price of soya protein is low partly because the oil, which until recently was the main reason for growing the crop in the U.S.A., covers production overheads and the soy meal was sold at an almost nominal price so that it would all find a market. The demand for meal and the protein extracted from it is now coming up to the same level as the sale of oil so that this situation could change in the near future. In terms of land area required, the production of soya or of beef demands a much greater area of land than does the same weight of algal protein. The land areas involved may be judged from Table 7, which takes the data of Table 3 converted into the land surface required to support one person in protein assuming an intake of 80 g./day and an equal protein efficiency ratio for all the sources. This gives a protein requirement of 29·2 kg./year.

Table 7. *Area needed to provide protein for one person using different sources*

Protein	Yield (kg./ha./yr)	Area yielding 29·2 kg./yr (m.^{2})
Meat from cattle on grassland	60	4,870
Wheat	300	970
Clover leaf protein	1,680	174
Chlorella pyrenoidosa	15,700	17·5
Spirulina platensis	24,300	12

Components of the overall cost of algae

The land surface requirements and also the labour requirements of the algal cultivation systems are considerably diminished but the other cost items must be brought down if adoption of algae as a staple feed is to be made commercially acceptable whether in developed or in developing countries. Fisher (1955) gives a cost analysis of microalgal production in lagoons, including a breakdown of the investment necessary on the 100-acre scale (Table 8).

Table 8. *Investment in microalgal pilot plant as percentage of the total*
Data of Fisher (1955)

Area preparation and growth tubes (polyethylene)	6·85
Circulation equipment	13·7
Cooling facilities	36·6
Preharvesting equipment	7·45
Harvesting equipment	7·25
Central gas preparation and distribution equipment	5·00
General facilities, equipment and buildings	6·57
Engineering and contingencies (20%)	16·6

Harvesting costs

These figures show that a large proportion of the investment is in circulation, harvesting and cooling equipment. This is probably true for installations culturing *Chlorella* or *Scenedesmus* wherever they be because of the difficulties of harvesting small algae. Golueke & Oswald (1965) give a very thorough analysis of the costs of harvesting, and consider autoflocculation, flotation, microstraining, electrolysis, sonic vibration, filtration, chemical flocculation and other methods in addition to centrifugation. Autoflocculation was a cheap harvesting method but depends on weather conditions. Flocculation with organic or inorganic electrolytes is the next cheapest method, but it gives a contaminated product. Centrifugation is reliable and clean but needs heavy initial, maintenance and running costs. At a culture density of 4 g. dry weight/l., one needs to centrifuge at least 125 litres of culture for harvesting each pound of algae and proportionally more if the culture is more dilute or the extraction is not complete. Fisher's (1955) costing gives 13 % as the capital contribution of the purchase of centrifugation equipment and it is likely that the share in running costs is higher than this. Golueke & Oswald (1965) considered that centrifugation was the most expensive harvesting method but that it was not generally possible to find an alternative to it.

Easy-to-harvest species would avoid most of the difficulties listed above. The filamentous algae such as *Spirulina, Spirogyra, Vaucheria, Porphyra, Hormidium, Stigeoclonium* and *Uronema* spp. would all be cheap to grow and harvest compared to the true microalgae. The filamentous algae do not seem to suffer from protozoan browsing to the same extent as the unicellular species, probably simply due to their much larger size and thicker walls. The thick walls would not be a great disadvantage if the filamentous algae were used for ruminant feeding. They would all be simple to harvest, and could be collected by raking, coarse filters, and mowers, and the concept of 'culture volume handled' would not apply since the alga would be removed from the culture medium at the start of harvesting instead of during a secondary operation. The use of these algae enables us to adjust the population density to that which is optimum for light utilization without needing to attempt agitation to induce turbulence and intermittent lighting effects. By contrast the installation at Trebon in Czechoslovakia which has been used for growing *Scenedesmus* (described by Enebo, 1969) uses a layer of turbulent culture 50 mm. deep with a concentration of $1 \cdot 5$–2 g. of algal dry weight per litre. This would have an optical density value of about 2 at 660 nm. (W. A. Vincent, personal observations). One supposes therefore that the region of the culture with an optimum range of illumination intensity extends from about 5 mm. deep, where the 4,000 foot-candles of midsummer midday sunlight will be decreased to the saturation intensity of 400 foot-candles, down to about 15 mm. deep where the light intensity would be about 4 foot-candles, which is roughly the compensation intensity at 20°. Thus 90 % of the culture is in fact under non-ideal conditions, and 70 % is losing weight by respiration faster than it is gaining it by photosynthesis. In the laboratory, rapid stirring in order to get the cells in and out of the well-illuminated volume of culture has given an increase in productivity of about 60 %, but it is not possible to agitate as quickly as this on a large scale. The whole basis of turbulent agitation is to get algae to share illumination of greater than saturation intensity, but as the critical illumination times are of the order of a few milliseconds it is seen that very great turbulence would be needed, far more than is possible in the Trebon installation.

At a surface illumination intensity of 4,000 foot-candles we need the equivalent of a 30 mm. layer of algal suspension of optical density $1 \cdot 0$ to decrease the illumination to the minimum permissible without decreasing the net yield of photosynthetic product. To predict the optical density of a suspension of filamentous algae is very difficult, but Emerson & Lewis (1943) calculated that *Chlorella* absorbs about 30 % of the red light (660 nm.) incident on one cell. If we assume that our filamentous alga is about ten times the thickness of a *Chlorella* cell and that it is optically similar, we should expect it to absorb 97 % of the incident red light. This means that a layer of filaments two deep will give nearly optimum conditions. This is obviously an extremely rough approximation but it is unlikely to be in error by two orders of magnitude, so we could not have, ideally, either one layer more or less of filaments. The validity of such calculations must be suspect when made in general without specifying the species of alga and its 'absorption coefficient per cell' which will depend not only on species but on nutrition, light intensity during growth and other factors. What I hope to have shown is that it is possible to perform such calculations, and that any particular result would be of the same order of magnitude.

We have considered the cost contribution of harvesting and have shown that a good deal of the harvesting overheads can be avoided by using filamentous species of algae. Other expenses which could be avoided are those of cooling, circulating the culture violently, enclosing the culture, and also enriching the atmosphere with carbon dioxide.

Cooling and circulating the culture

With mesophilic algae, the necessity for cooling is introduced inevitably by the use of a covered culture. In the absence of evaporative cooling and with the 'greenhouse effect' introduced by any cover not transparent to long-wavelength infrared radiation (i.e. all covers, for all practical purposes), the temperature in summer sun must be many degrees above the local shade temperature. The necessity to cool the culture may involve refrigeration equipment or a lavish use of cool water. This could use more water than would be used to irrigate the same area for conventional farming, but of course the cooling water need not be of 'agricultural' purity; it could be brackish or salty or suffering from natural or man-made pollution. Circulation of the culture is necessary if the cooling apparatus is not to be very extensive and therefore costly, and the circulation itself can be expensive in terms of capital expenditure on equipment and in the running costs created by the energy requirement.

Poor designs have relied upon the rapid circulation of the culture to get any added carbon dioxide into solution and distributed throughout the culture, and upon heat dissipation through a conventional heat exchanger with mains water as the heat sink. An example is the design used by Tamiya for a pilot plant which ran for a considerable time near Tokyo. This plant gave about $1 \cdot 7-4$ g. dry wt/m.^{2}/day in winter. The weak points are the absolute dependence on the pumps, the poor mixing with carbon dioxide, and the heat exchange isolated in one part of the circuit. However one operates such an apparatus, there must be gradients of pH value, carbon dioxide concentration, temperature and of nutrients other than carbon dioxide. Conditions could not therefore be optimal everywhere, and probably were not optimal anywhere.

Open cultures with evaporative cooling and with direct conduction to a convectionally replaced air layer will generally have a lower temperature than enclosed cultures, and they will also have a reasonably rapid exchange of gases with the atmosphere. The rate of loss of water by this route could probably be decreased by the use of suitable agents such as long-chain aliphatic alcohols to obtain some sort of balance between water economy and temperature control. This system of temperature control has the advantage of eliminating gross temperature gradients in the system. The fall of temperature at night, which is unavoidable with open systems, is an advantage since this decreases the rate of respiration and hence the loss of photosynthetic product. Davis *et al.* (1953) showed slightly higher yields of *Chlorella* under otherwise constant conditions, including day temperature, if the night temperature was low.

Changes in the overall ambient temperature with season would be better accommodated by using a series of algal strains or species with suitable temperature optima for growth than by trying to stabilize the culture temperature at a value different from the ambient. It may be added in passing that well-designed installations can do much to smooth the diurnal ambient temperature excursion if they have an appreciable thermal capacity. The algae listed in Table 5 (p. 65) have a wide range of optimum growth temperatures and would allow a seasonal succession of species to follow the temperature change. In winter, with suboptimal temperatures as well as low light intensity, growth rates are very low and would probably not cover the running costs at any latitude higher than 50°, i.e. north of the English Channel in Europe. The best solution here would be to accept the fact that algal, like conventional, farming has a close season and that we must operate on a seasonal basis. This does not mean that algal cultivation will not work in high latitudes. We are comparing it with conventional farming which is also basically an application of harnessing photosynthesis, and in general seems less efficient, and that seems economic even to much higher latitudes than our own. In general, the algal culture system has a more marked advantage the shorter the season, because of the shorter lag period compared to that of higher plant crops.

This acceptance of decreased algal production costs by eliminating the covers over the culture and circulation machinery imposes another limitation, namely that it will not be possible to use supplementary carbon dioxide. This tends to make the pH value of the culture critically important. The concentration of dissolved carbon dioxide in water in equilibrium with air is very low. The carbon dioxide transport inside the body of water depends on liquid movement, and we want to minimize the energy expended on agitation. The use of more alkaline growth media creates a 'carbon dioxide reserve' which makes less critical the circulation rate of liquid. The same considerations lead us also to decrease the thickness of the layer of liquid as far as possible.

General conclusions on culture methods

Several guide-lines emerge from the above considerations.

(a) The potential productivity per unit area is very high. Capital investment and running costs can be minimized by simplifying the installation and perhaps accepting a compromise lower yield.

(b) A seasonal temperature variation can be accommodated by changing the species under cultivation. The high cost of cooling can thus be avoided and the yield can be kept up throughout the season.

(c) An uncovered growth area is much cheaper and minimizes the cooling problem by day and night. It also makes best use of the incident light.

(d) The harvesting costs can be cut drastically by using filamentous or mat-forming species which often grow well in open culture and resist protozoan browsing.

(e) The carbon dioxide transport problem is decreased by the use of alkaline media and by thin layers of culture medium. Calculations of illumination requirement show that only two layers of algal filaments should be superimposed.

(f) Harvesting with simple tools could be done by unskilled personnel: the shortage of technically trained assistants is acute in many countries and the lack of sophistication of such a process is a definite advantage.

An ideal culture system therefore seems to be a sort of lawn of algal filaments covered by only a thin film of medium, perhaps seeping down a slope. Some circulation must exist to distribute the mineral nutrients necessary, and pumping of the alga-free fluid to allow its passage through the layer of algae under gravity seems the cheapest way to effect this circulation. Harvesting from such a culture could be done by raking or clipping but some care must be taken to leave the algal filaments as evenly distributed as possible in the wake of the harvesting device. Cutting only the optimum proportion of the filaments at any one pass gives a sort of continuous culture, with the frequency and intensity of cut adjusted to give the optimum yield per unit area and yield per unit effort or a compromise between the two.

CONCLUSIONS

Conventional agriculture relies on the ability of higher plants to fix carbon. Algal cultures could fulfil the same role in our future crowded biosphere with the possibility of increased production. Direct use of algae as human food is at first sight attractive because it offers the shortest food chain and therefore the highest yield of food protein per unit of light energy or per unit area of growing space. The psychological, technical and nutritional problems involved in direct use are immense and would involve very sophisticated technology to produce what would probably never be a very satisfactory product. The psychological and nutritional problems can probably be solved by accepting that algal cells are not that generally suitable for use as human food but that they are probably ideal for feeding ruminants which in their turn can provide more balanced and acceptable proteins, albeit in lower yield.

Ruminant animals can thus be fed filamentous algae, the cultivation of which instead of unicellular microalgae permits the installation to be simplified and the running costs to be minimized. Possible improvements of beef yields, comparing algal and pastoral primary food sources, are of the order of 50-fold. This figure is derived from data in Table 7 (p. 67), assuming a 15 % conversion of *Spirulina* protein into beef protein. Large increases could similarly be made in the yields of other products from ruminant animals. Examples are from any domesticated and presently wild ruminants such as the cow, sheep, goat, water buffalo, kudu, giraffe, and elephant; milk from any of these animals as well as wool and hide from some of them. The intensive fodder consumption would naturally produce an equally intensive waste disposal problem as in present-day factory farming where the food input is, in effect, concentrated from a wide area. Recycling of wastes would probably be extremely advantageous in such a system and would decrease the size of the mineral inputs otherwise necessary, as well as minimizing the growth of algae in the effluent water, where they are undesirable.

I would like to mention humbly the mental stimulus I have received from past colleagues including Jiri Bartos, C. Giddey, J. F. Gordon, A. T. James, I. Malek, J. D. A. Miller, the late Peter Ochsenbein, N. W. Pirie, S. Pribil, G. Priestley and Brian J. B. Wood, without which this paper would not have been written.

REFERENCES

ANON (1968a). Space food from bacteria. *Spaceflight* **10** (4), 124.

ANON (1968b). Regenerative life support. *Nature, Lond.* **224**, 112.

BECKER, M. J. & SHEFNER, A. M. (1963). Technical Documentary Report, No. AMRL-TDR-63-115.

BOYER, R. A. (1954). U.S. Patent 2,682,466, 29 June.

CALVERT, J. G. & PITTS, J. N. (1966). *Photochemistry*. New York: Wiley.

CHAPMAN, D., CHRISTENSEN, G., PILGRIM, A., STERN, J. & ZOMMERS, I. (1965). U.S. Patent 3,197,309, 27 July.

CLEMENT, G., GIDDEY, C. & MENZI, R. (1967). Amino acid composition and nutritive value of the alga *Spirulina maxima*. *J. Sci. Fd Agric.* **18**, 497.

DAVIS, E. A., DEDRICK, J., FRENCH, C. S., MILNER, H. W., MYERS, J., SMITH, J. H. C. & SPOEHR, H. A. (1953). Laboratory experiments on *Chlorella* culture at the Carnegie Institute of Washington Department of Plant Biology. In *Algal Culture from Laboratory to Pilot Plant*, pp. 105–153. Ed. J. S. Burlew. Carnegie Institute of Washington Publication 600.

DUCKHAM, A. N. (1968). Biological efficiency of food producing systems in A.D. 2000. *Chem. Ind.* 903.

EMERSON, R. & LEWIS, C. M. (1943). The dependence of the quantum yield of *Chlorella* photosynthesis on the wavelength of light. *Am. J. Bot.* **30**, 165.

ENEBO, L. (1969). Growth of algae for protein: state of the art. In *Engineering of Unconventional Protein Production. Chem. Eng. Prog. Ser.* **65**, p. 80.

FINK, H. (1955). *World Symposium on Applied Solar Energy*. Tucson: Arizona.

FISHER, A. W., JR. (1955). Engineering for Algae Culture. *Proc. World Symp. Appl. Solar Energy*.

GOLUEKE, C. G. & OSWALD, W. J. (1965). Harvesting and processing sewage-grown planktonic algae. *J. Water Pollution Control Fed.* p. 471.

HEDENSKOG, G., ENEBO, L., VENDLOVA, J. & PROKES, B. (1969). Investigation of some methods for increasing the digestibility *in vitro* of microalgae. *Biotechnol. Bioengng* **11**, 37.

HINDAK, F. & PRIBIL, S. (1968). Chemical composition, protein digestibility and heat of combustion of filamentous green algae. *Biologia Plantarum* **10** (3), 234.

HINTZ, H. F., HEITMAN, H., JR., WEIR, W. C., TORELL, D. T. & MEYER, J. H. (1966). Nutritive value of algae grown on sewage. *J. Anim. Sci.* **25**, 675.

HINTZ, H. F. & HEITMAN, H., JR. (1967). Sewage-grown algae as a protein supplement for swine. *Animal Prod.* **9**, 135.

HOOGENHOUT, H. & AMESZ, J. (1965). Growth rates of photosynthetic organisms in laboratory cultures. *Arch. Mikrobiol.* **50**, 10.

INSTITUT FRANÇAIS DE PETROLE (1967). *Une Nouvelle Algue Alimentaire*. Publication No. 14.237 F. Rueil-Malmaison.

JOHNSON, M. I. (1967). Horizons of Industrial Microbiology. *Impact of Science on Society (UNESCO)* **17** (3), 269.

KING, M. E. & SHEFNER, A. M. (1962). Technical Documentary Report No. AMRL-TDR-62-91.

KYOWA HAKKO KOGYO CO. LTD. (1968). Method for lysing the cells of micro-organisms. British Patent 1,134,351.

LACHANCE, P. A. (1968). Single-cell protein in space systems. In *Single Cell Protein*, Eds. S. R. Tannenbaum and R. I. Matales. Cambridge, Mass.: M.I.T. Press.

LACHANCE, P. & BERRY, C. A. (1967). *Nutrition Today* **2**, 2.

LOVERN, J. A. (1966). General outlook for fish protein concentrates. In *World Protein Resources*, pp. 37–51. Ed. A. A. Altschul. American Chemical Society.

MacCabe, F. & King, H. K. (1952). *Edinb. med. J.* **58**, 377.

Maynard Smith, J. (1969). Limitations on growth rate. In *Microbial Growth*, pp. 1–14. 19*th Symp. Soc. gen. Microbiol.* Eds. P. Meadow and S. J. Pirt. Cambridge University Press.

Meffert, M. E. & Pabst, W. (1963). Possibility of using the substance of *Scenedesmus obliquus* as source of protein in rat balance trials. *Nutr. Diets* **5**, 235.

Mitsuda, H. (1965). *First Int. Congr. Fd Sci. & Techn., London*, 1965. *Proc., vol.* 2. Ed. J. M. Leitch. New York: Gordon and Breach.

Mitsuda, H., Yasumoto, K. & Nakamura, H. (1969). A new method for obtaining protein lysates from *Chlorella* algae, *Torula* yeasts and other microbial cells. In *Engineering of Unconventional Protein Production*. Proceedings of a Conference at Santa Barbara, California, 1969. Chemical Engineering Progress Series **65**, No. 93.

Mustakas, G. C., Griffin, E. L., Jr. & Sohns, V. E. (1966). Full-fat soybean flours by continuous extrusion cooking. In *World Protein Resources*, pp. 37–51. Ed. A. A. Altschul. America Chemical Society.

Myers, J. (1969). Genetic and adaptive physiological characteristics observed in the *Chlorellas*. In *Proceedings, IBP/PP Meeting on the Productivity of Photosynthetic Systems*, pp. 195–203. Czechoslovakia: Trebon.

Northcote, D. H., Goulding, K. J. & Horne, R. W. (1958). The chemical composition of the cell wall of *Chlorella pyrenoidosa*. *Biochem. J.* **70**, 391.

Novotny, P. (1964). A simple rotary disintegrator for microorganisms and animal tissues. *Nature, Lond.* **202**, 364.

Pirie, N. W. (1961). Progress in biochemical engineering broadens our choice of crop plants. *Econ. Bot.* **15**, 302.

Pirie, N. W. (1969). *Food Resources, Conventional and Novel*. London: Penguin.

Powell, R. C., Nevels, E. M. & McDowell, M. E. (1961). *J. Nutr.* **75**, 7.

Ragonese, P. & Williams, J. A. (1968). A mathematical model for the batch reactor kinetics of algal growth. *Biotechnol. Bioengng* **10**, 83.

Reynolds, L. W. (1966). U.S. Patent 3,288,613.

Shapira, J. & Mandel, A. D. (1968). Nutritional evaluation of bacterial diets in growing rats. *Nature, Lond.* **217**, 1061.

Slack, C. R., Hatch, M. D. & Goodchild, D. (1969). Distribution of enzymes in mesophyll and parenchyma sheath chloroplasts of maize leaves in relation to the 4-carbon dicarboxylic acid pathway of photosynthesis. *Biochem. J.* **114**, 489.

Tamiya, H. (1955). *Proc. World Symp. Appl. Solar Energy*. Arizona: Phoenix.

Tamiya, H. (1959). Role of algae as food. *Proc. Symp. Algology*, UNESCO, New Delhi.

Theriault, R. J. (1965), Heterotrophic growth and production of xanthophylls by *Chlorella pyrenoidosa*. *Appl. Microbiol.* **13**, 402.

Toyama, Nobuo (1969). Saccharification of woody wastes with *Trichoderma viride* cellulase. *Appl. Microbiol. Lab. Reports* **22**.

Vincent, W. A. (1969). Algae for food and feed. *Process Biochemistry* (June), p. 45.

Verduin, J. (1953). A table of photosynthetic rates under various conditions. *Am. J. Bot.* **40**, 675.

Verduin, J. & Schmid, W. E. (1966). Evaluation of algal culture as a source of food supply. *Devs ind. Microbiol.* **9**, 205.

Westphal, O., Luderitz, O. & Bister, F. (1952). Über die Extraktion von Bakterien mit Phenol/Wasser. *Z. Naturforschung* **7**b, 148.

WINITZ, M., GRAFF, J., GALLAGHER, N., NARKIN, A. & SEEDMAN, D. A. (1965). Evaluation of chemical diets as nutrition for man in space. *Nature, Lond.* **205,** 741.

WISEMAN, A. (1969). Enzymes for breakage of microorganisms. *Process Biochem.* May, p. 63.

WOODHAM, A. A. (1969). New Horizons for chemistry and industry in the 1990's. *S.C.I. Symposium at Lancaster,* 7–11 July.

MICROBIAL PRODUCTION
OF FOOD ADDITIVES

ARNOLD L. DEMAIN

Department of Nutrition and Food Science
Massachusetts Institute of Technology
Cambridge, Massachusetts 02139, U.S.A.

INTRODUCTION

For many years microbial products have been used to enhance the quality, appeal, and availability of food and drink. A list of products which have been or are being used in the food and feed industries is shown in Table 1. The reason for this impressive use of microbes is simple: microorganisms have an amazing ability to utilize cheap sources of carbon and nitrogen to overproduce valuable low and high molecular weight metabolites. One may ask what is meant by the term 'overproduction'. Some define it as merely the excretion of an intracellular metabolite into the surrounding medium. I would propose a more stringent definition, at least for primary metabolites: overproduction is the accumulation of a metabolite, either intracellularly or extracellularly, to a level where (based on the total volume of broth) it exceeds by at least one order of magnitude the normal concentration required by fastidious microorganisms for optimal growth. The data in

Table 1. *Fermentation products used in the food or feed industries*

Class	Examples
Alcohols	Ethanol
Amino acids	Glutamic acid, lysine, threonine
Antibiotics	Nisin, tylosin, tetracyclines, penicillin, streptomycin, bacitracin
Antioxidants	Isoascorbic acid
Colouring agents	β-Carotene
Enzymes	Amylase, glucamylase, invertase, lactase, maltase, pectinase, cellulase, hemicellulase, protease, lipase, glucose oxidase, catalase
Nucleotides	5′-Guanylic acid, 5′-inosinic acid, 5′-xanthylic acid
Organic acids	Acetic acid, propionic acid, succinic acid, fumaric acid, lactic acid, malic acid, tartaric acid, citric acid, gluconic acid
Plant growth regulators	Gibberellic acid
Polyols	Glycerol, mannitol
Polysaccharides	Xanthan polymer
Protein	Single-cell protein
Sugars	Fructose, sorbose
Vitamins	Riboflavin (B_2), cyanocobalamin (B_{12})

Table 2. *Overproduction of some primary microbial metabolites*

Product	Growth requirement* (mg./litre)	Production† (mg./litre)	Ratio: $\dfrac{\text{Required}}{\text{Produced}}$
Glutamic acid	300	60,000	$2\cdot0 \times 10^2$
Lysine	250	42,000	$1\cdot7 \times 10^2$
Inosinic acid	25	13,000	$5\cdot2 \times 10^2$
Riboflavin	0·5	5,000	$1\cdot0 \times 10^4$
Cyanocobalamin	0·001	30	$3\cdot0 \times 10^4$

* These concentrations are generally used to give maximum growth of many micro-organisms. They probably are somewhat in excess.

† These concentrations approximately represent the maximum amounts reported in the scientific literature.

Table 2 emphasize the fantastic abilities of some overproducing micro-organisms.

The microorganism derives no benefit from overproduction; indeed it is a waste of effort from the cell's point of view. Any self-respecting microorganism uses its normal control mechanisms to prevent over-production and its permeability barriers to prevent leakage of meta-bolites from the cell. However, there do exist in Nature microbial 'freaks' that have quirks in their regulatory or membrane properties; these are the organisms that are screened out by applied microbiologists. The metabolic pathways of these organisms are usually identical to those of organisms already in captivity. Their looser regulatory mechanisms or less effective permeability barriers, however, become the target of development microbiologists who, by manipulating environment and genetics, further break down the protective mechanisms, converting these freaks of nature into efficient chemical factories capable of turning out, for example, 250 million pounds of monosodium glutamate per year.

In the following pages, I shall describe fermentations that produce certain important food additives. I have chosen to restrict myself to primary metabolites, i.e. amino acids, nucleotides, and a vitamin. In each instance, I shall emphasize the key biochemical factors which have led or should lead to development of successful economical fermentations.

THE GLUTAMIC ACID FERMENTATION

By far the most important commercial amino acid is monosodium glutamate, a potent flavour enhancer, and 90 % of the 250 million pounds of monosodium glutamate produced annually is made by fermentation.

The glutamic acid fermentation was discovered by Kinoshita, Udaka & Shimono (1957). Although many genera and species are included in the group of 'glutamate overproducers', e.g. species of *Micrococcus*, *Corynebacterium*, *Brevibacterium* and *Microbacterium*, all are taxonomically similar (Abe, Takayama & Kinoshita, 1967) and should be included in a single genus. Two common characteristics of these organisms are the basic biochemical keys to their glutamate overproduction, i.e. a deficiency of α-ketoglutarate dehydrogenase and a nutritional requirement for biotin. The enzymic block in the tricarboxylic acid cycle ensures the shunting of carbon to glutamate, while the block in biotin biosynthesis seems to result in an altered cellular permeability allowing the cell to excrete glutamate.

The major route of glutamate production from glucose appears in Fig. 1 (Tanaka, Iwasaki & Kinoshita, 1960; Shiio, Otsuka & Tsunoda, 1960*b*). Although the hexose monophosphate pathway is used to a limited extent, the major path of carbon from glucose to pyruvate is the Embden–Meyerhof pathway (Shiio, Otsuka & Tsunoda, 1960*a*; Oishi & Aida, 1963). Glutamate production is therefore inhibited by the glycolysis inhibitors, fluoride and iodoacetate (Birnbaum & Demain, 1969*a*). *Alpha*-ketoglutarate, which cannot be converted to succinyl-CoA because of the deficiency of α-ketoglutarate dehydrogenase, is reductively aminated to glutamate by the nicotinamide adenine dinucleotide phosphate (NADP) specific glutamate dehydrogenase. The source of the $NADPH_2$ is the isocitrate dehydrogenase reaction. The glutamate overproducers use the glyoxylate pathway rather than the tricarboxylic acid cycle to supply energy and intermediates for biosynthetic reactions. Routine molar yields of glutamate from sugar are 50 % of the theoretical yields, and broth concentrations exceed 60 g./litre. With such efficient production, monosodium glutamate only costs about U.S. $0·45 per pound.

During growth on glucose, the overproducers accumulate glutamate intracellularly until the cells become saturated at about 50 mg./g. dry weight (Matsuo *et al.* 1966). Then, presumably because of feedback regulation, accumulation ceases until the permeability barrier is altered to facilitate exit of the amino acid. This modification of permeability is effected by biotin limitation or by addition of agents such as penicillin or fatty-acid derivatives. Addition of penicillin during the logarithmic phase of growth triggers glutamate excretion, and the intracellular level of the amino acid rapidly drops to 5 mg./g. cells. Although viability decreases over 90 %, the cells continue to excrete glutamate for 40–50 h.; no lysis is evident as judged from optical density measurements or total

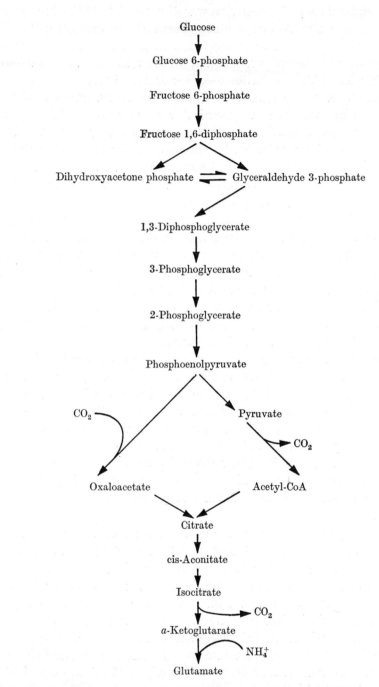

Fig. 1. Major reactions involved in glutamate production from glucose.

microscopic counts (Demain & Birnbaum, 1968). The increase in glutamate permeability appears to be in the outward direction only. Glutamate-producing (biotin-deficient) cells take up external glutamate at only 10 % the rate of normal cells (Oishi, 1967).

The evidence that permeability alteration is a key factor in glutamate fermentation may be summarized as follows:

(a) Cells grown with a concentration of biotin sufficient for growth engage in little or no glutamate overproduction; a suboptimal concentration of biotin is required. Cells grown in media containing sufficient biotin fail to excrete glutamate in resting-cell experiments even in the absence of biotin. Low-biotin cells, on the other hand, excrete glutamate under these conditions and addition of biotin to the resting-cell medium does not inhibit overproduction (Shiio, Otsuka & Takahashi, 1962). Resting cells, previously grown with sufficient biotin do, however, excrete glutamate in the presence of penicillin or fatty-acid derivatives (Shiio, Otsuka & Katsuya, 1963). The biotin concentration in the growth medium has no major influence on the level of many cellular enzymes or on the use of the major pathways of carbon metabolism.

(b) Biotin limitation or addition of fatty-acid derivatives alters the cellular morphology of glutamate-overproducing species.

(c) Crypticity of the cellular enzyme, α-ketoglutarate-aspartate transaminase, is eliminated or decreased markedly by biotin limitation or by addition of either fatty-acid derivatives or penicillin to cells sufficient in biotin.

(d) Biotin-limited cells have a smaller amino acid pool than biotin-sufficient cells. Whereas cells sufficient in biotin lose only neutral amino acids when washed with buffer, cells grown in media containing a limiting concentration of biotin lose practically their entire amino acid pool. Treatment of biotin-sufficient cells with penicillin or fatty-acid derivatives renders the charged amino acids susceptible to removal by buffer.

(e) Glutamate excretion triggered by penicillin results in a 60–80 % decrease in the volume of centrifugally packed cells, despite the absence of lysis, indicating some change in the cell surface (Demain & Birnbaum, 1968).

Studies on *Escherichia coli* by Broda (1968) have also confirmed the role of permeability alteration in glutamate excretion.

The saturating level of biotin in cells of *Brevibacterium lactofermentum* is 20 μg./g. dry cell weight. Only when the level drops to 0·5 μg./g. (1,300 molecules per cell) is glutamate excreted (Takinami, Yamada & Okada, 1966). The mechanism by which biotin controls permeability apparently is related to its role in fatty-acid synthesis. Biotin deficiency (as well as addition of fatty-acid derivatives or penicillin) causes marked changes in the fatty-acid composition of the envelopes of glutamate over-producers. However, there is almost no agreement as to the precise changes which take place in various species (Okabe, Shibukawa & Ohsawa, 1967; Otsuka & Shiio, 1968; Kanzaki *et al.* 1967, 1969), and it is therefore virtually impossible to decide which of the many fatty-acid changes that occur is responsible for permeability alteration. Nevertheless, studies on an oleate-requiring mutant of *Brevibacterium thiogentalis* leave no doubt that biotin acts through its effect on lipid composition (Kanzaki *et al.* 1967, 1969). The parent culture can use either biotin or oleate; glutamate production is triggered by growth in suboptimal levels of either nutrient. The mutant, on the other hand, is totally insensitive to biotin. It must have oleate for growth since it is blocked in an enzyme beyond the biotin-requiring step of fatty-acid biosynthesis. It overproduces glutamate, no matter what the concentration of biotin, as long as oleate is limiting. Since the mutant, unlike its parent, can overproduce glutamate in the presence of excess biotin, the sole function of biotin limitation in the normal fermentation is to provide the type of fatty-acid distribution in the membrane which will permit glutamate excretion. As expected, the oleate content of cellular lipids in the mutant is controlled by the oleate, but not by the biotin, concentration in the medium. On the other hand, either oleate or biotin can exercise this control in the parent.

The fatty-acid derivatives which best elicit glutamate excretion are polyoxyethylene sorbitan-monostearate (Tween 60), polyoxyethylene sorbitan-monopalmitate (Tween 40), the sucrose ester of stearic acid, and certain straight-chain (C_{12}—C_{18}) saturated fatty acids. Like penicillin, they are most effective when added during exponential growth.

Prior to the discovery of the penicillin effect (Somerson & Phillips, 1961), only pure carbon sources could be used for the glutamate fermentation because of the need to control carefully the biotin concentration. The use of penicillin (and the later use of fatty-acid derivatives) is of great commercial importance since crude materials such as molasses can serve as substrate. In fact, blackstrap molasses leads to greater glutamate production than an equivalent concentration of pure glucose or mixtures of glucose, fructose, and sucrose. This is apparently due to the presence in molasses of organic acids that have a dual beneficial effect. First, acetate or citrate can be converted to glutamate (Tsunoda, Shiio & Mitsugi, 1961; Kimura, 1964) after the cells adapt to such carbon sources; at least in the case of citrate, this adaptation involves the induction of citrate permease (Birnbaum & Demain, 1969 b). Secondly, growth in the presence of citrate leads to a more rapid rate of glucose consumption and hence to faster glutamate production than growth on glucose alone (Birnbaum & Demain, 1970). The mechanism is unknown but it might be a reversal of glucose (catabolite) repression of an early enzyme of the tricarboxylic acid cycle.

Production of glutamate from hydrocarbons and from ethanol has sparked interest recently. Special strains usually are isolated for these purposes. Glutamate production from ethanol by *Brevibacterium* sp. amounts to 60 g./litre, and 66 % of the theoretical yield based on consumed ethanol (Oki *et al.* 1969). Ethanol-utilizing strains require biotin and respond in the usual way to biotin limitation. Hydrocarbon utilizers, on the other hand, require thiamine; glutamate excretion can be triggered by thiamine limitation (Imada & Yamada, 1969) or by penicillin addition (Shiio & Uchio, 1969). It is possible that the thiamine requirement has nothing to do with permeability, but that it acts by restricting the production of α-ketoglutarate dehydrogenase (assuming that the hydrocarbon-utilizing strains are not deficient in this enzyme), thiamine pyrophosphate being the coenzyme of this enzyme. Glutamate overproduction upon addition of penicillin to cultures of *Corynebacterium hydrocarboclastus* R-7 amounts to 20 g./litre with a yield (w/w) of 72 % of consumed *n*-hexadecane.

THE LYSINE FERMENTATION

The bulk of the 1,400 million tons of cereals consumed in the World annually is deficient in the essential amino acid, L-lysine. Lysine supplementation converts such cereals into balanced food or feed, and several large-scale supplementation trials with human food are underway in various parts of the world. Thanks to the discovery by Kinoshita, Nakayama & Kitada (1958) that homoserine-requiring mutants of the glutamate-overproducer, *Corynebacterium glutamicum*, produce large amounts of lysine when grown under the proper conditions, an efficient fermentation is available which produces over 40 g. lysine/litre at a weight yield of 25 % of consumed glucose. Thus, L-lysine hydrochloride sells for about U.S.$1.00 per pound. Cultures which efficiently convert acetate to glutamate can be used to produce lysine from acetic acid after mutation to homoserine auxotrophy (Seto & Harada, 1969).

The glutamate overproducers, when mutated to loss of homoserine dehydrogenase, have proved to be the best overproducers of lysine. Before discussing the mechanism of lysine overproduction, let me first emphasize that concentrations of biotin optimal for growth must be provided to the culture. If suboptimal concentrations are provided, glutamate rather than lysine is excreted. Similarly, if penicillin is added to homoserine-less *C. glutamicum* growing in medium containing optimal biotin, glutamate is produced (Nara, Samejima & Kinoshita, 1964). The reason is that nitrogen assimilation by the glutamate-overproducers occurs only through the reductive amination of α-ketoglutarate to glutamate, the reaction being catalysed by the NADP-linked glutamate dehydrogenase (Kimura, 1962). Thus, the nitrogen of all of the natural amino acids is derived from internal glutamate by transamination. When

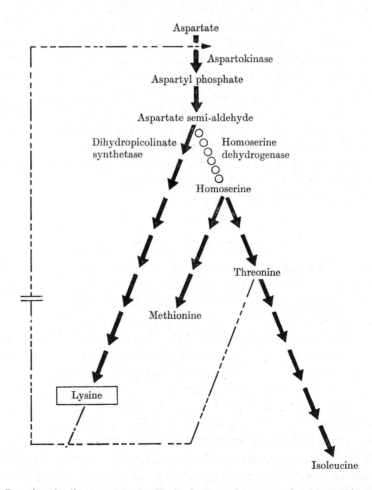

Fig. 2. Reactions leading to synthesis of lysine in *Corynebacterium glutamicum*. The enzymic block (○ ○ ○) is in homoserine dehydrogenase and results in a nutritional requirement for threonine plus methionine. Limited feeding of threonine bypasses (=) concerted feedback inhibition (— — — —) of aspartokinase, yielding lysine as the accumulated product.

the permeability mechanism is altered by limitation of biotin or addition of penicillin, the glutamate is lost from the cell and is no longer available as an intracellular nitrogen donor for lysine synthesis.

The key to lysine overproduction is the avoidance of feedback inhibition by: (a) limitation of a feedback inhibitor of aspartokinase; and by (b) possession of a feedback-insensitive dihydropicolinate synthetase. Lysine belongs to the aspartic-acid family, being produced from aspartate in a branched pathway along with threonine, methionine and isoleucine. The pathway (Fig. 2) is the same as in *Escherichia coli*, but the metabolic controls are different.

Whereas *E. coli* possesses three isozymic aspartokinases controlled separately by lysine, methionine, and threonine, *C. glutamicum* and *Brevibacterium flavum* apparently possess a single aspartokinase regulated by concerted (multivalent) feedback inhibition requiring the simultaneous presence of excess lysine and threonine (Nakayama & Kinoshita, 1966; Shiio & Miyajima, 1969). Since lysine producers lack homoserine dehydrogenase and require methionine plus threonine (or homoserine) for growth, the level of internal threonine is kept low by restriction of threonine in the medium. Thus, feedback inhibition of aspartokinase cannot occur even in the presence of extremely high lysine concentrations. As one would expect, addition of threonine to the medium severely inhibits lysine production. This inhibition is reversible by methionine (Daoust & Stoudt, 1966), which presumably competes with threonine at its allosteric binding site on aspartokinase. The threonine/methionine ratio in the medium is therefore quite important to lysine production.

In branched pathways, the end product usually controls both the initial enzyme of the common early sequence and the first enzyme of its specific branch. Thus, in *E. coli*, lysine inhibits dihydropicolinate synthetase. A second idiosyncracy of the lysine overproducers, however, is that dihydropicolinate synthetase is insensitive to lysine (Nakayama & Kinoshita, 1966). Thus, the metabolic flow from aspartate to aspartate semi-aldehyde proceeds smoothly into the lysine branch.

THE THREONINE FERMENTATION

L-Threonine is another nutritionally important essential amino acid. As shown in Fig. 2 (p. 84), it is a member of the aspartate family. At present, there is no direct fermentation that can produce threonine for as little as U.S. $1·00 per pound, but cheaper processes are being developed, especially in Japan.

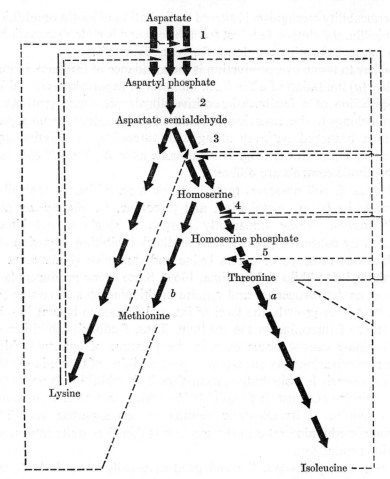

Fig. 3. The aspartic acid pathway in *Escherichia coli*. Feedback regulation of only the five aspartate → threonine reactions is shown. Note that there are three iso-enzymes for step 1 and two iso-enzymes for step 3. – – – indicates feedback repression; – – – – feedback inhibition; *a* and *b* refer to auxotrophic blocks in a high threonine producer (see p. 87).

The early literature on the microbial production of threonine was reviewed by Daoust (1966). The most successful threonine fermentations described to date have used mutants of *E. coli*. The keys to threonine overproduction have been: (*a*) a decreased intracellular concentration of inhibitors and co-repressors (such as methionine, lysine, and isoleucine); and (*b*) mutation of certain enzymes leading to resistance to feedback regulation. The regulatory pattern of the five enzymic steps leading from aspartate to threonine appears in Fig. 3. Certain enzymes which catalyse reactions between aspartate and threonine are repressed and inhibited by lysine, repressed by methionine, inhibited by threonine, and repressed in a concerted (multivalent) manner by threonine plus isoleucine. One of the first threonine fermentation processes was developed by Huang (1961) who employed a double mutant auxotrophic for both methionine and lysine. Because of the auxotrophic blocks, the feeding of low concentrations of methionine and lysine limited the intracellular concentration of these inhibitors and repressors, thereby allowing threonine overproduction to the extent of 4 g./litre. As one would expect, addition of excess lysine or methionine interfered with threonine production.

A recent process (Shiio & Nakamori, 1969) combines the auxotrophic mutation technique with the use of a threonine analogue to select mutants having a feedback-resistant enzyme(s). Shiio & Nakamori (1969) first isolated a threonine-excreting mutant (1·9 g./litre) of *E. coli* by selection in the presence of α-amino-β-hydroxyvalerate. Earlier, Cohen & Patte (1963) had shown that, in such a mutant, the threonine-sensitive homoserine dehydrogenase (step 3 in Fig. 3) was desensitized to threonine inhibition. The resistant mutant was next mutated to isoleucine auxotrophy (by elimination of threonine deaminase; *a* in Fig. 3), resulting in production of 4·7 g. threonine/litre. This increase evidently resulted from the blockage of threonine metabolism and (by feeding a limiting concentration of isoleucine) the elimination of normal concerted feedback repression of several enzymes by isoleucine plus threonine (Freundlich, 1963). Finally, the organism was mutated to methionine auxotrophy by removal of methionine synthetase (*b* in Fig. 3). This final mutant produced over 6 g. of threonine/litre.

PURINE NUCLEOTIDE FERMENTATIONS

The recent intense interest in nucleotide fermentations is due to the ability of three purine ribonucleoside 5'-monophosphates, namely, guanylic acid (guanosine 5'-monophosphate; GMP), inosinic acid (inosine 5'-monophosphate; IMP), and xanthylic acid (xanthosine 5'-monophosphate; XMP), to enhance flavour (in order of decreasing potency). Although the corresponding deoxyribonucleotides are also active, the following compounds are not: adenosine 5'-monophosphate (AMP), 2' and 3' isomers, nucleosides, free bases, and pyrimidine derivatives.

Direct fermentation

Much of the early research on nucleotide production dealt with excretion of nucleotide derivatives that arise from breakdown of ribonucleic

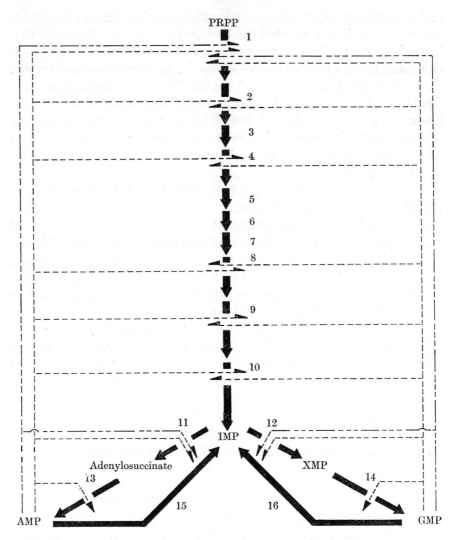

Fig. 4. Pathways leading to synthesis of purine nucleotides in microorganisms.
– – – indicates feedback repression; – – – – feedback inhibition.

acid during stress situations (Demain, 1968). At the same time, studies
were proceeding on the production of free bases and nucleosides by
auxotrophic mutants grown in the presence of limiting concentrations
of purines.

Figure 4 is a composite scheme for several bacteria which have been
studied, e.g. *Bacillus subtilis*, *E. coli*, *Aerobacter aerogenes* and *Salmonella typhimurium*. It shows the nucleotide biosynthetic pathway in microorganisms and some of the feedback regulatory circuits. Of course, the

minor differences between these microorganisms are not shown in the figure. Both AMP and GMP exert feedback repression and feedback inhibition on the initial common pathway and on their respective branches. An important site of control is enzyme 1, phosphoribosyl-pyrophosphate (PRPP) amidotransferase. In *A. aerogenes*, AMP and GMP can each inhibit the enzyme completely, but together they cause a synergistic inhibition at low concentrations (Nierlich & Magasanik, 1965). They apparently combine with the enzyme at different but interacting inhibitory sites. No such synergistic interaction is seen in *B. subtilis*, where AMP alone can cause complete inhibition but GMP cannot (Shiio & Ishii, 1969). Guanosine 5'-triphosphate (GTP) is a more potent inhibitor than GMP but it also cannot cause complete inhibition when present alone in excess.

The key to effective purine accumulation is the limitation of intra-cellular AMP and GMP. This limitation is best effected by restricted feeding of purine auxotrophs. Thus, adenine-requiring mutants lacking adenylosuccinate synthetase (enzyme 11; Fig. 4) accumulate hypoxan-thine that results from breakdown of intracellularly accumulated IMP to inosine and then to hypoxanthine. Mutant strains of *B. subtilis* that are altered in riboside breakdown accumulate inosine to concentrations as high as 11 g./litre (Yamanoi, Konishi & Shiro, 1967). Adenine limita-tion allows metabolites to flow rapidly to IMP, but most of the flow stops there due to inhibition and repression of IMP dehydrogenase by GMP. Guanine auxotrophs that lack XMP aminase (enzyme 14; Fig. 4) excrete over 5 g. of xanthosine/litre; a *B. subtilis* double mutant requir-ing adenine and guanine produces 9 g./litre (Fujimoto *et al.* 1966). For further information on feedback regulation in *B. subtilis*, the reader is referred to the works of Momose and his group (Momose, 1967).

Until 1964, no successful direct fermentation leading to production of nucleotides had been reported; nucleotidases and nucleosidases destroyed intracellular accumulated nucleotides and, when mutants lacking nucleotidases were isolated, they surprisingly produced little or no extracellular purine nucleotides (Fujimoto *et al.* 1965). Apparently in organisms like *B. subtilis*, the permeability barrier does not allow nucleotide excretion, and purines are excreted only after degradation (Demain & Hendlin, 1967). Nucleotidase-deficient mutants thus ceased production when the intracellular nucleotide level reached a concentra-tion capable of causing feedback regulation. An important advance, however, came with the discovery that the glutamate-overproducers could excrete intact nucleotides. When grown in media containing optimal concentrations of biotin, adenine auxotrophs of *C. glutamicum* produce IMP (Nakayama *et al.* 1964). Double mutants of either *C. glutamicum* or *Brevibacterium ammoniagenes* requiring both adenine and xanthine (blocks at adenylosuccinate synthetase (enzyme 11) and IMP dehydrogenase (enzyme 12)) show increased IMP formation since intracellular levels of both AMP and GMP can now be limited (Demain *et al.* 1965; Misawa, Nara & Kinoshita, 1969). As much as 13 g. of IMP/litre have been produced to date (Furuya, Abe & Kinoshita, 1968). Guanine-requiring mutants (blocked at enzyme 14) or adenine-guanine auxotrophs (blocked at enzymes 11 and 14) similarly accumulate XMP (Misawa, Nara & Kinoshita, 1964; Demain *et al.* 1965); concentrations of XMP in broths have reached almost 7 g./litre (Misawa *et al.* 1969).

Development of a direct fermentation to GMP has been more difficult since GMP, being an end product rather than an intermediate of the purine pathway, is a much more potent feedback regulator than is IMP or XMP. One of the main obstacles was believed to be the inhibition and repression of IMP dehydrogenase (enzyme 12; Fig. 4) by GMP. By mutating *C. glutamicum* to loss of this enzyme and then mutating it back to an active enzyme, a revertant was obtained that overproduced GMP in low concentrations (Demain *et al.* 1966). Further mutation increased productivity to 1 g./litre. A good enzyme to block in this organism would be GMP reductase (enzyme 16; Fig. 4), an enzyme which converts GMP back to IMP to complete a small interconversion cycle. This approach is suggested by recent studies with *B. subtilis* in which inosine-producing strains were mutated to guanosine producers. Konishi & Shiro (1968) first mutated an adenine-less parent (inosine producer) to 8-azaguanine resistance. The mutant obtained excreted xanthosine and a smaller amount of inosine, accompanied by hyperproduction (three-to six-fold) of IMP dehydrogenase. Apparently, the resistance phenotype was due to mutation to IMP dehydrogenase constitutivity. A spontaneous mutant was then isolated that produced guanosine (4 g./litre) and inosine (3 g./litre) instead of xanthosine plus inosine. This mutant was found to be devoid of GMP reductase. Momose and Shiio (1969) found that mutation to 8-azaxanthine resistance was even better than 8-azaguanine resistance for derepressing (or desensitizing) IMP dehydrogenase. Thus, by constructing a *B. subtilis* mutant lacking adenylosuccinate synthetase and GMP reductase, which was resistant to azaxanthine, they were able to obtain production of 4 g. guanosine/litre plus 9 g. inosine/litre. It is clear, then, that the keys to the ultimate GMP direct fermentation are: (*a*) a nucleotide-excreting organism such as *C. glutamicum*; (*b*) a lack of adenylosuccinate synthetase (to eliminate feedback regulation by AMP); (c) a derepressed or desensitized IMP dehydrogenase; and (d) a lack of GMP reductase.

Salvage synthesis

While investigating what appeared to be a direct IMP fermentation by an adenine-less mutant of *Brevibacterium ammoniagenes*, workers at the Kyowa Laboratories (Nara, Misawa & Kinoshita, 1967) noticed that, during the early part of the fermentation, only hypoxanthine was excreted. The excreted hypoxanthine was then consumed while extra-cellular IMP was being formed. Investigation showed that the pheno-menon resulted from intracellular accumulation of IMP, degradation to inosine and then to hypoxanthine, excretion of hypoxanthine, and finally a remarkable extracellular 'salvage synthesis' of IMP from the excreted hypoxanthine. The 'salvage synthesis' reaction:

$$purine + PRPP \rightarrow purine\ nucleotide + pyrophosphate$$

had been known for years to proceed inside cells of microbes that are provided with preformed purines; it had been demonstrated with cell-free extracts and with purified enzyme but had not been known to occur extracellularly with intact cells not given PRPP.

Investigators soon found that, as long as certain fermentation require-ments were met, several purine bases could be added to wild-type cultures of *B. ammoniagenes* for conversion to their respective nucleo-tides (Nara, Misawa & Kinoshita, 1968). The requirements, besides a high concentration of inorganic phosphate needed as a precursor, were a high Mg^{2+} concentration, addition of thiamine and pantothenate, and addition of a growth-limiting concentration of Mn^{2+}. Under normal conditions of cultivation in a medium containing a low concentration of phosphate, none of these factors is needed for growth. When the concentration of phosphate necessary to function as a nucleotide pre-cursor is added, however, growth does not occur unless Mn^{2+}, a high concentration of Mg^{2+}, thiamine, and pantothenate are added (Nara *et al.* 1969*a*). Under these conditions of growth, the concentration of Mn^{2+} is crucial to salvage synthesis; it must be low. Addition of growth-optimal concentrations of Mn^{2+} eliminates nucleotide synthesis. Low concentrations of Mn^{2+} not only trigger nucleotide formation; they inhibit DNA synthesis, cause a decrease in cell viability (although cell mass increases), shift cell morphology from normal small rods and ellipsoidal cells to elongated and swollen forms, and alter cellular permeability so that the cells excrete ribose 5-phosphate, PRPP, PRPP kinase, and purine nucleotide pyrophosphorylases for adenine, hypoxanthine, and guanine (Oka, Udagawa & Kinoshita, 1968). Thus, addition of a purine allows the following reactions to take place extracellularly:

$$\text{ribose 5-phosphate} + \text{ATP} \xrightarrow[\text{PRPP kinase}]{\text{Mg}^{2+}, \text{PO}_4{}^{3-}} \text{PRPP} + \text{AMP},$$

$$\text{PRPP} + \text{added purine} \xrightarrow[\substack{\text{purine nucleotide} \\ \text{pyrophosphorylase}}]{\text{Mg}^{2+}, \text{PO}_4{}^{3-}} \text{5'-purine nucleotide} + \text{pyrophosphate}$$

Although the investigators have not reported on the excretion of ATP, I assume that it is likely. If no purine is added, ribose 5-phosphate and the enzymes are found in the broth. The parallel between the effect of biotin limitation in glutamate excretion and limitation of Mn^{2+} in salvage synthesis is quite striking. Manganous ion limitation causes the fatty-acid composition of the cells to shift (Nara *et al.* 1969*b*) and can be replaced by penicillin or surfactants (Nara *et al.* 1969*c*).

The salvage synthetic route to nucleotides provides a practical means of producing GMP from guanine, as well as AMP from adenine. The additional excretion of transphosphorylases makes the organism also capable of forming GDP and GTP from guanine, ADP and ATP from adenine (Tanaka *et al.* 1968), and nicotinamide adenine dinucleotide (NAD) from adenine and nicotinamide (Nakayama *et al.* 1968).

THE RIBOFLAVIN FERMENTATION

Riboflavin (vitamin B_2) is produced commercially by both fermentation and chemical synthesis. Whereas the chemically synthesized material is generally used for pharmaceuticals and food, riboflavin produced by fermentation is used in animal feed.

The vitamin is generally present in microbial cells as the coenzymes flavin mononucleotide (FMN) and flavin adenine dinucleotide (FAD), which are bound to protein. Riboflavin overproducers contain as much FMN and FAD as normal microorganisms but they also have a large intracellular pool of free riboflavin. The flavin excreted by the overproducers is predominantly riboflavin. Normal microorganisms excrete less than 10 mg. riboflavin/litre. The overproducers can be divided into three groups. The low overproducers include the clostridia, which can produce, at best, about 100 mg./litre. Yeasts, especially some *Candida* species, are moderate overproducers that can be made to yield about 600 mg. riboflavin/litre. The high overproducers are two yeast-like moulds, *Eremothecium ashbyii* and *Ashbya gossypii*, which synthesize riboflavin in concentrations as high as 5,000 mg./litre.

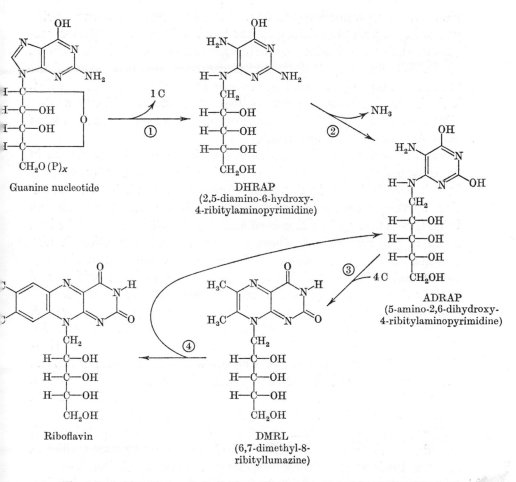

Fig. 5. Probable pathway of riboflavin production from a guanine derivative.

The biosynthetic path to riboflavin is fairly well understood, and Fig. 5 shows the pathway as it is known today (Oltmanns *et al.* 1969). The remaining questions concern: (*a*) whether GTP, GDP, GMP, guanosine, or guanine is the immediate purine derivative entering the pathway (Bacher & Lingens, 1969); (*b*) the actual number of reactions represented as steps 1 and 3 in Fig. 5; (*c*) the origin of the ribityl group (it does not appear to be ribose from guanine nucleotide; probably it comes from glucose through the pentose cycle (Plaut & Broberg, 1956)); and (*d*) the origin of the 4-carbon moiety that adds to 5-amino-2,6-dihydroxy-4-ribitylaminopyrimidine (ADRAP). Once thought to be acetoin, this 4-carbon precursor appears to be derived from the pentose cycle (Ali & Al-Khalidi, 1966). As would be expected, purines stimulate riboflavin overproduction (McNutt, 1954), and glycine, a known purine precursor, is very stimulatory except in *E. ashbyii* where glycine precursors such as threonine and serine are active (Goodwin & Horton, 1960). Riboflavin synthetase, which catalyses the final step of the sequence, is the only enzyme that has been purified and carefully studied. This interesting enzyme converts two molecules of 6,7-dimethyl-8-ribityllumazine (DMRL) to one molecule of riboflavin and one of ADRAP (Harvey & Plaut, 1966).

Practically nothing is known about the biochemical key to riboflavin overproduction, but it most likely involves iron. Ferrous ion severely inhibits riboflavin production by low overproducers; it inhibits formation by the moderate overproducers to a lesser extent, and has no inhibitory action against *E. ashbyii* and *A. gossypii*. The site of the iron effect is not riboflavin synthetase, since DMRL formation is also inhibited. Some workers feel that high concentrations of iron stimulate production of iron-containing enzymes, such as xanthine oxidase and urate oxidase, which (with the help of guanine deaminase) destroy guanine and thus deplete the cell of its purine precursor (Shavlovskii & Logvinenko, 1967; Schlee, Reinbothe & Fritsche, 1968). Cells containing sufficient Fe^{2+} do contain twice as much xanthine oxidase as Fe^{2+}-deficient cells, but no one has ever reported that additional purines can reverse inhibition of riboflavin formation by iron. I would favour a hypothesis that iron-flavoprotein is a repressor of riboflavin synthesis. Several enzymes (for example, dihydro-orotic dehydrogenase, xanthine oxidase, aldehyde oxidase, succinate dehydrogenase, and $NADH_2$ dehydrogenase) contain both flavin and iron as coenzymes (Miller & Massey, 1965). If one of these enzymes or a non-enzymic iron-flavoprotein were the repressor of riboflavin synthesis, growth in iron-deficient media would produce cells with little or no repressor, and flavin synthesis would be derepressed. Wilson & Pardee's (1962) results with *E. coli* suggest repression rather than feedback inhibition. To explain overproduction by *A. gossypii* and *E. ashbyii*, one might postulate that they are constitutive with respect to riboflavin-forming enzymes, or that the conditions used for the riboflavin fermentation inhibit formation of the repressor. One such condition could be temperature. Tanner, Vojnovich & Van Lanen (1949) reported that, although *A. gossypii* grows abundantly over a wide temperature range, riboflavin production was limited to a narrow range below the optimum temperature for growth. The highest yields were obtained between 26° and 28°. Pfeifer *et al.* (1950) compared cultures grown at 35° with cultures grown at 29° and found much faster sugar utilization but much poorer riboflavin

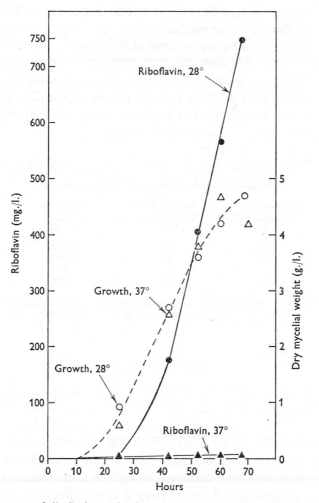

Fig. 6. Time-course of riboflavin production and growth of *Ashbya gossypii* at 28° and 37°.

production at the higher temperature. With these observations in mind, Kaplan & Demain (1970) studied riboflavin overproduction by *A. gossypii* in a semi-defined medium at 28° and 37°. They found the same rate of growth, but only at the lower temperature was riboflavin over-produced (Fig. 6). Temperature-shift experiments revealed that incubation at 37° was detrimental only during the growth phase; therefore a culture grown for 48 h. at 37° and then shifted to 28° could not overproduce riboflavin. Washed-suspension experiments confirmed these observations. Table 3 shows that suspensions prepared from mycelia grown at 37° made little or no riboflavin at 28° or at 37°; mycelium grown at 28°

produced riboflavin equally well at either temperature. These results suggest that riboflavin overproduction by *A. gossypii* is controlled by a repressor that fails to be synthesized at low temperatures. It would be interesting to study the production of riboflavin by *A. gossypii* grown at a temperature between 28° and 37° in the presence of different concentrations of iron. It is possible that riboflavin overproduction by cells grown at a less favourable temperature would be controlled by the concentration of iron in the growth medium.

Table 3. *Effect of growth temperature and suspension temperature on riboflavin production by Ashbya gossypii**

Growth temperature (°)	Suspension temperature (°)	Riboflavin produced (μM)
28	28	107·0
28	37	130·0
37	28	6·7
37	37	4·3

* Cultures were grown for 24 h. in a semi-defined medium. The mycelia were washed, harvested by centrifugation and suspended for 18 h. in a mixture containing glucose, mineral salts, glycine, tyrosine, hypoxanthine, histidine, and Tween 80.

CONCLUSIONS

The record of microbes in producing compounds that improve the nutritional quality, taste, and availability of food is one in which all applied microbiologists can take pride. It is remarkable that microbiologists are still largely responsible for the production of simple molecules such as glutamic acid and lysine. The Dutch synthetic L-lysine process no longer threatens to replace fermentation lysine, and synthetic L-glutamate still serves only a minor fraction of the world market. Synthetic riboflavin, made by a very efficient chemical process for many years, has yet to eliminate fermentation riboflavin. Indeed, the market for the natural product continues to expand much more rapidly than that for synthetic riboflavin. Clearly, the ability of microbes to overproduce metabolites will continue to be a major economic factor in the production of food additives.

For a completely different reason, I am proud of the record of applied microbiology. Applied research, especially in the areas of food additives and nutrition, has uncovered a number of extremely interesting findings significant to basic microbiology. Some of these discoveries have been discussed above, such as: (*a*) the alteration of permeability by biotin limitation, by penicillin addition, and by addition of fatty-acid derivatives; (b) the ability of certain bacteria to excrete phosphate esters such as nucleotides without alteration of permeability; (c) the ability of Mn^{2+} limitation to so alter permeability that normally intracellular enzymes and intermediates are excreted and participate in biosynthetic reactions outside the cell. Additional examples are the discovery of mevalonic acid, the key intermediate of isoprenoid biosynthesis which also turned out to be a growth factor for lactobacilli, the finding that bacteria can excrete high molecular-weight RNA during growth, and the discovery that *Brevibacterium liquefaciens* excretes massive amounts of cyclic 3',5'-AMP. Many more examples could be cited, but I believe it is clear that mission-oriented microbiological research, because its approach and pressures differ from those of non-oriented research, spins off exciting findings of general microbiological significance which might not have been uncovered for many years in laboratories interested in more basic problems. This is not to deny that the fantastic advances in industrial microbiology have borrowed heavily from the discoveries of basic geneticists, biochemists, and microbiologists. It is merely to point out that modern microbiology is a two-way street.

The preparation of this manuscript was supported by U.S. Public Health Service Research Grant AI-09345 from the National Institute of Allergy and Infectious Diseases.

REFERENCES

ABE, S., TAKAYAMA, K. & KINOSHITA, S. (1967). Taxonomical studies on glutamic acid producing bacteria. *J. gen. appl. Microbiol. Tokyo* **13**, 279.

ALI, S. N. & AL-KHALIDI, U. A. S. (1966). The precursors of the xylene ring in riboflavine. *Biochem. J.* **98**, 182.

BACHER, A. & LINGENS, F. (1969). The structure of the purine precursor in riboflavin biosynthesis. *Angew. Chem., internat. Edit.* **8**, 271.

BIRNBAUM, J. & DEMAIN, A. L. (1969a). Reversal by citrate of the iodoacetate and fluoride inhibition of glutamic acid production by *Corynebacterium glutamicum*. *Appl. Microbiol.* **18**, 287.

BIRNBAUM, J. & DEMAIN, A. L. (1969b). Conversion of citrate to extracellular glutamate by penicillin-treated resting cells of *Corynebacterium glutamicum*. *Agric. biol. Chem., Tokyo* **33**, 1169.

BIRNBAUM, J. & DEMAIN, A. L. (1970). Effect of citrate on utilization of glucose by resting cells of *Corynebacterium glutamicum*. *Can. J. Microbiol.* **16**, 210.

BRODA, P. (1968). Ribonucleic acid synthesis and glutamate excretion in *Escherichia coli*. *J. Bact.* **96**, 1528.

COHEN, G. N. & PATTE, J. C. (1963). Some aspects of the regulation of amino acid biosynthesis in a branched pathway. *Cold Spring Harb. Symp. quant. Biol.* **28**, 513.

DAOUST, D. R. (1966). Microbial production of threonine – a review. *Devl. industr. Microbiol.* **7**, 41.

DAOUST, D. R. & STOUDT, T. H. (1966). The biosynthesis of L-lysine in a strain of *Micrococcus glutamicus*. *Devl. industr. Microbiol.* **7**, 22.

DEMAIN, A. L. (1968). Production of purine nucleotides by fermentation. In *Progress in Industrial Microbiology*, vol. 8, p. 35. Ed. D. J. D. Hockenhull. London: Churchill.

DEMAIN, A. L. & BIRNBAUM, J. (1968). Alteration of permeability for the release of metabolites from the microbial cell. *Current Topics in Microbiol. Immunol.* **46**, 1.

DEMAIN, A. L. & HENDLIN, D. (1967). Phosphohydrolases of a *Bacillus subtilis* mutant accumulating inosine and hypoxanthine. *J. Bact.* **94**, 66.

DEMAIN, A. L., JACKSON, M., VITALI, R. A., HENDLIN, D. & JACOB, T. A. (1965). Production of xanthosine-5'-monophosphate and inosine-5'-monophosphate by auxotrophic mutants of a coryneform bacterium. *Appl. Microbiol.* **13**, 757.

DEMAIN, A. L., JACKSON, M., VITALI, R. A., HENDLIN, D. & JACOB, T. A. (1966). Production of guanosine-5'-monophosphate and inosine-5'-monophosphate by fermentation. *Appl. Microbiol.* **14**, 821.

FREUNDLICH, M. (1963). Multivalent repression in the biosynthesis of threonine in *Salmonella typhimurium* and *Escherichia coli*. *Biochem. biophys. Res. Commun.* **10**, 277.

FUJIMOTO, M., MOROZUMI, M., MIDORIKAWA, Y., MIYAKAWA, S. & UCHIDA, K. (1965). Studies on 5'-nucleotidase-lacking mutants derived from *Bacillus subtilis*. 2. The influence of the loss of 5'-IMP-dephosphorylating activity on the extracellular accumulation of hypoxanthine derivatives in an adenine-requiring mutant. *Agric. biol. Chem., Tokyo* **29**, 918.

FUJIMOTO, M., UCHIDA, K., SUZUKI, M. & YOSHINO, H. (1966). Accumulation of xanthosine by auxotrophic mutants of *Bacillus subtilis*. *Agric. biol. Chem., Tokyo* **30**, 605.

FURUYA, A., ABE, S. & KINOSHITA, S. (1968). Production of nucleic acid-related substances by fermentative processes. 19. Accumulation of 5'-inosinic acid by a mutant of *Brevibacterium ammoniagenes*. *Appl. Microbiol.* **16**, 981.

GOODWIN, T. W. & HORTON, A. A. (1960). Studies in flavinogenesis. 6. The role of threonine in riboflavin biosynthesis in *Eremothecium ashbyii*. *Biochem. J.* **75**, 53.

HARVEY, R. A. & PLAUT, G. W. E. (1966). Riboflavin synthetase from yeast. Properties of complexes of the enzyme with lumazine derivatives and riboflavin. *J. biol. Chem.* **241**, 2120.

HUANG, H. T. (1961). Production of L-threonine by auxotrophic mutants of *Escherichia coli*. *Appl. Microbiol.* **9**, 419.

IMADA, Y. & YAMADA, K. (1969). Utilization of hydrocarbons by microorganisms. 14. Formation of α-ketoglutarate, L-glutamate and DL-alanine by *Corynebacterium hydrocarboclastus* S 10 B 1. *Agric. biol. Chem., Tokyo* **33**, 1326.

KANZAKI, T., OKAZAKI, H., SUGAWARA, A. & FUKUDA, H. (1967). L-Glutamic acid fermentation. 4. The relation between the cellular fatty acid contents and the productivity of L-glutamic acid. *Agric. biol. Chem., Tokyo* **31**, 1416.

KANZAKI, T., ISOBE, K., OKAZAKI, H. & FUKUDA, H. (1969). L-Glutamic acid fermentation. 7. Relation between biotin and oleic acid. *Agric. biol. Chem., Tokyo* **33**, 771.

KAPLAN, L. & DEMAIN, A. L. (1970). Nutritional studies on riboflavin overproduction by *Ashbya gossypii*. In *Recent Trends in Yeast Research*. Ed. D. G. Ahearn. (In the Press.)

KIMURA, K. (1962). The significance of glutamic dehydrogenase in glutamic acid fermentation. *J. gen. appl. Microbiol., Tokyo* **8**, 253.

KIMURA, K. (1964). Utilization of organic acids for the glutamic acid production by *Micrococcus glutamicus*. *J. gen. appl. Microbiol., Tokyo* 10, 23.

KINOSHITA, S., NAKAYAMA, K. & KITADA, S. (1958). L-Lysine production using microbial auxotroph. *J. gen. appl. Microbiol., Tokyo* 4, 128.

KINOSHITA, S., UDAKA, S. & SHIMONO, M. (1957). Studies on the amino acid fermentation. 1. Production of L-glutamic acid by various microorganisms. *J. gen. appl. Microbiol., Tokyo* 3, 193.

KONISHI, S. & SHIRO, T. (1968). Fermentative production of guanosine by 8-azaguanine-resistant *Bacillus subtilis*. *Agric. biol. Chem., Tokyo* 32, 396.

MATSUO, T., OYAMA, Y., TANIMOTO, H., HASHIDA, W. & TERAMOTO, S. (1966). Fermentative production of glutamic acid and its application. X. Effects of antibiotics on the glutamic acid fermentation. 3. Effects of penicillin and biotin. *Amino Acid Nucleic Acid, Tokyo* 12, 915.

McNUTT, W. S. (1954). The direct contribution of adenine to the biogenesis of riboflavin by *Eremothecium ashbyii*. *J. biol. Chem.* 210, 511.

MILLER, R. W. & MASSEY, V. (1965). Dihydroorotic dehydrogenase. 1. Some properties of the enzyme. *J. biol. Chem.* 240, 1453.

MISAWA, M., NARA, T. & KINOSHITA, S. (1964). Production of nucleic acid-related substances by fermentative processes. 6. Accumulation of 5'-xanthylic acid by guanine-requiring mutants of *Micrococcus glutamicus*. 2. Studies on cultural conditions. *Agric. biol. Chem., Tokyo* 28, 694.

MISAWA, M., NARA, T. & KINOSHITA, S. (1969). Production of nucleic acid-related substances by fermentative processes. 23. Derivation of IMP-producing mutants of *Brevibacterium ammoniagenes*. *Agric. biol. Chem., Tokyo* 33, 514.

MISAWA, M., NARA, T., UDAGAWA, K., ABE, S. & KINOSHITA, S. (1969). Production of nucleic acid-related substances by fermentative processes. 22. Fermentative production of 5'-xanthylic acid by a guanine auxotroph of *Brevibacterium ammoniagenes*. *Agric. biol. Chem., Tokyo* 33, 370.

MOMOSE, H. (1967). Genetic and biochemical studies on 5'-nucleotide fermentation. 3. Genetic derepression of enzyme formation in purine nucleotide biosynthesis in *Bacillus subtilis*. *J. gen. appl. Microbiol., Tokyo* 13, 39.

MOMOSE, H. & SHIIO, I. (1969). Genetic and biochemical studies on 5'-nucleotide fermentation. 4. Effect of GMP reductase and purine analogue resistance on purine nucleoside accumulation pattern in adenine auxotrophs of *Bacillus subtilis*. *J. gen. appl. Microbiol., Tokyo* 15, 399.

NAKAYAMA, K. & KINOSHITA, S. (1966). Some considerations on the mechanism in biosynthesis of lysine. *Devl. industr. Microbiol.* 7, 16.

NAKAYAMA, K., SUZUKI, T., SATO, Z. & KINOSHITA, S. (1964). Production of nucleic acid-related substances by fermentative processes. 5. Accumulation of inosinic acid by an adenine-auxotroph of *Micrococcus glutamicus*. *J. gen. appl. Microbiol., Tokyo* 10, 133.

NAKAYAMA, K., SATO, Z., TANAKA, H. & KINOSHITA, S. (1968). Production of nucleic acid-related substances by fermentative processes. 17. Production of NAD and nicotinic acid mononucleotide with *Brevibacterium ammoniagenes*. *Agric. biol. Chem., Tokyo* 32, 1331.

NARA, T., MISAWA, M. & KINOSHITA, S. (1967). Production of nucleic acid-related substances by fermentative processes. 13. Fermentative production of 5'-inosinic acid by adenine auxotroph of *Brevibacterium ammoniagenes*. *Agric. biol. Chem., Tokyo* 31, 1351.

NARA, T., MISAWA, M. & KINOSHITA, S. (1968). Production of nucleic acid-related substances by fermentative processes. 14. Fermentative production of 5'-purine ribonucleotides by *Brevibacterium ammoniagenes*. *Agric. biol. Chem., Tokyo* 32, 561.

NARA, T., SAMEJIMA, H. & KINOSHITA, S. (1964). Effect of penicillin on amino acid fermentation. *Agric. biol. Chem., Tokyo* **28**, 120.

NARA, T., KOMURO, T., MISAWA, M. & KINOSHITA, S. (1969 *a*). Production of nucleic acid-related substances by fermentative processes. 29. Growth responses of *Brevibacterium ammoniagenes*. *Agric. biol. Chem., Tokyo* **33**, 1030.

NARA, T., MISAWA, M., KOMURO, T. & KINOSHITA, S. (1969 *b*). Production of nucleic acid-related substances by fermentative processes. 31. Biosynthetic mechanisms involved in 5'-purine ribonucleotide production by *Brevibacterium ammoniagenes*. *Agric. biol. Chem., Tokyo* **33**, 358.

NARA, T., MISAWA, M., KOMURO, T. & KINOSHITA, S. (1969 *c*). Production of nucleic acid-related substances by fermentative processes. 30. Effect of antibiotics and surface active agents on 5'-purine nucleotide production by *Brevibacterium ammoniagenes*. *Agric. biol. Chem., Tokyo* **33**, 1198.

NIERLICH, D. P. & MAGASANIK, B. (1965). Regulation of purine ribonucleotide synthesis by end product inhibition. The effect of adenine and guanine ribonucleotides on the 5'-phosphoribosylpyrophosphate amidotransferase of *Aerobacter aerogenes*. *J. biol. Chem.* **240**, 358.

OISHI, K. (1967). Metabolic control of glutamate synthesis in glutamic acid bacteria. *J. agric. chem. Soc., Japan* **41**, 35.

OISHI, K. & AIDA, K. (1963). Studies on amino acid fermentation. 11. Effect of biotin on the Embden-Meyerhof-Parnas pathway and the hexose-monophosphate shunt in a glutamic acid-producing bacterium, *Brevibacterium ammoniagenes* 317-1. *Amino acid nucleic acid, Tokyo* **8**, 35.

OKA, T., UDAGAWA, K. & KINOSHITA, S. (1968). Unbalanced growth death due to depletion of Mn^{2+} in *Brevibacterium ammoniagenes*. *J. Bact.* **96**, 1760.

OKABE, S., SHIBUKAWA, M. & OHSAWA, T. (1967). L-Glutamic acid fermentation with molasses. 9. Relation between the lipid in the cell membrane from *Microbacterium ammoniaphilum* and the extracellular accumulation of L-glutamic acid. *Agric. biol. Chem., Tokyo* **31**, 789.

OKI, T. NISHIMURA, Y., SAYAMA, Y., KITAI, A. & OZAKI, A. (1969). Microbial utilization of non-carbohydrate substances. I. Fermentative production of L-glutamic acid from ethanol. 2. Influence of several factors on L-glutamic acid production from ethanol. *Amino acid nucleic acid, Tokyo* **19**, 82.

OLTMANNS, O., BACHER, A., LINGENS, F. & ZIMMERMANN, F. K. (1969). Biochemical and genetic classification of riboflavine deficient mutants of *Saccharomyces cerevisiae*. *Molec. gen. Genetics* **105**, 306.

OTSUKA, S. & SHIIO, I. (1968). Fatty acid composition of cell wall-cell membrane fraction from *Brevibacterium flavum*. *J. gen. appl. Microbiol, Tokyo* **14**, 135.

PFEIFER, V. R., TANNER, F. W., JR., VOJNOVICH, C. & TRAUFLER, D. H. (1950). Riboflavin by fermentation with *Ashbya gossypii*. *Ind. Engng Chem.* **42**, 1776.

PLAUT, G. W. E. & BROBERG, P. L. (1956). Biosynthesis of riboflavin. 3. Incorporation of ^{14}C-labelled compounds into the ribityl side chain. *J. biol. Chem.* **219**, 131.

SCHLEE, D., REINBOTHE, E. & FRITSCHE, W. (1968). Der Enifluss von Eisen auf den Purinstoffwechsel und die Riboflavinbildung von *Candida guilliermondii* (Cast.) Lang. et. G. *Z. allg. Mikrobiol.* **8**, 127.

SETO, K. & HARADA, T. (1969). Formation of L-lysine from acetic acid by a homoserine auxotroph of *Corynebacterium acetophilum* A51. *J. Ferment. Techn.* **47**, 558.

SHAVLOVSKII, G. M. & LOGVINENKO, E. M. (1967). Catabolism of xanthine in the presence of the fungus *Candida guilliermondii* and the role of metals in the process. *Biokhimya (USSR)* **32**, 59.

SHIIO, I. & ISHII, K. (1969). Regulation of purine ribonucleotide synthesis by end product inhibition. 2. Effect of purine nucleotides on phosphoribosylpyrophosphate amidotransferase of *Bacillus subtilis*. *J. Biochem., Tokyo* **66**, 175.

SHIIO, I. & MIYAJIMA, R. (1969). Concerted inhibition and its reversal by end products of aspartate kinase in *Brevibacterium flavum*. *J. Biochem., Tokyo* **65**, 849.

SHIIO, I. & NAKAMORI, S. (1969). Microbial production of L-threonine. 1. Production of *Escherichia coli* mutant resistant to α-amino-β-hydroxyvaleric acid. *Agric. biol. chem., Tokyo* **33**, 1152.

SHIIO, I., OTSUKA, S. & KATSUYA, N. (1963). Cellular permeability and extracellular formation of glutamic acid in *Brevibacterium flavum*. *J. Biochem., Tokyo* **53**, 333.

SHIIO, I., OTSUKA, S. & TAKAHASHI, M. (1962). Effect of biotin on the bacterial formation of glutamic acid. 1. Glutamate formation and cellular permeability of amino acids. *J. Biochem., Tokyo* **51**, 56.

SHIIO, I., OTSUKA, S. & TSUNODA, T. (1960a). Glutamic acid formation from glucose by bacteria. 3. On the pathway of pyruvate formation in *Brevibacterium flavum* No. 2247. *J. Biochem., Tokyo* **47**, 414.

SHIIO, I. OTSUKA, S. & TSUNODA, T. (1960b). Glutamic acid formation from glucose by bacteria. 4. Carbon-dioxide fixation and glutamate formation in *Brevibacterium flavum* No. 2247. *J. Biochem., Tokyo* **48**, 110.

SHIIO, I. & UCHIO, R. (1969). Microbial production of amino acids from hydrocarbons. 4. L-Glutamic acid production by *Corynebacterium hydrocarboclastus* R-7. *J. gen. appl. Microbiol., Tokyo* **15**, 65.

SOMERSON, N. & PHILLIPS, T. (1961). Procede d'obtention d'acide glutamique. Belg. Patent no. 593,807.

TAKINAMI, K., YAMADA, Y. & OKADA, H. (1966). Biochemical effects of fatty acid and its derivatives on L-glutamic acid fermentation. IV. Biotin content of growing cells of *Brevibacterium lactofermentum*. *Agric. biol. chem., Tokyo* **30**, 674.

TANAKA, K., IWASAKI, H. & KINOSHITA, S. (1960). Studies on the fermentation of L-glutamic acid. 5. Accumulation of glutamic acid by bacteria and biotin. *J. agric. chem. Soc., Japan* **34**, 593.

TANAKA, H., SATO, Z., NAKAYAMA, K. & KINOSHITA, S. (1968). Production of nucleic acid-related substances by fermentative processes. 15. Formation of ATP, GTP and their related substances by *Brevibacterium ammoniagenes*. *Agric. biol. Chem., Tokyo* **32**, 721.

TANNER, F. W., JR., VOJNOVICH, C. & VAN LANEN, J. M. (1949). Factors affecting riboflavin production by *Ashbya gossypii*. *J. Bact.* **58**, 737.

TSUNODA, T., SHIIO, I. & MITSUGI, K. (1961). Bacterial formation of L-glutamic acid from acetic acid in the growing culture medium. II. Growth and L-glutamate accumulation in a chemically defined medium and some other fermentation products. *J. gen. appl. Microbiol., Tokyo* **7**, 30.

WILSON, A. C. & PARDEE, A. B. (1962), Regulation of flavin synthesis by *Escherichia coli*. *J. gen. Microbiol.* **28**, 283.

YAMANOI, A., KONISHI, S. & SHIRO, T. (1967). Studies on inosine fermentation – production of inosine by mutants of *Bacillus subtilis*. 5. Specific requirement of adenine for regulation of inosine formation and adenine-sparing action of AICA and amino acid mixture. *J. gen. appl. Microbiol., Tokyo* **13**, 365.

MICROBIAL DISEASE AND PLANT PRODUCTIVITY

S. H. CROWDY AND J. G. MANNERS

Department of Botany, University of Southampton

INTRODUCTION

The microbes most deeply involved in plant disease are fungi and viruses. Bacteria cause a few serious diseases, but the overall losses which result from these are relatively small. There are a variety of ways in which these losses are caused, ranging from a general reduction in vigour associated with systemic infection such as is frequently caused by viruses, to losses which can be associated with specific symptoms. Thus leaf infections will reduce the area available for photosynthesis and the flow of nutrients to the plant, while root infections, or infections of the vascular tissues, by reducing the flow of water, can lead to wilting and death of the host. In most cases, the causes of loss are complex and additional physiological effects such as the influence of toxins produced by the parasite or in response to infection have to be considered. The effect of disease on productivity can be studied as the gross loss in quality as well as quantity of crop, in which case we are dealing with an economic assessment of the value of the disease to the community or the individual; this information is essential both to provide data for allocating research priorities and to make a proper assessment of the economics of disease control. In addition, to provide a basis for rational control measures, the way in which infection affects the production of the individual plant must be studied as a part of a wider investigation of the interaction between the parasite, its host and the environment.

ESTIMATION OF LOSSES IN INDIVIDUAL CROPS

Measuring the economic cost of a plant disease is difficult and uncertain and the necessary data are only rarely available even in highly developed countries; in developing countries, where the need to maximize production is most vital and it is most necessary to make the best use of men and materials by a proper assessment of priorities: data for making these assessments are almost non-existent. The basis on which these assessments should be made has been dealt with in a number of publications. Any method must be reasonably rapid, reproducible and independent of a possible bias related to different observers. The final result must be based on a truly representative sample of the crop in the area under consideration. Precise and highly reproducible estimates of loss can in some cases be derived from the results of experiments on the control of specific diseases. These provide accurate measures in the specific circumstances in which the experiment was performed. However, they are unlikely to provide a comprehensive picture, since the experiments are usually deliberately located in areas where the disease is likely to be serious. If no infection occurs, the experiment is a failure. This bias may sometimes be increased unconsciously, if the results are interpreted by a plant pathologist with a vested interest in demonstrating the economic importance of his work. Reliable estimates must be based on a random sample of the crop. In principle, the data could be derived from a large number of randomly distributed experiments, however these are cumbersome and require a large amount of time and labour. In practice an adequate sample would not be possible. This difficulty is probably best met by breaking the problem down into three distinct stages, namely estimating the intensity of disease in the field, relating disease intensity to yield loss and making an adequately extensive survey of disease intensity, from which can be derived an estimate of overall loss.

The whole process has been surveyed recently by Large (1966) and Moore (1969). The initial stage, estimating disease intensity, is probably the most reliable and straightforward and our information on this stage is the most extensive. The estimates must be numerical so that they can be treated by normal arithmetical and statistical processes, they must be repeatable, they must give comparable results in the hands of a number of different individuals and, finally, they must be rapid so that large areas can be surveyed in a short time. The actual criteria used will vary with the type of disease; leaf diseases are commonly recorded as the percentage of the leaf area affected and standardized diagrams are usually provided to guide the estimator. Systemic infections and root diseases, which affect the whole plant are more commonly estimated as the percentage of infected plants in a standardized sample, or by an overall rating of symptoms, though the last may introduce a subjective judgement which could affect the result. Virus diseases seem particularly difficult to deal with since they occur in a variety of strains of different virulence and the age of the plant at infection may greatly affect the result. Watson & Mulligan (1960) developed a symptom rating system for barley yellow dwarf virus based on intensity of chlorosis, necrosis and stunting; more recently Doodson & Saunders

(1970) have suggested a revised method of estimation based on decreases of height, leaf length and area of discoloration; the revised formula seems to need uninfected plants for comparison and is probably well suited to give guidance in breeding experiments: it may be too complicated to use in extensive surveys.

The theoretical considerations involved in the final stage, relating disease intensity to yield loss, have been discussed by Van der Plank (1963). A comprehensive model should take account of a variety of factors including the amount of the disease, its duration, the maturity of the plant when infection takes place and environmental factors such as soil fertility. In practice, lack of information prevents the construction of an adequate model and the relationship between disease intensity and yield loss is established empirically within defined statistical limits in a series of specially designed experiments in which yields of infected and uninfected plots are compared. If enough experiments are carried out, there is usually a reasonable spread of disease intensity and associated yield loss records; these records are transformed to give a linear relationship between symptoms and yield. This can be expressed as a linear regression, which is applied to the survey data. In these experiments the uninfected plots are maintained free of disease by spraying, or suitably weighted comparisons may be made between the performance of susceptible and resistant varieties. Regressions of this type and other empirical relationships have been reported for a number of leaf diseases; for example, cereal mildews caused by *Erysiphe graminis* (Large & Doling, 1962, 1963), sharp eyespot caused by *Corticium solani* (Pitt, 1966), yellow rust caused by *Puccinia striiformis* (Doling & Doodson, 1968), potato blight caused by *Phytophthora infestans* (Large, 1958) and some of the cereal root diseases, notably take-all caused by *Ophiobolus graminis* (Slope, 1967; Rosser & Chadburn, 1968). Similar equations have been applied to the virus diseases sugarbeet yellows (Hull, 1953) and, as noted above, barley yellow dwarf.

There are obvious errors inherent in this approach, since there is no certainty that the spray is completely without phytotoxicity or that it is controlling only the disease under investigation. One must also accept the assumption that all varieties react in the same way to the disease in all the locations examined. Some of these difficulties are illustrated by the assessment of the losses due to different levels of infection by barley mildew. This disease was examined by Large & Doling (1962) who related the yield loss, y, to the estimate of disease intensity, x, by the equation

$$y = 2 \cdot 5(x)^{-\frac{1}{2}}. \tag{1}$$

This relationship was based on a series of experiments in which the disease was controlled with a lime sulphur spray. Recently Doodson & Saunders (1969) have suggested that this estimate is too low. These authors do not suggest a revised equation, but the data they publish, in which the yield of the infected plots is compared with the yield in plots in which the disease was controlled by the modern fungicide ethirimol (active ingredient 5-*n*-butyl-ethylamino-4-hydroxy-6 methyl pyrimidine) indicate that this equation underestimates yield loss by about 25 %. Sulphur sprays are known to be toxic to some plants and reduce yield without producing well defined symptoms; in commercial use, they have been recorded as reducing yields of apples by as much as 53 % (Cobham, 1965) so a reduction of 25 % due to sulphur damage would not be entirely unexpected. However, we must also note that ethirimol may not be entirely safe on barley and may, on its own, cause losses of the order of 7 % (Wolfe, 1969). An allowance for this would require a further correction to the yield estimates based on equation (1); these may underestimate the losses by 34 %. This sequence of events illustrates some of the difficulties and uncertainties in such methods and underlines the need to review the basic relationships between disease intensity and yield as techniques and varieties change.

Estimates of losses on a regional basis

Even when estimates of regional yield losses are available, their presentation in terms which have a real meaning presents considerable difficulty. In the simplest case, which ignores the possibility of penalties associated with the loss of quality, regional losses are normally presented in terms of the weight of crop lost. A comprehensive survey, covering a number of diseases, will produce a number of such estimates for each crop. These estimates are clearly not always additive since, for example, an apple which is already valueless as the result of insect attack cannot contribute to the losses attributed to a disease or another insect. However, the integration of these different estimates is a question of arithmetic guided by a knowledge of the behaviour of the disease in the crop and a sound integration will indicate the relative importance of the various diseases and pests in the crop. It is very much more difficult to arrive at an estimate of the value of these losses to a region or a country yet this should be the basis on which research priorities are decided. This aspect of the question has been studied in some detail by Ordish (1952) and Ordish & Dufour (1969) and is properly the realm of the economist. A straightforward estimate of loss by multiplying the estimated loss by the current market price will frequently give a

Table 1. *Some crop returns in California*

	Year	Calamity	Crop (1,000 tons)	Return to growers (U.S. 10^6)	Result to growers group
Peaches	1928	—	414	6·9	—
	1929	Frost	179	12·2	Gain
Prunes	1928	—	220	22·0	—
	1929	Frost	103	16·0	Loss
Oranges	1927	—	12·2*	50·0	—
	1928	Fruitfly	9·1*	52·2	Gain
	1929	—	17·7*	47·7	—
	1930	Fruitfly	9·3*	54·6	Gain

* Millions of boxes.

very misleading figure since it ignores the impact on the price of the remaining product reaching the market; in some circumstances a 'calamity' will actually benefit the farming community as a whole as is illustrated by data presented by Smith *et al.* (1933) and considered again by Ordish & Dufour (1969) shown in Table 1. However, one should always remember that any overall gains to the growers as a group conceals disaster to the unfortunate individuals who contribute to the shortage which raises the price. In view of the uncertainty over price, Ordish (1952) prefers to estimate losses as the 'untaken acres'; 'area wasted' might be a better term, since 'acres' are about to disappear as a measure of area. This essentially represents the area devoted to cultivating the disease and includes the cost of seed, cultivation, pest control, labour and other expenses incidental to raising the crop which the disease uses. The area released by controlling the disease is available to grow more of the original crop or an alternative if this seems advisable. In current literature this estimate of area wasted is probably the most common means of presenting losses due to pests and diseases.

Adequate surveys which related estimates of disease intensity to regional yield losses are very rare and the process can be illustrated by two classical surveys carried out by the Ministry of Agriculture Food and Fisheries in England and Wales. The first, extending over about 10 years, on potato blight, a disease caused by *Phytophthora infestans*, has been summarized by Large (1958), who also describes how disease intensity was related to yield. Potato blight is a disease which is commonly controlled with protectant fungicides. Since the epidemiology of the disease is fairly well understood, it is possible to delay the start of spraying until conditions for infection have occurred. However, once spraying has started, prophylactic spraying must be repeated to maintain a protective cover necessary to prevent the spread of the disease. The surveys included inspections of unsprayed plots so that they revealed not only the effectiveness of the treatments used to control the disease, but also provided an estimate of the losses which would have occurred if no control measures had been employed; from these the value of the control treatments could be estimated. Further, since the results were assessed on a regional basis, it was possible to indicate the potential value of spraying in different regions. Strickland (1966) has integrated Large's data into a wider review of pest losses and estimated that the losses which would have been caused by *P. infestans*, if no control measures had been taken, would amount on the average to about 12 % of the crop occupying 20,900 wasted hectares. The second survey was devoted to the leaf diseases of spring barley in 1967 and the results were published by James (1969). Losses will vary considerably from year to year and a survey confined to a single year must be interpreted with reservations: the results have been summarized and presented in Table 2.

Table 2. *Estimated losses due to leaf diseases of barley*

Losses due to	Loss (%)	Wasted hectares (× 1,000)
Mildew	18	328
Rust	4	81
Others	2	45
Total	24	454

Area of barley grown, 2,141,000 ha.
Average yield, 3·54 metric tonnes per ha.

It is clear that in this year by far the most serious losses were caused by mildew and, since the observers did not consider the incidence of mildew exceptional, this probably represents a reasonable estimate of average losses. Since barley is not over-produced in this country and deficiencies have to be made good by importing, it is probably reasonable in this case to attribute, as James (1969) did, a cash value to the crop calculated from the actual market price; on this basis mildew cost the economy £40–50 million in a single year. However, this figure was based on Large and Doling's relationship (equation 1); if this is, in fact, an underestimation as has been suggested, a better measure of the wasted hectares would be about 400,000 (24 %) for mildew alone and a loss of revenue of about £60 million. This estimate of loss is considerably higher than the 4 % estimated by Doling (1963) from the best data available prior to the 1967 survey. This example has been considered in some detail because it illustrates three rather important points. First, it shows how much loss a disease, which until recently has not been considered very serious, can actually cause. Secondly, how susceptible the estimates of loss are to changes in the basic assumptions, and thirdly, how difficult it is to provide reliable estimates of disease loss without the basic survey data and experimental work to allow its interpretation.

Regional estimates of losses due to disease are few and their accuracy is suspect since the basic data are only rarely available and have to be replaced by intelligent approximations, frequently based on data derived for other purposes. However, even imperfect estimates are valuable if they indicate the order of magnitude of the problems to be investigated. Three sets of data are frequently quoted: estimates prepared by Ordish (1952), a survey by Strickland (1966) and a survey by Watts Padwick (1956). Ordish's estimates are based on published reports and relate mainly to 1939; Strickland has published a number of surveys, based on reports and data gathered by the Ministry of Agriculture Food and Fisheries; most of these refer to losses caused by pests, but the one referred to also includes data on diseases: where relevant, this survey includes estimates of disease losses which would have occurred if no control measures had been taken. Strickland's estimate antedates the survey of cereal leaf diseases published by James (1969) and probably seriously underestimates disease losses in these crops; Crowdy (1970) suggested a revised estimate of these losses taking account of this information. Watts Padwick (1956) presents a series of disease loss estimates for tropical countries which formerly formed part of the British Empire. These are based on a rather primitive Colonial Plant Disease Survey in which losses were classified in three classes: A, causing losses over 10 %; B, causing losses under 10 %; and C, causing very small losses. These three classes were arbitrarily rated as causing losses of 15 %, 5 % and 1 % respectively. This was excellent pioneering work and still stands on its own as a comprehensive survey of disease losses in tropical crops. However, in the light of our present knowledge of the relationship between symptoms and losses it would be wrong to regard these figures as giving more than a general indication of the order of magnitude of these losses. Strickland (1966), with much more reliable data than were available to Watts Padwick (1956), estimated disease losses in Britain at about 4 %; later survey work has indicated that this estimate should be revised to about 18–29 %. Watts Padwick estimated average overall losses as 12 %; if these are scaled proportionately, the revised figure would be about 50 %. There is little reason to believe that losses due to disease are actually less in the tropics than in Britain, equally a 50 % loss seems excessive and probably a fair estimate would be in the neighbourhood of 30 %, about half-way between the extreme estimates. Whatever level is chosen, the losses are immense simply because world agricultural output is immense.

MECHANISMS BY WHICH A PARASITE CAUSES LOSSES

The grower is not usually interested in the means by which a pathogen causes losses, but this aspect of the situation is important to the plant pathologist who, by a knowledge of the processes involved, hopes to put himself in a stronger position for devising control measures, whether they be genetical, chemical or cultural. In this section we shall, for the most part, be concerned with the relationship between the pathogen and the single host plant, but it must be emphasized that loss depends on the effect of the pathogen on the population, rather than on the individual. When scattered plants in a closely grown crop, e.g. most field or plantation crops, are killed, or have their growth checked, the total loss will depend on the extent to which the loss is compensated by increased growth of surrounding shoots or plants. In potato leaf roll (Tuthill & Decker, 1941), the yield of a healthy plant is slightly increased when it is next to a diseased one, so that as long as the infection rate is reasonably low, compensation occurs. Where all plants or shoots are equally affected, or where large patches die, as in some root diseases, the extent of compensation will be negligible. Even when conditions are such that compensation might be expected, it is not always found.

Pathogens may be divided into those which kill the host plant, or part of it, and live on the dead remains, and those which extract nutrients from the living tissues. The former may be called necrotrophs and the latter biotrophs (Wood, 1967).

Necrotrophic pathogens

When part of a plant is killed, the mechanism by which loss is caused is usually obvious. If a photosynthetic organ such as a leaf is killed, e.g. by attack by *Venturia inaequalis* in the case of apple, the effect is similar to that of defoliation: the assimilate normally exported from that leaf will be lost; if the stem is ringed, as in apple canker, caused by *Nectria galligena*, the part of the plant above the ring will die, and the corresponding part of the yield lost; if the roots are killed, the whole plant will die. But there are certain less obvious aspects. For example, when part of a plant is killed and the fungus continues to live, does it obtain all its nutrient from the dead tissues, or is some obtained from living parts of the plant? Little work appears to have been done on this problem: Shaw, Brown & Rudd Jones (1954), in experiments involving the use of ^{14}C, found little or no accumulation of assimilate around lesions of *Botrytis* sp. on broad beans, *Helminthosporium* on wheat or *Pseudomonas coronafaciens* on oats. The results obtained with these pathogens were similar to those obtained by mechanically wounding healthy tissues.

Another question which arises with certain necrotrophic pathogens is that of determining the extent of the dead tissue. This involves defining death in relation to plant cells. It is generally agreed that, for practical purposes, a plant cell can be regarded as dead when the cell membrane has lost its semi-permeable properties, a character which may be determined with the aid of suitable dyes. However, the advent of electron microscopy enables other criteria to be employed. Examination of electron micrographs of cells affected by the soft rot bacterium *Erwinia carotovora* var. *atroseptica* (Fox, 1970) has shown that a progressive series of degenerative changes occurs. At what stage the process becomes irreversible, i.e. at what stage the cell becomes effectively dead, is uncertain. Disintegration of the membranes to an extent visible in electron micrographs is the final stage, though staining in neutral red and caustic potash (Strugger, 1935) indicates that no cell behind the infection front has an intact tonoplast.

Not only the amount of dead tissue, but also the extent to which the diseased tissues are degraded by the pathogen may affect the extent of loss. For example, Fox (1970) found that at low temperatures (around 5°) disintegration at the ultrastructural level in soft-rotted potatoes was relatively slight, whereas at higher temperatures (around 25°) it was much greater. The mechanism by which cytoplasmic breakdown increased at high temperatures was uncertain. Microbodies giving a positive reaction to a histochemical test (Barka, 1960) for acid phosphatase were identified in potato tuber cells; these were around 0·5 μ diameter in healthy cells and much the same size in diseased cells kept at 5°. In diseased tissues kept at 25°, however, the microbodies had increased to around 5 μ diameter, suggesting that they possessed the properties of lysosomes (De Duve, 1959) and were responsible for much of the extensive cytoplasmic damage seen at high temperatures; the possibility of 'cytases' produced by the pathogen could not be excluded. A correlation was noted between the extensive cytoplasmic damage visible at the electron-microscope level and the general experience of growers that at low temperatures the rot is inconspicuous, even though often widespread, whereas at high temperatures the products of cell breakdown are evil smelling and run out of the diseased tubers, contaminating neighbouring ones, so that the whole consignment becomes tainted and unsaleable, even though only a limited proportion may be actually rotted.

Biotrophic pathogens

In the case of biotrophic pathogens one cannot assess loss by merely assessing the amount of dead tissue, because the premature death which occurs is only the last stage of the process whereby the host is damaged: most of the loss is due to reduced and/or abnormal growth of parts of the host plant which are still living. A good deal of attention has been directed to the production of abnormal structures such as galls, but it seems that losses due to the diversion of materials into such structures are relatively slight in relation to those caused by the pathogen reducing the growth of the host plant as a whole. Consideration will be confined to this latter form of effect. In such cases, the extent of loss, the relative damage to various parts of the plant and the dependence of the effect on the duration of attack can be demonstrated by classical growth analysis techniques as employed, for example, by Murphy (1935) on crown rust of oats. Such an approach is still valuable, forming, as it does, an essential basis, too often neglected, for studies on the means by which the pathogen causes the observed losses. Murphy (1935) showed that the degree of yield loss depended on the duration of pathogenic attack (as might be expected) and that the part of the plant bearing the pathogen was not necessarily the part where growth was most affected. In fact the greatest losses (on a percentage basis) were in the grain and the roots, not the leaves, to which the pathogen was largely confined. Similar results were obtained by Yarwood & Childs (1938) working with *Puccinia helianthi* on sunflower.

Doodson, Manners & Myers (1964) investigated the effect of attack by *Puccinia striiformis* (yellow rust) on the growth and yield of a spring wheat, and found that the extent of damage depended on the duration of infection. Furthermore, the nature of the damage varied according to the date of infection. Early infection reduced the number and weight of grain per ear, as well as the single grain weight. With later infection, single grain weight, but not grain number, was reduced. The number of florets was reduced by early infection, but not by late (an examination of the apex showed that by the time late infection was carried out, the floret initials had already been laid down). The number of spikelets was not affected by infection (the initiation of spikelets, though not visible at the onset of 'early infection', presumably commenced at a very early stage).

Rust attack at any stage, except very late in development, reduced the height of the plant and the amount of tillering, as well as the length of leaves not mature when infection developed. The dates of ear emergence and anthesis were delayed by approximately 14 days. There was a very striking reduction in root development – greater than that in the development of any other part of the plant. It had been thought that because nearly all the assimilate passing to the grain originates from the flag leaf, rachis and glumes (Boonstra, 1936; Mayer & Porter, 1960), damage to the lower leaves was unimportant. However, Doodson *et al.* (1964) found that if only the lower leaves were rusted, yield was still substantially reduced, largely due to a reduction in the number of grain per ear, though the single grain weight was also slightly reduced. In fact, rusting of any part of the plant at any time had an effect on yield.

The means by which the various effects revealed by growth analysis are brought about have been studied in a number of biotrophic pathogens, and it appears that the mechanisms in all investigated cases have something in common. As Sempio (1959) commented, plants attacked by such pathogens commonly exhibit the symptoms of chronic starvation. Such symptoms may be induced by one or more of a variety of means:

(*a*) The parasite may act as a metabolic sink.

(*b*) The parasite may produce toxins which damage the host, or may induce the production by the host of such toxins.

(*c*) The hormone balance in the host may be altered.

(*d*) The formation of abnormal tissue, at the expense of normal growth, may be induced.

All these processes have been studied in plants attacked by biotrophic pathogens, but attention has, in general, been concentrated on the nature of the physiological and biochemical processes involved, rather than on the extent of the damage to the host. There have been exceptions, however, notably in the classical work of Murphy (1935) already mentioned, and that of Allen (1942) on *Erysiphe graminis* on wheat. More recently Doodson *et al.* (1964, 1965) have correlated yield losses with changes in assimilation rates and translocation patterns.

The various means by which pathogens bring about such effects may be considered in turn.

Effect on assimilation

Allen (1942), Livne (1964) and Doodson *et al.* (1965) have all shown that when plants are attacked by obligate parasites, the assimilation rate is reduced, whatever criterion (assimilation per unit weight of tissue, per unit area of leaf, or per unit amount of chlorophyll) is employed.

Certain effects on the photosynthetic system may be observed anatomically. The whole leaf may become chlorotic; this is due to a reduction in the amount of chlorophyll per chloroplast, rather than to a reduction in chloroplast number, at least in *Puccinia recondita* on wheat (Manners & Gandy, 1954). Conversely, green islands may appear around a pustule when the remainder of the leaf is yellow. This has been reported by Allen (1942) for cereal mildew and by various workers in certain rusts. Bushnell & Allen (1962) found that *E. graminis* on barley caused the tissues around a pustule first to become chlorotic, then redevelop chlorophyll. Development of green islands might also be induced by extracts of conidia of *E. graminis* or uredospores of *Puccinia graminis*. Substances such as auxins or kinetin (Mothes, Engelbrecht & Kulajewa, 1959) may also induce green island formation. Kinetin induced islands contain more starch but less nitrogen than mildew induced islands in barley (Bushnell, 1967). The chlorophyll around the green islands associated with pustules of *Uromyces phaseoli* on bean is photosynthetically active (Wang, 1961). Electron micrographs show (Plumb, 1969) the grana and intergranal lamellae of chloroplasts in leaves infected with *E. graminis* become progressively more disorganized as infection progresses.

Physiologically, the initial effect of infection is often that the rate of photosynthesis rises during the 24–48 h. following inoculation. This was seen by Doodson *et al.* (1965) in wheat infected by *P. striiformis*, in barley infected by *E. graminis* by Allen (1942), and in dwarf beans infected by *Pseudomonas phaseolicola* (R. Whitbread & C. N. Hale, private communication). The mechanism by which this increase is brought about is obscure. In the later stages of infection the rate of photosynthesis falls and Doodson *et al.* (1965) found that in wheat plants infected with *P. striiformis* the rate of fall was greater than the rate of decline in chlorophyll content, i.e. that assimilation rate/unit chlorophyll drops. Allen (1942) found this to be true also in barley infected with *E. graminis*. Wynn (1963) showed that photosynthetic phosphorylation per unit chlorophyll was not affected, so there must be damage to some process subsequent to phosphorylation.

Livne (1964) found a marked compensation effect in beans (*Phaseolus vulgaris*) infected with *Uromyces phaseoli*. In unifected leaves of infected plants the rate of assimilation was as much as 50 % above that of corresponding leaves on healthy plants. This may have been due to the removal of carbohydrate stimulating photosynthesis in such leaves, since it has been shown that in such plants there is enhanced translocation out of uninfected leaves. On the other hand, it may have been due to some more specific effect of the rust. The question of compensation is a difficult one on which to generalize; for example, Last (1963) showed that 7–10 days after inoculation of barley plants with *E. graminis*, when less than 30 % of the leaf was infected, the reduction was more than that attributable to loss of leaf surface; when more than 30 % was infected, it was less, i.e. compensation was occurring.

Effect on respiration

In recent years many workers have studied the effects of pathogen attack on respiration (see the review by Millerd & Scott, 1962), so only certain salient points will be discussed here. The general pattern of the effects is that described by Allen (1942) for wheat mildew. The respiration rate rises quite soon after infection occurs and this early rise can scarcely be attributed to the respiration of the pathogen. As long as the host tissue is alive, the rate tends to be above that of corresponding healthy plants, but as necrosis sets in, the rate drops. The important point in relation to yield loss, as emphasized by Allen (1942), is that in biotrophic pathogens there is often a considerable period when the respiration rate of an infected leaf is high, but the assimilation rate low, so that the net amount of carbohydrate available to the plant is very low.

Table 3. *Effects of attack by Erysiphe graminis on photosynthesis and respiration rates in barley (Allen, 1942)*

The results are expressed as moles C/unit wt. leaf/day.

	Age of infection (days)			
	0	4	8	12
Photosynthesis	45·3	30·2	6·7	3·0
Respiration	2·1	3·4	5·4	2·9

In the later stages at least, it is impossible in a straightforward comparative experiment to determine how much of the additional respiration is attributable to the host and how much to the pathogen, but this is irrelevant in relation to effects on yield, since in either case the energy used in respiration is lost to the host. If it is desired to distinguish between the respiration of the host and that of the pathogen, one of the techniques listed by Wood (1967) must be employed. The means by which such increases in respiration are brought about is debatable, and different factors may be involved in different host–parasite combinations. Increased availability of substrates (Allen, 1942; Shaw & Samborski, 1956) is probably involved. The production of toxins uncoupling phosphorylation and thus causing inhibition of the Pasteur effect has been postulated (Allen, 1953; Farkas & Kiraly, 1955), but more recent evidence (Shaw & Samborski, 1957) suggests that uncoupling, if it occurs at all, is unimportant. It seems more likely that the increased accumulation of metabolites, often in a mobile form, around lesions is the main cause of the increased respiration observed. For instance, this is seen when ^{14}C fed to the plant as CO_2 or carbohydrate accumulates there. In some cases hypertrophy occurs, and where this is not seen the dry weight per unit area of infected leaves is often higher than that of corresponding control leaves. Thus Inman (1962) found that starch and sugar contents per gram fresh weight in leaves of *Phaseolus vulgaris* infected with *Uromyces phaseoli* were double those in corresponding leaves of control plants at certain stages of the development of the infection. Such accumulations do not necessarily imply increased synthesis: they have also been reported in rusted cereals (Samborski & Shaw, 1956), where the evidence suggests little or no diversion of assimilates produced elsewhere into infected leaves (Doodson *et al.* 1965; Siddiqui, 1966).

In rusts and mildews, at least, increased respiration is associated with an increased importance of the hexose monophosphate (HMP) pathway (Farkas & Kiraly, 1955; Daly, Sayre & Pazur, 1957). More recent work by Daly, Bell & Krupka (1961) suggests that much of the HMP respiration is carried out by the parasite rather than by the host, since in the rust concerned (*Puccinia carthami* on safflower) there is a large amount of rust mycelium and in their experiments the rise in respiration coincided with sporulation. Furthermore, the respiration of rust uredospores seems to be largely via the HMP pathway (Shaw, 1964). Much of the increased respiration in diseased plants, however, occurs before the pathogen sporulates, and must be due to stimulation by the parasite of host respiration, which still appears to proceed at least in part by the EMP route (Daly, 1967).

Effect on transport

Most of the information available is on the translocation of assimilates (Smith, Muscatine & Lewis, 1969) as revealed by the growth of plants in $^{14}CO_2$, but the effect of infection on the transport of water and mineral salts may also be important. Since many substances transported in the phloem appear to move down a sugar concentration gradient, information on the movement of soluble carbohydrate is likely to be applicable to other phloem-transported substances. One of the most frequently reported effects of attack by biotrophic pathogens on translocation is that involving the retention of assimilate in an infected organ. This has been demonstrated with many pathogens, e.g. *Erysiphe graminis* (Allen, 1942; Shaw & Samborski, 1956), *Puccinia graminis* (Shaw & Samborski, 1956), *P. striiformis* (Doodson *et al.* 1965), in contrast to the situation in necrotrophic pathogens where, as already noted, there is little or no accumulation. Much of the accumulation is in the fungal tissues rather than in host (Thrower, 1965), though some retention in host tissues also occurs, depending partly on the state of the remainder of the plant. For example Doodson *et al.* (1965) found that when a single leaf of a wheat plant was infected, the proportion of the assimilate exported was only 0·4 % compared with 20 % in a corresponding healthy leaf on a control plant. When the whole plant was infected, however (Siddiqui, 1966), the proportion exported, in a comparable experiment, fell only from 35 % to 15 %.

The form in which the carbohydrate accumulates varies according to the host. In cereals infected with rust or mildew, much of the insoluble fraction is starch, as shown by Bushnell & Allen (1962) for *E. graminis* on barley. The same holds true for legumes as shown by Thrower (1965), working on rust (*Uromyces trifolii*) of *Trifolium subterraneum*. In a composite, *Tussilago farfara*, infected by *Puccinia poarum*, the insoluble material was mostly inulin (Yuen, 1969). Daly, Inman & Livne (1962) and Livne (1964) showed that when [^{14}C]glucose was fed to safflower infected by *Puccinia carthami* the soluble fraction of the accumulated carbohydrate contained not only host sugars such as glucose, fructose and sucrose, but also compounds tentatively identified as trehalose, mannitol and arabitol, which between them form nearly half the radioactivity in the soluble carbohydrate fraction at the sporulating stage. These three compounds have subsequently been identified in host–parasite combinations as diverse as *Erysiphe graminis* on barley (Edwards & Allen, 1966), *Puccinia poarum* on *Tussilago farfara* (Yuen, 1969) and *Ustilago nuda* on wheat (R. Gaunt, private communication). In infection by *Ustilago nuda* these compounds accumulate in the nodes and heads, where the fungus is concentrated, but not elsewhere. These fungal metabolites appear to be poorly utilized by higher plants unless they are normal constituents of the species concerned (Lewis & Smith, 1967), hence conversion of carbohydrate to these forms will remove it from the carbohydrate pool of the host, and will thus have a 'sink' effect.

In some cases there is evidence that a pathogen not only reduces export of carbohydrate from an infected organ, but also that it promotes import into such an organ. Livne & Daly (1966) and Thrower & Thrower (1966) showed that in legumes, assimilates in the translocation system could be diverted to an infected leaf from elsewhere, in an otherwise uninfected plant. Doodson *et al.* (1965), in comparable experiments on wheat, were unable to demonstrate any such movement into an infected, mature leaf. This may, as Thrower & Thrower (1966) suggested, be due to some special physiological or anatomical property of monocotyledonous leaves. It does not, however, apply to wheat stem apices, as R. Gaunt (private communication) found that there was a massive diversion to the apex in a smutted plant at a growth stage earlier than that at which massive flow to the apex normally occurs. This influence was effective over long distances; assimilate could be attracted from a main stem to a tiller or vice versa.

The attraction of assimilate to a diseased plant organ, together with the decreased assimilation and increased respiration found, will starve the plant of carbohydrate. In such circumstances the distribution of carbohydrate in the plant as a whole may be altered, quite apart from any specific flow to, or retention in, diseased organs. This aspect of translocation effects has been much neglected, partly because many workers have employed plants with only a single infected leaf, so that the plant as a whole was not starved. In such circumstances the translocation pattern of the plant as a whole may be almost unaffected as found by Doodson *et al.* (1965), working on wheat infected by *Puccinia striiformis*.

However, the greatly decreased root growth of rusted cereal plants, already mentioned, led Siddiqui (1966) to suspect that when an entire wheat plant was infected with *P. striiformis* and each leaf fed in turn (in a different replicate), as soon as it was fully grown the translocation pattern might be distorted. He found that this was so: the percentage of the total translocate moving to the leaves was enhanced in infected plants, at the expense of the roots, when the fed leaf was the second, third or fourth leaf, at the expense of both roots and tillers when the fifth, sixth or seventh leaf was fed. By the time that the flag leaf (leaf eight) was fully developed, the predominance of the ear as a sink was even more pronounced in the infected than in the control plants.

Earlier workers (see Sempio, 1959) considered that the observed effects of most biotrophic pathogens on the growth of the host could be explained by assuming that the former were acting as metabolic sinks. But this was not entirely satisfactory as a theory, since it could scarcely explain the massive accumulation of carbohydrates metabolized by the host in the neighbourhood of the lesions of the parasite. Undoubtedly the pathogen acts as a sink, both by using carbohydrate for its own structure, by using it as an energy source, and by secreting it in the form of compounds not available to the host plant, but this is not a complete explanation.

In some of the earliest work on this aspect of host–parasite relations Shaw & Samborski (1956) drew attention to the resemblances between accumulations at meristems and that around disease lesions, and suggested that diffusible growth regulators might be responsible. As pointed out by Smith *et al.* (1969), it is difficult in many experiments where hormone is exogenously applied to determine whether the effect observed is due to the hormone itself or increased metabolic activity at the site of application. However, evidence is accumulating that alteration in hormone balance may frequently accompany attack by biotrophic pathogens and it seems likely that at any rate some part of the observed alterations in translocation patterns is due to them. Evidence for the involvement of such substances is of four types. First, there is evidence of increased hormone content in infected tissues. Increased β-indolyl acetic acid (IAA) levels have been found in many cases (Sequeira, 1963). Enhanced gibberellin levels have been reported in the thistle *Cirsium arvense* when attacked by *Puccinia punctiformis* (Bailiss & Wilson, 1967). Kiraly, El Hammady & Pozsar (1967) found increases in cytokinin levels in bean (*Phaseolus vulgaris*) attacked by *Uromyces phaseoli* and in broad bean (*Vicia faba*) attacked by *U. fabae*. Secondly, the flow of assimilates into leaves, rather than out of them, as in healthy leaves, can be induced by removal of the terminal bud (Pozar & Kiraly, 1966). This reversal can be prevented by applying kinetin or benzyladenine, but not IAA, to the cut apex. Thirdly, the production of green islands around rust pustules, often coincident with increased rates of senescence elsewhere, can, as already mentioned, be induced by local application of certain growth factors. Fourthly, the amount of mycelium present in the early stages of some infections, e.g. those of cereal smuts (R. Gaunt, private communication), is so small that it is not plausible to postulate that the substantial diversions of assimilates observed is due solely to the pathogens acting as a metabolic sink. The exact means by which such substances may affect the growth of the plant is still not fully understood despite much work during the past few years. The diversion of food materials to infected leaves is clearly one possible mechanism, but this is not the whole picture, as various workers, e.g. Yao & Canvin (1969), have shown that even when supplies of soluble carbohydrate are very adequate, an inbalance of growth factors such as gibberellin may inhibit growth in wheat.

CONCLUSION

In the case of some crop diseases we now know by what means the pathogen reduces the yield of the host. Usually there is a suggestion that more than one mechanism is involved, and the relative importance of the various mechanisms thought to be concerned is not known. In any case, external as well as internal factors have to be taken into consideration. It is also probably true to say that so far, it has not been possible to apply our knowledge of the mechanism by which such losses are caused to the reduction of those losses. Up to now disease control has been dependent mainly on the breeding of resistant varieties and on the use of fungicidal chemicals, and provided that control was achieved, a knowledge of the exact mechanism by which it was attained was of little importance. However, the frequent appearance of pathogen races overcoming major gene resistance in new varieties, or even developing resistance to chemicals, suggests that in future more sophisticated methods, involving minor genes conferring 'horizontal' resistance (Van der Plank, 1968), and chemicals whose mode of action is known may become necessary. The development of such methods will inevitably be expensive, and this will make it all the more necessary to obtain accurate information, not merely of the amount of disease present, but also of the extent of loss in yield, and of the loss in financial and economic terms.

REFERENCES

ALLEN, P. J. (1942). Changes in the metabolism of wheat leaves induced by infection with powdery mildew. *Am. J. Bot.* **29**, 425.

ALLEN, P. J. (1953). Toxins and tissue respiration. *Phytopathology* **43**, 221.

BAILISS, K. W. & WILSON, I. M. (1967). Growth hormones and the creeping thistle rust. *Ann. Bot.* N.S. **31**, 195.

BARKA, T. (1960). A simple azo-dye method for histochemical demonstration of acid phosphatase. *Nature, Lond.* **187**, 248.

BOONSTRA, A. E. H. R. (1936). Der Einfluss der verschiedenen assimilierenden Teile auf den Samenertrag von Weizen. *Z. Zücht.* A **21**, 115.

BUSHNELL, W. R. (1967). Symptom development in mildewed and rusted tissues. In *The Dynamic Role of Molecular Constituents in Plant–Parasite Interaction.* Ed. C. J. Mirocha and I. Uritani. St Paul, Minnesota: American Phytopathological Society.

BUSHNELL, W. R. & ALLEN, P. J. (1962). Induction of disease symptoms in barley by powdery mildew. *Pl. Physiol., Lancaster* **37,** 50.

COBHAM, R. O. (1965). *Fernhusrt Orchard.* Haslemere, Surrey: Plant Protection Ltd.

CROWDY, S. H. (1970). Future technical developments in control of pests and diseases. In *Potential Crop Production in Britain.* (In the Press.)

DALY, J. M. (1967). Some metabolic consequences of infection by obligate parasites. In *The Dynamic Role of Molecular Constituents in Plant–Parasite Interaction.* Ed. C. J. Mirocha and I. Uritani. St Paul, Minnesota: American Phytopathological Society.

DALY, J. M., BELL, A. A. & KRUPKA, L. R. (1961). Respiratory changes during development of rust diseases. *Phytopathology* **51,** 461.

DALY, J. M., INMAN, R. E. & LIVNE, A. (1962). Carbohydrate metabolism in higher plant tissues infected with obligate parasites. *Pl. Physiol., Lancaster* **37,** 531.

DALY, J. M., SAYRE, R. M. & PAZUR, J. H. (1957). The hexose monophosphate shunt as the major respiratory pathway during sporulation of rust of safflower. *Pl. Physiol., Lancaster* **32,** 44.

DE DUVE, C. (1959). Lyzosomes: a new group of cytoplasmic particles. In *Subcellular Particles.* Ed. T. Hayashi. New York: Ronald Press.

DOLING, D. A. (1963). The incidence and economic importance of cereal disease. *Proc. 2nd Br. Insectic. Fungic. Conf.* p. 27.

DOLING, D. A. & DOODSON, J. K. (1968). The effect of yellow rust on the yield of spring and winter wheat. *Trans. Br. mycol. Soc.* **51,** 427.

DOODSON, J. K., MANNERS, J. G. & MYERS, A. (1964). Some effects of yellow rust (*Puccinia striiformis*) on the growth and yield of a spring wheat. *Ann. Bot.* N.S. **28,** 459.

DOODSON, J. K., MANNERS, J. G. & MYERS, A. (1965). Some effects of yellow rust (*Puccinia striiformis*) on [14]carbon assimilation and translocation in wheat. *J. exp. Bot.* **16,** 304.

DOODSON, J. K. & SAUNDERS, P. J. W. (1969). Observations on the effects of some systemic chemicals applied to cereals in trials at the N.I.A.B. *Proc. 5th Br. Insectic. Fungic. Conf.* p. 1.

DOODSON, J. K. & SAUNDERS, P. J. W. (1970). Some effects of barley yellow-dwarf virus on spring and winter cereals in glasshouse trials. *Ann. appl. Biol.* **65,** 317.

EDWARDS, H. H. & ALLEN, P. J. (1966). Distribution of products of photosynthesis between powdery mildew and barley. *Pl. Physiol., Lancaster* **41,** 683.

FARKAS, G. L. & KIRALY, Z. (1955). Studies on the respiration of wheat infected with stem rust and powdery mildew. *Physiologia Pl.* **8,** 877.

FOX, R. T. V. (1970). *Mechanisms for the infection of potato tubers by the soft rot organism Erwinia carotovora var. atroseptica (van Hall) Holland and associated defence mechanisms.* Ph.D thesis, University of Southampton.

HULL, R. (1953). Assessments of losses in sugar beet due to virus yellows in Great Britain, 1942–52. *Pl. Path.* **2,** 39.

INMAN, R. E. (1962). Disease development, disease intensity and carbohydrate levels in rusted bean plants. *Phytopathology* **52,** 1207.

JAMES, W. C. (1969). Survey of foliar diseases of spring barley in England and Wales in 1967. *Ann. appl. Biol.* **63,** 253.

KIRALY, Z., EL HAMMADY, M. & POZSAR, B. I. (1967). Increased cytokinin activity of rust-infected bean and broad bean leaves. *Phytopathology* **57,** 93.

LARGE, E. C. (1958). Losses caused by potato blight in England and Wales. *Pl. Path.* **7,** 39.

LARGE, E. C. (1966). Measuring plant disease. *A. Rev. Phytopath.* **4,** 9.

LARGE, E. C. & DOLING, D. A. (1962). The measurement of cereal mildew and its effect on yield. *Pl. Path.* **11**, 47.

LARGE, E. C. & DOLING, D. A. (1963). Effect of mildew on yield of winter wheat. *Pl. Path.* **12**, 128.

LAST, F. T. (1963). Metabolism of barley leaves inoculated with *Erysiphe graminis* Mérat. *Ann. Bot.* N.S. **27**, 685.

LEWIS, D. H. & SMITH, D. C. (1967). Sugar alcohols (polyols) in fungi and green plants. I. Distribution, physiology and metabolism. *New. Phytol.* **66**, 143.

LIVNE, A. (1964). Photosynthesis in healthy and rust-affected plants. *Pl. Physiol., Lancaster* **39**, 614.

LIVNE, A. & DALY, J. M. (1966). Translocation in healthy and rust-affected beans. *Phytopathology* **56**, 170.

MANNERS, J. G. & GANDY, D. G. (1954). A study of the effect of mildew infection on the reaction of wheat varieties to brown rust. *Ann. appl. Biol.* **41**, 393.

MAYER, A. & PORTER, H. K. (1960). Translocation from leaves of rye. *Nature, Lond.* **188**, 921.

MILLERD, A. & SCOTT, H. J. (1962). Respiration of the diseased plant. *Ann. Rev. Pl. Physiol.* **13**, 559.

MOORE, F. J. (1969). The development of disease assessment methods. *Proc. 5th Br. Insectic. Fungic. Conf.* p. 711.

MOTHES, K., ENGELBRECHT, L. & KULAJEWA, O. (1959). Über die Wirkung des Kinetins auf Stickstoff-verteilung und Eiweissynthese in isolierten Blatten. *Flora, Jena* **147**, 445.

MURPHY, H. C. (1935). Effect of crown rust infection on yield and water requirement of oats. *J. agric. Res.* **50**, 387.

ORDISH, G. (1952). *Untaken Harvest.* London: Constable.

ORDISH, G. & DUFOUR, D. (1969). Economic basis for protection against plant diseases. *A. Rev. Phytopath.* **7**, 31.

PITT, D. (1966). Studies on sharp eyespot disease of cereals. III. Effects of disease on the wheat host and the incidence of disease in the field. *Ann. appl. Biol.* **58**, 229.

PLUMB, R. T. (1969). *Host–parasite inter-relationships in Erysiphe graminis infection of wheat.* Ph.D. thesis, University of Southampton.

POZSAR, B. I. & KIRALY, Z. (1966). Phloem-transport in rust infected plants and the cytokinin-directed long-distance movement of nutrients. *Phytopath. Z.* **56**, 297.

ROSSER, W. R. & CHADBURN, B. L. (1968). Cereal diseases and their effects on intensive wheat cropping in the East Midland Region, 1963–65. *Pl. Path.* **17**, 51.

SAMBORSKI, D. J. & SHAW, M. (1956). The physiology of host parasite relations. II. The effect of *Puccinia graminis tritici* Erikss. and Henn. on the respiration of the first leaf of resistant and susceptible species of wheat. *Can. J. Bot.* **34**, 601.

SEMPIO, C. (1959). The host is starved. In *Plant Pathology, an Advanced Treatise*, vol. 1. Ed. J. G. Horsfall and A. E. Dimond. New York: Academic Press.

SEQUEIRA, L. (1963). Growth regulators in plant disease. *A. Rev. Phytopath.* **1**, 5.

SHAW, M. (1964). The physiology of rust uredospores. *Phytopath Z.* **50**, 159.

SHAW, M., BROWN, S. A. & RUDD JONES, D. (1954). Uptake of radioactive carbon and phosphorus by parasitized leaves. *Nature, Lond.* **173**, 768.

SHAW, M. & SAMBORSKI, D. J. (1956). The physiology of host–parasite relations. I. The accumulation of radioactive substances at infections of facultative and obligate parasites including tobacco mosaic virus. *Can. J. Bot.* **34**, 389.

SHAW, M. & SAMBORSKI, D. J. (1957). The physiology of host–parasite relations. III. The pattern of respiration in rusted and mildewed cereal leaves. *Can. J. Bot.* **35**, 389.

SIDDIQUI, M. Q. (1966). *The translocation pattern of assimilates in wheat, with special reference to factors affecting translocation to the roots and root growth.* Ph.D. thesis, University of Southampton.

SLOPE, B. D. (1967). Disease problems of intensive cereal growing. *Ann. appl. Biol.* **59**, 317.

SMITH, D., MUSCATINE, L. & LEWIS, D. (1969). Carbohydrate movement from autotrophs to heterotrophs in parasitic and mutualistic symbiosis. *Biol. Rev.* **44**, 17.

SMITH, H. S., ESSIG, E. O., FAWCETT, H. S., PETERSON, G. M., QUAYLE, H. J., SMITH, R. E. & TOLLEY, H. R. (1933). The efficiency and economic effects of plant quarantines in California. *Univ. Calif. Bull.* no. 553.

STRICKLAND, A. H. (1966). Some costs of insect damage and crop protection. *Proc. F.A.O. Symposium on Integrated Pest Control*, **1**, 75.

STRUGGER, S. (1935). Beiträge zur Gewebephysiologie der Wurzel. Zur analyse und Methodik der Vitalfarbung pflanzlicher Zellen mit neutral rot. *Protoplasma* **24**, 108.

THROWER, L. B. (1965). On the host–parasite relationship of *Trifolium subterraneum* and *Uromyces trifolii*. *Phytopath. Z.* **52**, 269.

THROWER, L. B. & THROWER, S. L. (1966). The effect of infection with *Uromyces fabae* on translocation in broad bean. *Phytopath. Z.* **57**, 267.

TUTHILL, C. S. & DECKER, P. (1941). Losses in yield caused by leaf roll in potatoes. *Am. Potato J.* **18**, 136.

VAN DER PLANK, J. E. (1963). *Plant Diseases: Epidemics and Control.* New York: Academic Press.

VAN DER PLANK, J. E. (1968). *Disease Resistance in Plants.* New York: Academic Press.

WANG, D. (1961). The nature of starch accumulation at the rust infection site in leaves of Pinto bean plants. *Can. J. Bot.* **39**, 1595.

WATSON, M. A. & MULLIGAN, T. E. (1960). Comparison of two barley yellow-dwarf viruses in glasshouse and field experiments. *Ann. appl. Biol.* **48**, 559.

WATTS PADWICK, G. (1956). Losses caused by plant disease in the colonies. *Phytopath. Pap.* no. 1.

WOLFE, M. S. (1969). Pathological and physiological aspects of cereal mildew control using ethirimol. *Proc. 5th Br. Insectic. Fungic. Conf.* p. 8.

WOOD, R. K. S. (1967). *Physiological Plant Pathology.* Oxford: Blackwell.

WYNN, W. K. (1963). Photosynthetic phosphorylation by chloroplasts isolated from rust-infected oats. *Phytopathology* **53**, 1376.

YAO, Y. Y. & CANVIN, D. T. (1969). Growth responses of Marquillo × Kenya Farmer wheat dwarf 1 and 2 to gibberellic acid, kinetin and indolebutyric acid under controlled environmental conditions. *Can. J. Bot.* **47**, 53.

YARWOOD, C. E. & CHILDS, J. F. L. (1938). Sone effects of rust infection on the dry weight of host tissues. *Phytopathology* **28**, 723.

YUEN, C. K. L. (1969). *The movement of carbohydrates from green plants to fungal symbionts.* Ph.D. thesis, University of Sheffield.

MICROBIAL DISEASES AND ANIMAL PRODUCTIVITY

J. B. DERBYSHIRE

Institute for Research on Animal Diseases, Compton, Newbury, Berkshire

INTRODUCTION

In the context of this contribution, microbial disease is defined as any infection with bacteria, viruses or protozoa which decreases the productive capacity for an animal. The animals involved are domesticated farm livestock maintained primarily for food production. Animals maintained essentially for sport or recreational use are excluded from consideration. At the level of the individual animal, productivity is best measured in terms of efficiency of food conversion. From the point of view of the livestock producer, additional factors include a high return on the capital and labour costs of his operation. At the national level, animal productivity is concerned primarily with the efficient utilization of national resources, and with import saving or exports.

It will be apparent from this definition that the effect of microbial disease on animal productivity is invariably unfavourable. The only minor qualification that this statement requires is to indicate that, on occasion, measures which are introduced in order to control animal disease may at the same time create generally more favourable husbandry conditions. There is no doubt that the presence of disease in a population is a potent stimulus to a critical appraisal of the husbandry system. Any possible directly beneficial effects of microbial infection on animal productivity are excluded from consideration in this paper. The influence of the intestinal microflora on animal productivity is dealt with in another contribution to this Symposium (see p. 149). However, the border between health and disease is not always clearly defined, and the important influence of subclinical microbial disease on productivity will be emphasized.

It is hardly necessary to stress the importance of maintaining and improving the productive efficiency of farm livestock. While estimates of world food supply in relation to population increase range from Malthusian gloom to cautious optimism, it is an inescapable fact that animal protein is becoming increasingly scarce in relation to the expanding human population. According to the Food and Agriculture

Organization statistics, in 1968 world food production increased by 3 %, but the production of meat increased by only 2 %. Food conversion via the animal is economically wasteful, although nutritional and genetical research on farm animals is minimizing this wastage, and it should be kept in mind that the ruminant species share with plants the ability to convert non-protein nitrogen into high quality protein. It is clear, however, that wastage due to disease should not be allowed to widen further the food conversion gap involved in animal protein production. Economic pressures in this context are likely to increase as manufactured goods of vegetable origin compete with natural foods, such as meat and milk, for the household budget. Another factor of importance is competition for land. Four cows maintained under non-intensive husbandry conditions occupy more than enough land to house four families. In recent years the shortage of land and labour has led to increasingly intensive systems for livestock production, but these new systems of management have created new disease problems.

Consideration will be given first to the ways in which animal productivity is decreased by microbial disease. It will be shown that wastage due to disease may range from the total loss represented by death of the animal, through production losses which may occur in clinically sick animals, to the more subtle wastage in subclinical infections which may only be detected as a fractional decrease in efficiency of food conversion. In addition to the above direct effects of disease, consideration will also be given to some of the indirect consequences. These losses may be very high indeed and the productivity of an area may be limited by diseases in the indigenous population which preclude the introduction of high-yielding animals.

The measures required to control animal disease will depend upon the nature of the problem in the particular area. In the technically advanced countries, many of the major epidemics of farm livestock have been controlled or eradicated, and the measures necessary to achieve this success will be discussed. In less developed areas, this kind of infection remains as a significant source of loss and a major barrier to productive expansion of the local animal industry. Some of the efforts which are being made to deal with the situation will be outlined, and an indication will be given of the problems which are associated with the implementation of adequate disease control measures.

The potential benefits of genetical improvement of livestock and improved nutrition are lost in the absence of efficient disease control. Conversely, in order to make a major contribution to productivity, it is essential that successful disease control be accompanied by improved

dietary and breeding practices, or an expanding population of unproductive animals will result. Thus it is essential to distinguish between
increased production and increased productivity, for only the latter
makes a net contribution either to the world food supply or the national
economy. These considerations are particularly relevant to certain
developing countries with relatively primitive systems of animal
husbandry.

Current problems in the agriculturally developed countries are of a
somewhat different order. Continual vigilance is required to exclude
epidemics, but in addition new problems have arisen. Increasingly
intensive husbandry systems have led to the emergence of diseases which
are unrecognized under more natural conditions. Infections of this kind,
which are particularly important in young animals, range from severe
fatal diseases to conditions in which the losses are measured in terms
of lowered production rather than clinical disease. The control of this
kind of infection is currently one of the most important areas of
veterinary microbiological research. The merits and hazards of dietary
supplementation with antibiotics, and the use of specific pathogen-free
(S.P.F.) animals will be discussed in this context. A further problem
associated with intensive animal husbandry is the disposal of increasing
quantities of farm effluent in such a manner as to avoid the spread of
disease to the human and animal population in the area.

It will be apparent that the scope of this contribution is extremely
broad. Adequate consideration of the subject requires an attempt to
cover the situation on a global basis and calls for reference to political,
social and economic factors in addition to the microbiological aspects
of the problem, for in some parts of the world the administrative
difficulties in disease control are greater than the biological problem.
Therefore it will be possible only to outline the general principles involved, with specific examples drawn from a wide field.

PRODUCTIVITY LOSSES DUE TO MICROBIAL DISEASES

It is impossible to quote a figure for world-wide losses caused by microbial diseases in animals. Inadequate reporting of all diseases in certain
countries, and on certain diseases in all countries, severely limits the
data which are available, although through the efforts of the Food and
Agriculture Organization (F.A.O.) and L'Office International des
Epizooties (O.I.E.) data collection is improving. In the Animal Health
Yearbook for 1962, the F.A.O. (1963) attempted to assess the economic
losses caused by animal diseases. An assessment of total losses was

available from only a few countries, but even where intensive veterinary activity had been in progress for many years, losses still ranged between 15 % and 20 % of the total value of annual animal production. Where veterinary activities were less intensive, losses were estimated at 30–40 %. The F.A.O. publication (F.A.O., 1963) made it clear that losses due to the major epidemic diseases, such as foot-and-mouth disease, swinefever and Newcastle disease, were most impressive. Losses from chronic and intercurrent diseases were less dramatic but still high.

Apart from inadequate reporting and data collection, estimates of disease losses are difficult to make because of the problem of defining and assessing invisible and consequential losses. The main advantage of accurately assessing animal disease losses is in defining their impact on the economy of a country, as a stimulus to the introduction of appropriate control measures. Statistics on the relative importance of different diseases indicate where veterinary services and research are most urgently required. Individual livestock owners are also stimulated to invest more in disease control and health insurance programmes.

More detailed consideration will now be given to the ways in which microbial diseases in animals cause loss and thereby decrease the productivity of the livestock industry.

Mortality

Death is the most obvious and dramatic effect of the introduction of a disease into a population, as it involves the total and immediate loss of the animals involved. In Britain during the last century, mortality in cattle due to rinderpest was as high as 90 % in some of the infected areas, and the situation due to the same disease in Southern Africa at the beginning of this century was similar. Some diseases produce a high mortality only in young animals. Thus porcine transmissible gastroenteritis virus kills all infected piglets under 10 days of age, so that losses in large breeding units can be considerable. Overall neonatal mortality in farm animals tends to be very high; about 20 % of piglets born die before weaning, and a considerable proportion of neonatal mortality in both calves and piglets is associated with enteric bacterial infection.

Production losses

This kind of loss, although less dramatic than the death of an animal, can be equally wasteful, as the maintenance of non-productive stock represents a continual drain on valuable feedingstuffs. Thus foot-and-mouth disease is rarely fatal, but may involve major losses in meat and milk production because of its debilitating effect on infected animals.

Bovine mastitis, due to streptococcal and staphylococcal infection, is rarely a killing disease but acute infection may result in total cessation of lactation, and in the insidious chronic form milk-yield losses of at least 10 % are common. In 1960 the average annual total loss due to bovine mastitis in France was estimated at 100 million U.S. dollars (F.A.O., 1963). This figure included the loss in milk production, the costs of premature replacement of incurable cows, and losses in the manufacture of dairy products. In the poultry industry, depression of egg production is a widely recognized result of a variety of mild viral and mycoplasma infections. In this kind of disease, depressed production is often the first and sometimes the only indication of the presence of infection. Calf pneumonia, an increasing problem in intensive beef production, is important because of the diminished productivity of an infected herd. The mortality associated with the disease is rarely more than 2·5 %, but to this loss must be added the chronic depression of growth rate, which may be only 50 % of the normal and may necessitate culling of unproductive animals (MacLean, 1969). In uncomplicated enzootic pneumonia of pigs deaths are extremely rare, but growth and food conversion rates are adversely affected to such an extent that the national economic loss from this disease in Britain was estimated at 10 million pounds sterling annually (Beveridge, 1960). Livestock production may be seriously impaired by infections of the genital tract which lead to abortion, stillbirth and infertility. Examples of such diseases include brucellosis, vibriosis and trichomoniasis in cattle, and vibriosis and enzootic abortion in sheep and parvovirus infection in swine. Harnett (1956) estimated that, in Ireland in 1954, 10 % of potential calf production was lost, amounting to some 123,000 calves. It was calculated that half of this figure was due to abortion or stillbirth, and the remainder represented potential production lost due to infertility.

Consequential losses

The strain on the national economy associated with some of the indirect effects of animal disease may far exceed the immediate effects on the industry. There is much variation among different infections and in different countries in the kinds of consequential loss which are important. In the case of the zoonoses, the economic loss from human disease may be considerable, but this form of consequential loss is slightly beyond the scope of this paper, as animal productivity is not directly involved. Disease in draught animals is also important from the point of view of its indirect effect on agricultural productivity. In countries which lack alternative means of traction, an infection such as the 1959/60 epidemic of African horsesickness in the Near East may result in the total loss of field crops. In those countries in which livestock and livestock products are important in the export trade, foreign-trade restrictions due to the presence of a contagious disease may severely affect the economy. In the foot-and-mouth disease outbreak in Canada in 1951–2, the one million Canadian dollars paid in compensation was considered a small sum compared with the losses suffered due to the closure of foreign markets. In Ireland, the annual cost of the bovine tuberculosis eradication scheme was estimated in 1959 at close to 10 million U.S. dollars, but this expenditure was justified by the threatened export market in the United Kingdom, valued at 110–140 million U.S. dollars annually (F.A.O., 1963). In some examples, the disease may be relatively unimportant locally as a cause of loss. In certain African countries, rinderpest and foot-and-mouth disease are regarded as most important because of the loss of potential export markets, and this also applies to foot-and-mouth disease in South America. The local losses due to the infections are less well appreciated.

Perhaps the most far-reaching consequences of the presence of microbial disease in an indigenous animal population in an undeveloped country are the restrictions which this places on the development of the livestock population. At the extreme, the prevalence of carriers of infection in the wild population, and of the vectors of infection, may totally preclude the development of large areas of land for livestock production, and in the less extreme situation, the presence of infection in the domesticated native stock may make it impossible to introduce safely high-yielding animals to replace unproductive but resistant stock. The economic importance of this form of loss is impossible to assess in monetary terms, but its significance has long been recognized in some countries such as South Africa (Jansen, 1969) which pioneered veterinary preventive medicine in the African continent from the beginning of this century. Elsewhere the position has unfortunately been slow to improve.

Finally, it is necessary to make some reference to the costs of the control and prevention of microbial disease in relation to productivity. While control measures are intended to minimize losses due to disease, it is usually necessary to include their cost when an assessment is made of the total impact made by the disease on the national economy. Control measures may be expensive, since they may include widespread vaccination, vector control, preventive chemotherapy and quarantine measures. Where control measures include the payment of compensation to livestock owners, a distinction has to be made between that paid for animals which have died from a disease and that paid for healthy animals slaughtered to prevent spread of the disease. In the former case, the payment made merely transfers the loss from the owner to a government agency or some kind of insurance programme, while in the latter, compensation constitutes an additional consequential loss due to the disease. The unique feature of expenditure on the control of disease is that this portion of the consequential loss of the disease may justifiably be regarded as an investment to prevent a greater loss in productivity, provided the control measures are effective.

PRODUCTIVITY AND THE CONTROL OF MAJOR EPIDEMICS

The effective control of disease does not necessarily lead to increased productivity. This will only result if the increased number of livestock can be fed adequately and if they are genetically capable of a high level of production. Otherwise the result may be overstocking with unproductive animals which will require to be culled in order to avoid starvation. In some areas, culling is not acceptable, as the owner's social standing may be reflected by the number rather than the quality of his animals. Elsewhere, religious taboos may prevent the culling of unproductive animals.

The ultimate in microbial disease control is eradication, rigorously defined by Cockburn (1963) as 'the eradication of the pathogen that causes the infectious disease; so long as a single member of the species survives then eradication has not been accomplished'. This objective is highly desirable, since control measures are no longer required once complete eradication has been achieved. Eradication is very difficult to achieve on a global basis, but national and regional eradication programmes for certain infections have achieved considerable success. Reid (1969) listed the general requirements for the effective eradication of disease as: (1) a stable system of government; (2) the support of the local population; and (3) financial and technological assistance. The specific requirements are: (1) identification and treatment of cases and carriers; (2) removal of reservoirs of infection; (3) quarantine arrangements; and (4) immunization.

General requirements for epidemic disease control
Political stability

A stable form of government is an important basis for national eradication programmes. Such programmes are usually long-term endeavours which are vulnerable to political policy change. Reid (1969) pointed out that in times of political upheaval funds and manpower may be diverted to other activities. Preventive measures can only be efficiently applied when the civil authority is able to enforce the appropriate regulations. Also of major importance is the frequent need of international cooperation in the successful application of control measures. Cockrill (1965) has reviewed the existing international machinery relating to disease control measures in animals. The main organizations involved are the F.A.O. and the World Health Organization (W.H.O.) of the United Nations. The former covers all animal infectious diseases, and the latter only the zoonoses. A more senior body is O.I.E., which collects and disseminates information on livestock disease and advises on control methods, but does not carry out field programmes. Many other national and regional organizations operate on a less wide basis, such as the Pan American Foot and Mouth Disease Centre which provides technical assistance and support for the prevention and control of foot-and-mouth disease in the Americas, the European Commission for the Control of Foot and Mouth Disease, and the Inter-African Bureau for Animal Health.

Support of the local population

Local support is vital to the success of any disease-control programme, since the latter is dependent upon rapid notification of outbreaks of disease and a high level of cooperation in dealing with outbreaks, and in some instances with the collection of animals for examination and immunization. Livestock owners may be encouraged to cooperate by compensation for disease losses, by bonus payments in recognition of successful eradication at the local level, or by the enforcement of penalties when irregularities come to light. In spite of rigid precautions, the illegal movement of livestock has been responsible for the spread of infectious diseases. Todd (1958) gave several historical examples, dating from the introduction of contagious bovine pleuropneumonia into England by illegally imported Dutch cattle in 1839 to the appearance of foot-and-mouth disease in Germany as a result of smuggling immediately after the Second World War. More recently, Cockrill (1963) pointed to the well-established smuggling routes in many of the Latin American countries, along which thousands of cattle may pass illegally across long and ill-defined borders which are impossible to guard. In the same paper, Cockrill suggested the possibility of intercontinental smuggling of livestock from India to South America, which could lead to epidemics on a devastating scale.

Financial and technological assistance

The need for financial and technological assistance in controlling the major epidemics is apparent from a study of the global distribution of these diseases, which continue to cause the greatest losses in undeveloped countries. The knowledge required for their effective control is frequently available, but the resources in materials and personnel required to apply this knowledge may be lacking. One of the most important epidemic diseases is rinderpest, and Scott (1964a) indicated that the persistence of this disease was related inversely to the resources in staff and finance in the countries involved. The 23 countries plagued by rinderpest in 1962 were populated by 261 million cattle and 6,000 veterinarians, while the United States, for example, contained 96 million cattle and 20,000 veterinarians. The 'emerging diseases', such as African swinefever, African horsesickness and the South African Territories (S.A.T.) types of foot-and-mouth disease have emerged most frequently in underdeveloped countries where veterinary services may be inadequate. For this reason, Cockrill (1965) proposed the establishment of an International Emergency Fund to finance the early containment of outbreaks of epidemic diseases in areas where assistance is required. All countries should prepare contingency plans for action in dealing with outbreaks of exotic diseases. Cockrill (1964) also discussed some of the requirements for technical assistance in animal health, and indicated the importance of such assistance from the point of view of the technically advanced countries as an investment in the protection of their own livestock. The provision of supplies of vaccine alone is inadequate. Assistance in the building, equipping and staffing of diagnostic and production laboratories, and the establishment and staffing of veterinary field services are needed. Assistance in disease control is steadily increasing, both by the international agencies and on a country-to-country basis. Cockrill (1965) warned against too disproportionate an emphasis on research in technical assistance programmes, pointing out that in many instances attention might be given to the practical utilization of existing knowledge as an alternative to the initiation of ambitious, institutionalized research activities.

Specific requirements for epidemic disease control

Diagnosis and treatment of infected animals

In some infections the disease may be so characteristic that diagnosis does not present a problem. Indeed, rinderpest and contagious bovine pleuropneumonia were eradicated from Great Britain before the causes were known. This was only possible because infected animals could be identified on the basis of history, symptoms and post-mortem findings. Today, rinderpest may be confused with the mucosal disease complex in cattle (Scott, 1964*b*). In other diseases symptomatic diagnosis is impossible. Thus tuberculosis and brucellosis in cattle may not produce symptoms, but in both instances reliable diagnostic tests of infection are available. In the absence of reliable diagnostic procedures, or in the absence of the trained personnel required to carry out the tests involved, control is impossible. Even in the technically advanced countries there is a limit to the availability of veterinary resources and a scale of priorities has to be established for disease eradication programmes.

When infected animals have been identified, it is necessary to eliminate the infection. It is seldom possible to achieve this object by therapy, and the procedure of choice is usually slaughter. In the short term this is clearly unproductive, and may be very costly indeed in the case of valuable breeding stock. Major financial resources are required to operate this rigorous approach because some form of compensation may be required where the infection does not decrease the profitability of the infected animals. In some infections it is possible to recover some or all of the value of the animal if the carcase can be utilized for human food, but the risk of disseminating infection may preclude this action. This may occur during transportation, or indirectly by the feeding of contaminated uncooked garbage to susceptible animals, as in swinefever and vesicular exanthema.

In some countries the system of husbandry renders the identification and treatment of infected animals extremely difficult. Discussing the control of East Coast fever (Theileriosis), Wilde (1967) pointed out that there are large areas in East Africa which are used as grazing grounds by nomadic pastoralists, and in which no effective control of the disease can be practised. Considerable losses are accepted as normal, and under these conditions it was expected that East Coast fever would be prevalent for many generations, until stock movement could be effectively controlled. Wilde (1967) regarded East Coast fever as one of the major influences retarding livestock development in East Africa, and it is clear that the problem is sociological as much as scientific.

Removal of reservoirs of infection

In addition to carrier host animals, other reservoirs of infection may occur. Pastures and buildings may remain contaminated following outbreaks of disease. A pasture may frequently be regarded as safe following cultivation, although contamination with anthrax spores may persist for many years. The efficiency with which buildings may be disinfected depends partly on the nature of the infection and partly on the state of the building. Preliminary cleansing is essential, but this may be impossible in poorly constructed accommodation which may have to be replaced, adding further to the cost of control. Many microorganisms are readily destroyed by mild chemical disinfectants. Others, such as some of the smaller viruses, are highly resistant to many disinfectants, and are only destroyed by highly corrosive substances such as the chlorine-containing compounds. It is particularly important to disinfect vehicles that have been used to transport infected livestock or livestock products, or a method of actively spreading the infection will remain.

Wild animals may become carriers of infectious diseases, and they may thus be a hazard to domestic stock locally, and in other parts of the world if they are exported to zoological collections. Thus the introduction of wild ruminants which may be infected with the virus of rinderpest or foot-and-mouth disease constitutes a serious hazard. Wart hogs and bush pigs are especially dangerous as carriers of African swinefever. Effective game eradication appears to be extraordinarily difficult to achieve. Du Toit (1959) described two such attempts in Zululand, which were intended to control the hosts of tsetse flies, vectors of trypanosomiasis. In 1929 and 1930 an attempt was made to create a buffer zone, and although 37,861 animals were killed the results were most disappointing and the epidemic of trypanosomiasis persisted. A second more extensive effort was made later, designed to eliminate game outside certain reserves, but although 138,529 head of game were destroyed, the results were again disappointing and the campaign was finally abandoned in 1950. The restriction of wild animals to game reserves by fencing has been more successful. Jansen (1969) described the erection of a fence on the western side of the Kruger National Park, a vast undertaking which proved highly successful.

Arthropods play a vital part as vectors of some important infections, and consideration has to be given to their control. The chemical control of tsetse flies was described in some detail by Du Toit (1964). The procedure outlined included surveys to determine the breeding grounds of the flies, insecticide applications, bush clearing and dipping, and resulted in the eradication of one important species. Costs were discussed in detail, and it was concluded that the aerial application of insecticide was the least expensive method of eradication, estimated at some 10 shillings per acre for eight applications. This cost has to be related to the fact that, in tsetse-fly areas, livestock production is virtually impossible, and that eradication of the fly can open up vast areas of hitherto unproductive country to livestock production. The spraying of tsetse-fly breeding sites in the Eastern Caprivi strip began in 1964, and already the number of cattle in the area has trebled (Jansen, 1969). Yeoman (1964) surveyed the part played by ticks in the economy of East Africa. The control of tick-borne protozoal infections, of which the most important is East Coast fever, is vital to the agricultural economy of this area. Wilde (1967) indicated that, in Kenya, the recent replacement of large herds in which organized tick control was carried out, with small holdings of few cattle, might represent a considerable hazard in relation to the control of these diseases. Haig (1955) reviewed the tick-borne rickettsial infections, such as heartwater, in South Africa and pointed out that the regular dipping of cattle and sheep could never destroy all the ticks on a farm since the amblyommae feed on a variety of animal species. Some of the important arbovirus diseases of farm livestock were reviewed by Haig (1965). In some of these diseases the virus can be maintained in wild species of animals in addition to the arthropod vector. Thus Nairobi sheep disease, which is suspected of being transmitted by ticks, can be maintained in wild ruminants such as the blue duiker, and Rift Valley fever and Wesselsbron disease, for which mosquito transmission is suspected, have a wide host range including rodents and man. Bluetongue and horsesickness are transmitted by the almost universally present *Culicoides*, and antibodies against bluetongue virus were demonstrated in the sera of ten common game species in Kenya by Walker and Davies (1971). Eradication of such infections in endemic areas is thus virtually impossible.

Cockrill (1966) drew attention to the potential hazard of stored semen as a reservoir of infection which can transmit some of the important diseases of breeding such as brucellosis, vibriosis and trichomoniosis. The use of artificial insemination can be an enormous asset in the rapid genetical improvement of unproductive stock, but it is necessary to appreciate the disease hazard of semen, as the preservation techniques used would also preserve a wide range of contaminating infective agents. Biological products may also be vehicles for extraneous infective agents. Todd (1958) quoted examples ranging from Newcastle disease vaccine containing *Salmonella pullorum*, to foot-and-mouth disease virus in smallpox vaccine, and the development of scrapie disease in sheep vaccinated against louping-ill. Carelessly handled imported laboratory cultures may also spread disease. In this context, Todd (1958) cited an outbreak of fowl plague in the United States caused by virus illegally imported for experimental use.

Quarantine and sanitary measures

Major strides have been made in disease eradication and control, particularly in the developed countries, by the rigorous application of quarantine and sanitary measures, including compulsory slaughter. The number of successes is impressive, and a few examples may be given. Great Britain eradicated rinderpest and contagious bovine pleuropneumonia at the end of last century, and swinefever much more recently. Foot-and-mouth disease presents a greater problem in Britain; eradication has been interspersed with repeated reintroductions of the virus. Australia eradicated foot-and-mouth disease in 1872, and has subsequently remained free from the infection; and rinderpest was eradicated in 1923. There is no doubt that the relative isolation of the Australian continent has contributed to this success, but her rigid quarantine measures have to be vigilantly maintained. Japan has been free of rinderpest, contagious bovine pleuropneumonia and foot-and-mouth disease for many years. In North America one of the most signal successes in disease control has been the stamping out of vesicular exanthema, which spread widely from California in 1952 but is now believed to have been totally eradicated.

Cockrill (1963) discussed the changing status of animal quarantine at the present time, and expressed the cautionary view that some of the past successes, such as those mentioned above, are increasingly at risk today, and that the emerging infections such as bluetongue, African horsesickness, exotic types of foot-and-mouth disease and lumpyskin disease, constitute an ever-increasing challenge to the efficiency of national quarantine measures. Rapid air travel can transport infected stock or infected vectors over a wide area, and infection may also be widely disseminated in exported frozen meat. The increases in world travel and world trade call for increased vigilance in the enforcement of quarantine measures, and increase the need for international cooperation in disease control. The risk of major epidemics will only be diminished by effective control measures which decrease the incidence of these diseases in the endemic areas.

Vaccination

The control of some infections in certain areas is impracticable or impossible solely by the elimination of infected animals and the application of quarantine and sanitary measures. It is under these circumstances that vaccination is indicated. The objective of vaccination may be merely to decrease, in the individual animal, the risk of loss, or it may be designed to so diminish the incidence of the disease in the population over a wide area that eradication by a stamping-out policy may eventually become economically feasible. A further use of vaccination is to create buffer zones of resistant animals in an attempt to prevent or limit the spread of infection.

The range of vaccines available is very wide. In a recent review, Rweyemamu (1970) listed 23 viral diseases of farm livestock for which vaccines are used. To this list must be added those vaccines for use in the control of bacterial disease, such as anthrax, brucellosis, the clostridial diseases, salmonellosis, swine erysipelas and haemorrhagic septicaemia. Generalizations about the value of vaccination in terms of productivity are difficult to make because of the wide range of products available and widely varying circumstances. At the outset it is necessary to indicate that the immunizing power of certain vaccines which are available commercially is open to question, and the contribution which they make to increased productivity is doubtful. This category might include certain vaccines marketed as prophylactics for bovine mastitis, coliform infections in young animals, and respiratory infections in calves. For many of the major epidemic diseases, however, efficient vaccines are available and it is necessary to give some consideration to the ways in which they may be used in relation to animal productivity. As far as vaccination of the individual animal is concerned, the effectiveness of the procedure is assessed by balancing the cost of producing and administering the vaccine against the value of the animal and risk of exposure to infection. In this situation it may be justified to vaccinate valuable breeding stock, but not animals destined for slaughter at an early date.

Vaccination as a part of a national or regional control or eradication programme for epidemic diseases is a more complex matter. To be effective in these circumstances it is essential to immunize as large a proportion of the population as possible. In an assessment of campaigns against foot-and-mouth disease, Henderson (1970) regarded vaccination of less than 80 % of the cattle in an area as unsatisfactory. In the same paper Henderson discussed some of the other technical aspects of foot-and-mouth disease vaccines, many of which are applicable to other infections. He emphasized the importance of repeated vaccination at appropriate intervals, the necessity of considering vaccination of all susceptible species, and the need for quality control of vaccine production and adequate facilities for the distribution and storage of vaccine. Clearly, adherence to these criteria, although essential in order to make a successful impact on a disease, can be extremely costly. Ferguson (1969) warned that, although the efficacy of many vaccines has increased in recent years, their application in the field may be limited by the cost of mass vaccination, and suggested that in the case of rinderpest consideration should be given to relating mass vaccination to the periodicity of epidemics. MacFarlane (1969) discussed this question further in relation to the joint campaign against rinderpest (J.P. 15) in Africa, in which more than 80 million head of cattle were vaccinated, and the incidence of the disease dropped dramatically in those areas in which the campaign was operated. Lepissier & MacFarlane (1967), commenting on the success of the J.P. 15 campaign, considered that the technique of mass vaccination under coordinated control could be adapted and used in the control of other major epizootics in Africa. The development of attenuated live virus vaccines, such as the rinderpest vaccine, has been a major factor in the success of control programmes against many diseases.

Jansen (1969) gave some data on the control of diseases in South Africa by means of vaccination. The figures are impressive from the point of view of the scope of the exercise, and of its effectiveness in terms of increased productivity. In 1927, 1,121 outbreaks of anthrax occurred in South Africa. Following successful research on the development of a vaccine, compulsory vaccination in affected areas decreased the incidence of anthrax to less than 18 outbreaks in 1967; about five million animals are immunized annually. Vaccination against bluetongue is more complicated because of the multiplicity of serotypes of the virus which occur. The present vaccine protects against 14 types of virus, and no less than 25 million doses are issued annually. A polyvalent vaccine has also been developed against horsesickness and through its general use the disease, which used to kill up to 40 % of horses in one season, has virtually disappeared from South Africa. Jansen stated in his paper that as many as 105 million doses of vaccine, directed against 27 diseases, were currently being manufactured annually at Onderstepoort.

The use of vaccination in the face of an epidemic is well exemplified by the international campaign against the S.A.T. 1- and A-type foot-and-mouth disease epidemics in the Near East during the last decade, a feature of which has been the success of the policy of frontier vaccination (Cockrill, 1965). Repeated vaccination was carried out on the Turkish frontier from 1962, and this was a factor of great importance in confining the disease. The cost of maintaining such buffer zones is extremely heavy, and is usually only possible, as in the example quoted, with the support of several interested countries. A further prerequisite for the success of this policy is continual monitoring of the types and subtypes of virus to which the population is at risk, in order to maintain the appropriate antigens in the vaccines used in the buffer zone.

PRODUCTIVITY AND DISEASES
OF INTENSIVE HUSBANDRY

A feature of agricultural development in the technically advanced countries has been the trend towards increasingly intensive systems of animal husbandry, culminating in the concept of 'factory farming', which has received a certain amount of undesirable publicity in recent years. Intensivism in livestock production connotes the use of genetically improved animals which are capable of the highest levels of production, fed a diet which is scientifically designed to give optimum production at minimum cost, and continuous housing of large groups of animals in order to minimize the labour requirements of the operation. This trend towards larger units of highly productive animals has been most apparent in the poultry industry, but similar developments are occurring in the dairy and pig industries, and in beef production. These changes would be impossible in the absence of efficient epidemic disease control, but they have unfortunately created some new disease problems which were not apparent under less intensive husbandry systems. Moreover, the profit margins in operations of this kind tend to be so narrow that mild, subclinical infections assume major importance, and have to be controlled in order to maintain profitability. Many of the important diseases of intensively produced stock are of multiple aetiology, in which a variety of infectious agents are involved, together with various management factors. Productivity can only be increased and maintained under these conditions by high standards of preventive medicine.

General principles of preventive medicine in intensive units

Isolation

In order to exclude the entry of new infections, as high a degree of isolation of the unit as possible should be maintained. Where possible, the animal population should be self-contained, and if it is necessary to introduce new stock these should be purchased from a source of known health history. This requirement is particularly difficult to fulfil at present in beef production, where the tendency has been for calves to be purchased from a variety of sources, with the attendant risk of importing disease. This practice places severe limitations on disease control in the beef industry. Foodstuffs are a constant source of danger because of the possibility of contamination with pathogenic organisms. Heard (1969) recommends steam-pelleting of food and the use of paper sacks to minimize the risk of contamination to some extent. Animal feedingstuffs can be sterilized by gamma irradiation, but this might double their cost and render the procedure uneconomic. Efficient rodent control is required, and visitors must be excluded from the unit.

Building construction and ventilation

The design of animal houses is a compromise between the conflicting demands of low-capital costs, minimum labour requirements and disease control. Only too frequently the first of these factors has precedence over the others, sometimes with disastrous results. Multiple small units in which the animals can remain from birth to slaughter are desirable for disease control but may be expensive to construct and manage. Large buildings with communal air space and effluent collection may be cheaper to operate, but increase the risk of cross infection. Adequate ventilation is essential for the prevention and control of respiratory infections, which are important in poultry, pig and beef units. Detailed recommendations on the modern housing of livestock are given by Sainsbury (1967).

Records

Extensive and detailed records of production are essential in order to indicate where subclinical infection might be adversely affecting productivity, and to provide economic justification for preventive procedures. Records of fertility, births, feed consumption and growth rates are required, together with details of illness, post-mortem findings and carcase evaluation. The capital and labour costs of the operation must also be detailed. The need for accurate recording cannot be over-estimated in any attempt to cost the wastage from disease in a herd, and to assess the cost effectiveness of any exercise designed to minimize microbial infection and increase productivity.

Health control schemes

Appreciation of the importance of preventive medicine in maintaining and increasing productivity in intensive livestock husbandry has led to various attempts to develop organized animal health programmes. These have been most abundant in two main areas, namely the attempted control of bovine mastitis, and health schemes in the pig industry. Organized disease control is also very much a feature of the operation of the larger poultry companies who, having achieved major genetical improvements in production, now look increasingly to improved health as a basis for greater productivity (Powell, 1970). In many respects, mastitis in the dairy cow typifies the disease problems of intensive livestock production. A wide range of bacterial infections may produce mastitis, and although the disease is not unknown in range cattle, it is essentially a problem of the high-producing dairy animal. While the disease may be severe, it is most commonly a mild or subclinical infection which decreases the milk yield of the cow by at least 10 %. The disease has not proved amenable to vaccination and the value of treatment is rather limited. Control schemes at various levels have proliferated in almost all of the major dairying countries; some of these were reviewed by Blackburn (1962). One of the oldest schemes was the New York State mastitis programme which was started in 1946. Subsequently a National Mastitis Council was formed in the United States, with the object of promoting educational, research and field activities to combat the disease. The basis of all mastitis control programmes is the identification of problem herds by means of indirect tests for mastitis on milk samples, followed by the implementation of control measures at the herd level. These measures are aimed at the culling of intractably infected cows and improvements in management practices, particularly relating to hygiene at milking which are intended to minimize the spread of infection within a herd. A major advance in mastitis control was the discovery of the value of intramammary infusions of suitable antibiotics during the non-lactating period in decreasing the incidence of udder infection (Smith *et al.* 1967). There is considerable hope that the routine application of this procedure, together with rational sanitary precautions, could profoundly diminish the very considerable losses currently associated with mastitis, and make a major contribution to productivity in the dairy industry.

Preventive medicine in the pig industry was reviewed by Melrose (1965), who drew attention to programmes of varying scope in Holland, Norway, Sweden, Canada, Western Germany, the United States and Great Britain. Some of these schemes are directed towards the control of a single disease or group of diseases, frequently enzootic pneumonia, an important cause of diminished performance in intensively reared pigs. Such a scheme, which is privately operated by the Pig Health Control Association in Britain, was fully described by Goodwin & Whittlestone (1960). Considerable success had been achieved in the control of enzootic pneumonia, but unexplained breakdowns had occurred in carefully controlled herds, and further progress along these lines will depend upon the development of a satisfactory specific diagnostic procedure for enzootic pneumonia. Of broader scope is the Pig Health Scheme operated in Britain by the Ministry of Agriculture. This is an official attempt to bring combined private and state veterinary services into the field of preventive medicine in the pig industry. The scheme is at an early stage of development, but it indicates an imaginative approach in relating veterinary services to current problems by instituting routine advisory visits to farmers.

Repopulation with specific pathogen-free (S.P.F.) animals

Of great potential value in microbial disease control is the use of S.P.F. animals to repopulate herds. The principle involved is to obtain new-born animals free of infection by hysterectomy or hysterotomy, and to rear them in strict isolation. Although certain infections are capable of crossing the placental barrier, the procedure is theoretically capable of producing animals free from a very wide range of microbial infections. The technique was pioneered under field conditions in pigs in Nebraska, U.S.A. (Young, 1964), and subsequently extended nationally on a considerable scale. Some fairly extravagant claims have been made for the value of repopulation of swine herds with S.P.F. animals, both from the point of view of improved productivity and of the control of infectious diseases. Interpretation of production benefits is frequently difficult because of the lack of strict comparability of data, and some workers quote depressing disease statistics in repopulated herds (Goodwin, 1965). From the economic aspect, the initial cost of the primary S.P.F. foundation stock is high, and there may in addition be a production loss due to the running down of an existing herd. The major limitation of the technique in relation to disease control is the difficulty in maintaining adequate standards of isolation of the herd following repopulation, as the animals are highly susceptible to any infection which may be introduced. The standards required to maintain S.P.F. pig herds from infection have not been defined, but experimental studies with this object in view are currently in progress at my institution. Difficulties encountered in the maintenance of S.P.F. poultry were discussed recently by Cooper (1970); S.P.F. calf production has not yet been described commercially. In the pig, however, S.P.F. techniques have been of some value in microbial disease control, and the potential value is such that further investigations would appear to be worth while. Heard & Jollans (1967) gave a detailed description of an S.P.F. pig herd which was under their control, and their most encouraging finding in terms of productivity was a food conversion rate of 2·8 compared with the national average of 3·95. There were in addition significant improvements in terms of lower mortality and disease incidence.

Feed medication

A preventative procedure of some importance in intensive animal production units is the inclusion of suitable concentrations of antibiotic or chemotherapeutic substances in the diet as a prophylactic against certain specific microbial infections. Examples are to be found in the poultry industry, in which the use of coccidiostats has successfully controlled coccidiosis and proved itself to be an economically sound procedure, and where the use of antimycoplasmal drugs has dramatically lowered financial losses due to respiratory diseases (Oakley, 1970). In the pig industry, feed medication has been widely used in the control of enteric disease due to bacterial infection, particularly in association with the early weaning of piglets.

A distinction has to be made between the strategic use of food medication in order to combat a specific microbial disease, and the use of food additives as 'performance improvers'. In the latter case, antibiotics are added to diets in order to increase the growth rate and food-conversion efficiency of livestock. The mechanism of the improved performance is not fully understood, although it may be postulated that the animals are suffering from subclinical infections whose effect is diminished by the feeding of low concentrations of antibiotics. Feed medication has been practised most widely in the poultry and pig industries, and the antibiotics most widely used have been penicillin and the tetracyclines. The contribution to the productivity of these industries has been considerable. In 1969 the Swann Committee quoted a figure of one million pounds sterling as a conservative estimate of the annual profit gained by the current use of antibiotic growth promotants within the broiler, turkey and pig industries (Joint Committee on the Use of Antibiotics in Animal Husbandry and Veterinary Medicine, 1969). However, largely because of fears of the consequences of the transference of infectious drug resistance among the bacterial flora of animals and man, the Swann Committee recommended that the use of penicillin, chlortetracycline and oxytetracycline as food additives should no longer be permitted without prescription. The use of other antibiotics such as bacitracin, virginiamycin, nitrovin and flavomycin as growth promotants may be permitted for poultry (Smith, 1970) and, in the pig industry, copper sulphate has already largely supplanted the antibiotics as a growth promoter.

A further important recommendation of the Swann Committee was the expansion of epidemiological investigations of infectious diseases of animals. This is highly desirable on a broad basis, not only in relation to infections of obvious economic and public health importance such as salmonellosis. Further study is urgently required on the spread of infectious diseases in herds and flocks, particularly in relation to the increasingly intensive husbandry systems which are being developed.

Modification of husbandry methods in the light of epidemiological findings is likely to contribute much more in the long term to animal productivity than the empirical, non-specific use of antibiotics to control disease.

Effluent disposal

The intensification of livestock production has created a microbiological hazard to health in the form of large volumes of farm effluent. On many farms 'slurry', a fluid mixture of faeces and urine with a high water content, has replaced the traditional farmyard manure. Rankin & Taylor (1969) indicated that a 100-cow dairy can produce 20,000 gallons of slurry per week, and the farmer usually has to dispose of this material by irrigation of his own land. These authors reviewed the possible hazards of this procedure from the point of view of the distribution of pathogens. A further problem arises because, in some instances, the physical difficulties of handling slurry may limit the productive expansion of an intensive unit. Work is now being developed which is aimed to devise procedures whereby farm effluent can be treated to make it acceptable to river authorities or to municipal sewage plants, or to render its disposal on the farm efficient and safe. The main difficulties in achieving this are economic. Many traditional sewage treatments are relatively expensive to instal and operate, and could make intensive livestock production unprofitable.

CONCLUSIONS

1. Animal productivity is greatly decreased by microbial diseases. The losses are greatest in the underdeveloped areas, but are also considerable in the technically advanced countries with intensive animal husbandry.

2. In the underdeveloped countries, additional financial and technical assistance are required in order to combat epidemic disease on a realistic scale.

3. In these countries, education is required to remove certain sociological barriers to productive expansion and microbial disease control.

4. The further development of cooperative disease control programmes on a global basis is required in relation to the increasing risks of major epidemics.

5. In the technically advanced countries it is essential to maintain quarantine barriers against epidemic diseases.

6. Intensive livestock production demands an increased epidemiological research effort in order to develop management systems which will minimize the effects of microbial infection.

I wish to thank Dr W. Ross Cockrill of F.A.O. and Mr W. G. Beaton of the Centre for Tropical Veterinary Medicine for the kind provision of reference material.

REFERENCES

BEVERIDGE, W. I. B. (1960). Economics of animal health. *Vet. Rec.* **72**, 810.

BLACKBURN, P. S. (1962). Reviews of the progress of diary science. Section E. Diseases of dairy cattle. Mastitis. *J. Dairy Res.* **29**, 329.

COCKRILL, W. R. (1963). The changing status of animal quarantine. *Brit. vet. J.* **119**, 338.

COCKRILL, W. R. (1964). The profession and the science: international trends in veterinary medicine. *Adv. vet. Sci.* **9**, 252.

COCKRILL, W. R. (1965). The principles and application of international disease control. *Vet. Rec.* **77**, 1438.

COCKRILL, W. R. (1966). Patterns of disease. *Vet. Rec.* **78**, 259.

COCKBURN, A. (1963). *The Evolution and Eradication of Infectious Diseases*. Baltimore: John Hopkins Press.

COOPER, D. M. (1970). Poultry: Principles of disease control. I. Production of specific pathogen-free stock by management-environment control. *Vet. Rec.* **86**, 388.

DU TOIT, R. M. (1959). The eradication of the tsetse fly (*Glossina pallidipes*) from Zululand, Union of South Africa. *Adv. vet. Sci.* **5**, 227.

DU TOIT, R. M. (1964). Trypanosomiasis and the control of tsetse flies by chemical means. *Onderstepoort J. vet. Res.* **26**, 317.

F.A.O. (1963). The economic losses caused by animal diseases. In *Animal Health Yearbook*, 1962, p. 284. Rome: F.A.O.

FERGUSON, W. (1969). Periodicity of epidemics and cost-benefit of vaccination programmes. *Vet. Rec.* **85**, 420.

GOODWIN, R. F. W. (1965). The possible role of hysterectomy and related procedures for the eradication and control of pig diseases in Britain. *Vet. Rec.* **77**, 1070.

GOODWIN, R. F. W. & WHITTLESTONE, P. (1960). Experiences with a scheme for supervising pig herds believed to be free from enzootic pneumonia (virus pneumonia). *Vet. Rec.* **72**, 1029.

HAIG, D. A. (1955). Tickborne rickettsioses in South Africa. *Adv. vet. Sci.* **2**, 307.

HAIG, D. A. (1965). The arboviruses. *Vet. Rec.* **77**, 1428.

HARNETT, P. (1956). The significance of veterinary science in the national economy. *Irish vet. J.* **10**, 130.

HEARD, T. W. (1969). Preventive medicine in pig practice. *Vet. Ann.* **10**, 72.

HEARD, T. W. & JOLLANS, J. L. (1967). Observations on a closed, hysterectomy-founded pig herd. *Vet. Rec.* **80**, 481.

HENDERSON, W. M. (1970). Foot and mouth disease: a definition of the problem and views on its solution. *Brit. vet. J.* **126**, 115.

JANSEN, B. C. (1969). Past, current and future control of epizootic diseases in South Africa. *Trop. Anim. Hlth Prod.* **1**, 96.

JOINT COMMITTEE ON THE USE OF ANTIBIOTICS IN ANIMAL HUSBANDRY AND VETERINARY MEDICINE (1969). *Report*. Cmnd. 4190. London: H.M.S.O.

LEPISSIER, H. E. & MACFARLANE, I. M. (1967). The technique of massive vaccination in the control of rinderpest in Africa – the joint campaign against rinderpest (J.P. 15). *Bull. Off. Int. Epizoot.* **68,** 681.

MACFARLANE, I. M. (1969). Periodicity of epidemics and cost-benefit of vaccination programmes. *Vet. Rec.* **85,** 725.

MACLEAN, C. W. (1969). Preventive medicine in intensive and semi-intensive beef units. *Vet. Ann.* **10,** 83.

MELROSE, D. R. (1965). Preventive methods in pig farming. *Vet. Rec.* **77,** 1350.

OAKLEY, R. G. (1970). Poultry: principles of disease control. IV. Medication. *Vet. Rec.* **86,** 429.

POWELL, D. G. (1970). Poultry: principles of disease control. II. Eradication. *Vet. Rec.* **86,** 397.

RANKIN, J. D. & TAYLOR, R. J. (1969). A study of some disease hazards which could be associated with the system of applying cattle slurry to pasture. *Vet. Rec.* **85,** 758.

REID, D. (1969). General aspects of disease eradication. *Vet. Rec.* **84,** 626.

RWEYEMAMU, M. (1970). Viral vaccines in veterinary medicine. *Vet. Bull.* **40,** 73.

SAINSBURY, D. (1967). *Animal Health and Housing.* London: Baillière, Tindall and Cassell.

SCOTT, G. R. (1964*a*). Rinderpest. *Adv. vet. Sci.* **9,** 114.

SCOTT, G. R. (1964*b*). The new pseudorinderpests. *Bull. epizoot. Dis. Afr.* **12,** 287.

SMITH, H. W. (1970). The 'cost' of Swann. *Vet. Rec.* **86,** 133.

SMITH, A., WESTGARTH, D. R., JONES, M. R., NEAVE, F. K., DODD, F. H. & BRANDER, G. C. (1967). Methods of reducing the incidence of udder infections in dry cows. *Vet. Rec.* **81,** 504.

TODD, F. A. (1958). Defense against imported animal diseases. *Adv. vet. Sci.* **4,** 1.

WALKER, A. R. & DAVIES, F. G. (1971). Preliminary survey of the epidemiology of bluetongue disease in Kenya. *J. Hyg., Camb.* (In the Press.)

WILDE, J. K. H. (1967). East Coast fever. *Adv. vet. Sci.* **11,** 207.

YEOMAN, G. H. (1964). Control of ticks in East Africa. *Outl. Agric.* **4,** 126.

YOUNG, G. A. (1964). S.P.F. swine. *Adv. vet. Sci.* **9,** 61.

THE ROLE OF INTESTINAL FLORA
IN ANIMAL NUTRITION

D. LEWIS AND H. SWAN

University of Nottingham

The contribution made by intestinal microflora to the nutrition of the host animal depends largely upon the location of the main sites of microbial activity within the alimentary tract. In the pig and the horse for example, the main site is the caecum, located posterior to the main areas of digestion and absorption; whilst in the ruminant the major microbial activity occurs in the rumen and reticulum, anterior to the small intestine. The main microbial fermentation in the ruminant is prior to gastric, duodenal and intestinal activity with the result that the bodies of the microorganisms can be digested and the products made available to the host animal. In the caecal digestion of fibre the microbial waste products can be absorbed but the bodies of the microorganisms are largely voided in the faeces.

In establishing the role of intestinal microflora in animal nutrition it is therefore desirable to compare such species as the pig, sheep and fowl. An essential difference may well be in the capacity to digest fibre: in the case of cereal grains the fowl probably digests less than 10% of the fibre, the pig about 20–30%: in the case of the ruminant animals probably half the fibre is digested. The digestibility of various forage celluloses and hemicelluloses by ruminants and non-ruminants was compared by Keys, Van Soest & Young (1969). They noted that there was a 40–60% digestibility of both cellulose and hemicellulose in the sheep but only 20–40% was digested in the pig and in most cases less than 15% in the rat (Table 1).

Though there is considerable information available for these species to identify the organisms present in the different areas of the alimentary tract and also on the composition of the digesta, for example, organic acids, there is remarkably little data available concerning the effective contribution of the microbial population in nutritional terms. Thus information is available on the volatile fatty acid (VFA) concentration within different regions of the alimentary tract (Table 2) and the proportions of individual fatty acids in the sites of major microbial activity (Table 3). However, it must be recognized that the concentration of the acid within the digesta is not an indication of the amount pro-

Table 1. *Comparison of hemicellulose and cellulose digestibilities in ruminants and non-ruminants fed alfalfa, brome grass and orchard grass* (*Keys, Van Soest & Young, 1969*)

		Digestibility (%)		
		Sheep	Pig	Rat
Alfalfa	Hemicellulose	49	42	46
	Cellulose	53	38	23
Brome grass	Hemicellulose	76	46	11
	Cellulose	72	38	4
Orchard grass	Hemicellulose	76	46	8
	Cellulose	68	42	4

Table 2. *Volatile fatty-acid concentrations in digesta from different parts of the digestive tracts of the fowl* (*Annison, Hill & Kenworthy, 1968*), *pig* (*Friend, Nicholson & Cunningham, 1964*) *and the sheep* (*Ward, Richardson & Tsein, 1961; Annison, 1954*)

	mmoles/kg.		
	Fowl	Pig	Sheep
Rumen	—	—	119
Stomach, gizzard or abomasum	5	55	13
Small intestine	10	3	19
Caecum	107	182	60
Colon	51	190	7

Table 3. *Volatile fatty acid (VFA) proportions within the sites of major microbial activity in the fowl* (*Annison, Hill & Kenworthy, 1968*), *pig* (*Friend, Nicholson & Cunningham, 1964*) *and the sheep* (*Annison, 1954*)

	Total VFA (%)		
	Fowl caecum	Pig caecum	Sheep rumen
Acetic acid	61	55	54
Propionic acid	27	37	25
Butyric acid	11	6	17
Higher acids	1	2	4

duced but represents the difference between that produced and absorbed. The amount metabolized by the gut wall and by bacteria must also be taken into account. Nor do the proportions in which organic acids are found in the digesta necessarily indicate their relative rates of production, as it is possible that the different acids are absorbed or metabolized at different rates. Such data qualitatively confirm the sites of major microbial activities but cannot supply quantitative information concerning the contribution made to the nutrition of the animal.

GENERAL COMPARATIVE ISSUES

Fowl

Though it is clearly recognized that there is a substantial microbial population in many areas of the alimentary tract of the bird (see Jayne-Williams & Coates, 1969) it is fully accepted that the main microbial action occurs within the caeca (Barnes & Shrimpton, 1957). The fact that the caeca are a site of major fermentative activity is confirmed by information on gas production, cellulose digestion and VFA production. It is, however, difficult to ascribe a substantial role to these organs since caecectomy appears to have no detrimental effect. Developing techniques in gnotobiology allow comparison of 'germ-free' and conventional chickens and show the influence of the gut flora upon digestion. It is obvious that there is a substantial synthesis of vitamins by the gut flora (see Jayne-Williams & Coates, 1969) but there is little other quantitative information available. The situation is made more difficult to interpret by the marked morphological and physiological differences between germ-free and contaminated animals: it is not possible to differentiate between the contributions that might be made by the microorganisms and the secondary physiological changes.

It was shown by Annison, Hill & Kenworthy (1968) that there was a much higher concentration of VFA's in the caeca than in other parts of the digestive tract (see Table 2). The VFA's were also shown to be present in portal and peripheral blood and it seemed at first that they originated from the caecum. When germ-free chickens were examined there was a similar pattern of VFA's in peripheral blood but only negligible amounts in the caeca. It seems therefore that the VFA's of peripheral blood are not of caecal origin. These findings again exemplify the difficulties in obtaining quantitative information on the nutritional significance of the alimentary microflora.

Pig

It is apparent that there is a substantial microbial breakdown of food constituents within the digestive tract of the pig (Cranwell, 1968). There is no digestion of cellulose in the stomach or small intestine but a substantial bacterial breakdown in the large intestine, particularly in the caecum and proximal colon. The VFA concentrations in these organs (see Table 2) indicate that there is an active fermentation, a conclusion confirmed by microscopic examination, but there is little information available on the quantitative contribution made to the nutrition of the host animal. A tentative estimate by Friend, Nicholson & Cunningham (1964) suggests, however, that approximately 20 % of the energy requirement for 'maintenance' of the young growing pig could be contributed by VFA's absorbed from the alimentary tract. There is little information available on vitamin synthesis within the pig digestive tract (Cranwell, 1968). Though Michel (1962) showed that analogues of vitamin B_{12} are synthesized by bacteria in the caecum he also showed that 95 % of these materials remained within the bacteria.

Ruminant

The particular features of the ruminant depend essentially upon the rumen (or proventriculus, Habel, 1965) which is a large organ with a remarkably dense population of bacteria and protozoa which ferment carbohydrates and produce VFA's as end-products that can serve as energy sources for the host animal. With respect to cellulose and related materials this represents a substantial advantage to the animal since they would otherwise be largely wasted. A similar fermentative process occurs with starch and sugars but since they would be effectively digested and absorbed in the small intestine this represents a net loss to the animal since the end-products of fermentation are of less nutritive value than the initial materials.

The microbial metabolism of nitrogenous compounds is complex and whether there is or not a net gain to the protein economy of the host animal depends upon the diet. There can be advantages as a result of a changed pattern of amino acids or a loss in terms of ammonia absorbed from the rumen. The influence upon the ruminant's vitamin economy is of outstanding importance, since the synthetic abilities of the microbial population provide the host animal with a complete supply of the B-vitamins. In the ruminant, in contrast to other species of farm animals, there is much known about the pathways whereby the alimentary microbial population contributes to the nutritional economy of the host and also about the magnitude of this contribution under different circumstances.

QUANTITATIVE NUTRITIONAL IMPLICATIONS

Because the ruminant can digest fibrous materials, it can be offered a diet composed of ingredients unsuitable for or of little value to other farm animals and man. On the other hand, rumen activity may exert a detrimental effect upon production standards by wasting energy-yielding nutrients. An indication of the relative efficiency of nutrient utilization at different dietary levels of energy-yielding nutrients can be obtained by comparing food conversion ratios for different species (Table 4). It is possible that the different species are within different physiological stages of, say, maturity but it is clear that there is a reduced efficiency in the pig as compared with the fowl whilst in the ruminant noticeably less satisfactory standards are achieved. On the other hand, the extent to which there is an improvement in food utilization as the dietary energy level is raised is probably of more significance. In the fowl a 25 % increase in dietary energy supply leads to an 18 % improvement in food conversion, the comparable value for the pig is 13 % whilst for the steer it is 25 %. It is, however, difficult to ascribe clear significance to such comparisons between species because of the difficulties in standardizing patterns of food intake and general levels of growth performance.

Table 4. *Relative efficiency of food utilization at different dietary energy levels for the young growing pig, chicken and ruminant*

DE = digestible energy; ME = metabolizable energy (both in kcal./kg.). FCR = food conversion ratio (units food per unit live-weight gain).

Pig, 25–60 kg.		Chicken, 0–56 days		Steer, 200 kg. live-weight	
DE	FCR	DE	FCR	ME	FCR
2,650	2·83	2,800	2·43	2,200	7·4
2,850	2·67	3,000	2·30	2,600	5·5
3,050	2·45	3,200	2·24	3,000	4·3
3,250	2·39	3,300	2·13	3,400	3·5
3,450	2·33	3,400	2·06		
3,650	2·31	3,500	1·99		

A more direct comparison of the relative efficiency with which different species are able to utilize the energy-yielding constituents of their diets is possible by observing the relative metabolizable energy values. Values for the pig, fowl and ruminant (Table 5) can be considered in terms of the possible effect of the presence of different alimentary microbial populations in different segments of the alimentary tract. It is again apparent that fibrous materials (e.g. lucerne or wheat bran)

Table 5. *Comparative metabolizable energy values (kcal./kg.) for standard food ingredients in the case of the pig, fowl and ruminant*

(See Lewis & Morgan, 1963; Robinson, Prescott & Lewis, 1965; Morgan, Cole & Lewis, unpublished; NRC-NAS, 1969; various personal communications.)

	Pig	Fowl	Sheep
Wheat	3,200	3,350	3,150
Maize	3,450	3,550	3,250
Barley	2,550	2,850	3,000
Oats	2,750	2,750	2,750
Milo (sorghum)	3,250	3,400	2,900
Soyabean meal	2,850	2,550	2,950
Cottonseed meal	2,800	2,750	2,750
Groundnut meal	2,800	2,800	2,750
Fish meal (white, 65% protein)	2,800	2,750	2,250
Meat meal (60% protein)	2,800	2,500	—
Lucerne	1,600	1,200	2,100
Wheat bran	1,550	1,250	2,500

in the ruminant show a considerable advantage in terms of digestibility whereas for the materials relatively low in fibre content (e.g. maize or fishmeal) the ruminant appears to be somewhat penalized.

METABOLISM IN THE RUMEN
Carbohydrates

In non-ruminant animals glucose is recognized to be the major end-product of carbohydrate digestion and it is essential for the maintenance of cellular activity especially within the blood and nervous systems. The presence of the microbial population in the rumen results in a different situation in ruminants where the contribution of glucose is somewhat overshadowed by VFA's. Though the process as a whole makes it possible for refractory fibrous materials to be utilized the penalty is an inevitable wastage of energy-yielding nutrients. In the microbial conversion of carbohydrates to VFA's the latter are waste products insofar as the microorganisms are concerned but for the host animal they are of value and in addition, the microbial cell material contributes to those substances passing along the alimentary tract as effective food.

The conversion of dietary carbohydrate to acetic, propionic and butyric acids emphasizes the precarious nature of glucose supply to the ruminant: it accentuates the importance of gluconeogenesis (the formation of glucose from materials other than sugars). The short-chain fatty acids probably represent two-thirds of the energy supply of the animal and are utilized in many ways that spare or form glucose. It was shown by Leng, Steele & Luick (1967) that about half of the glucose of the ruminant is formed from propionate produced in the rumen and about a fifth is converted to lactic acid during passage through the rumen wall which enters a gluconeogenetic pathway in the liver via phosphoenolpyruvate. The remainder of the propionic acid is directed via methylmalonyl CoA, succinate and oxaloacetate into the glucose forming route. Acetic and butyric acids on the other hand do not give rise to glucose. It has already been mentioned that glucose is vital for certain living processes and the substantial degradation of hexoses in the rumen results in a precarious carbohydrate economy – changes either in the diet or in the demands for carbohydrate are liable to alter the carbohydrate balance from the normal to the pathological (e.g. ketosis). It is thus especially important to recognize the importance of the propionic acid proportion of the VFA's produced in the rumen and also to be aware of the significance of the net passage of carbohydrate material out of the rumen (undegraded dietary carbohydrate plus that microbially synthesized).

It has often been blithely assumed that since there is a very rapid fermentation of starch and other related products in the rumen hardly any of it escapes undegraded to the duodenum. During recent years this assumption has been questioned and several observers have found that perhaps 10 or 20 % of dietary starch escapes microbial breakdown. In examining these conclusions it is important to ensure that satisfactory procedures have been employed – for the determination of α-linked glucose polymers, for the collection of suitable duodenal samples and for the recognition of the importance of overall digestibility. Data collected by Armstrong & Beever (1969) indicate the digestibility of starch when different diets are fed and also give values for the percentage of the starch digested that has been degraded in the rumen, i.e. prior to passage along the duodenum to the intestines. In the case of hay, the overall digestibility is low but some 20 % of that digested leaves the rumen intact: this is not of much quantitative importance since the content in the feed is so low. In the case of most of the other diets less than 10 % of the digested starch leaves the rumen intact. There is, however, a significant difference in the case of those diets containing ground maize when some 30 % of the starch escapes breakdown: much of the data on ground maize refers to cattle whereas for the other cereals sheep were usually used. The quantity of glucose that becomes available to the animal can be calculated from these data and in the case of diets containing ground maize it seems that the glucose requirement is likely to be met by that absorbed from the small intestine. In those cases where the glucose supply is in any way precarious, it would seem that the inclusion of ground maize within the diet might constitute a useful precaution.

There are innumerable observations showing that VFA production represents half to three-quarters of the effective digestible energy value of the diet. The significance of these acids to the economy of the animal can also be evaluated in other ways. It is possible for example to measure what is termed the entry rate, the rate at which a particular metabolite enters the blood stream of the animal from the alimentary tract. This has been evaluated by Annison *et al.* (1967) and a comparison can be made (see Table 6) between the effective entry of glucose, acetate and long-chain fatty acids. Sheep were used: those fed were offered food every hour whilst those fasted were not offered food for a period of 24 h. The particular significance of acetate can readily be recognized. Propionate was not included in this evaluation since its fate is inextricably associated with that of glucose and in fact its contribution is essentially covered under glucose. An alternative form of evaluation of the significance of VFA's was also used by Annison *et al.* (1967) by partitioning the carbon dioxide output. The fraction of the carbon dioxide output derived from a substance represents the ratio of the terminal value of the specific activity of the respired CO_2 to that of the traced substance. This again represents an effective measure of the contribution made by particular nutrients to the overall energy supply of the animal. The particular significance of acetate is obvious as also is that of the longer chain fatty acids, presumably mobilized from depot lipids, after a short period of fasting.

Table 6. *Entry rates and oxidation of metabolites serving as sources of energy in the sheep (Annison et al. 1967)*

Substance	Entry rate (mg./min./kg.$^{0.75}$)		CO_2 output (% total)	
	Fed	Fasted	Fed	Fasted
Glucose	5·0	3·8	9·1	11·2
Acetate	10·8	5·8	31·6	22·1
β-OH-butyrate	1·4	1·5	10·4	4·8
Palmitate	—	1·0	—	4·7
Oleate	—	0·9	—	4·0
Stearate	—	0·9	—	4·4

A ruminal fermentation carries with it the advantages of an improvement in the utilization of low quality fodder, the synthesis of certain B-vitamins and also in some cases of improving the protein quality of the diet. However, the price paid by the animal for these advantages in terms of energy utilization is high: it is in effect that of maintaining a microbial population in the rumen at the expense of degradation of carbohydrate to VFA's and methane plus the energy expenditure involved in reconverting fatty acids to the carbohydrate required for normal mammalian metabolism. The utilization of those fatty acids that are more directly oxidized also involves a probable penalty.

During the fermentation process in the rumen there is a substantial production of methane. Though this is by no means unique for the ruminant as methane is formed in much smaller quantities by other herbivores. Some of the methane is absorbed directly from the rumen and is subsequently excreted by the lungs but most of the gas is eructed. There is no evidence that the energy trapped as methane is to any extent available to the ruminant animal and this pathway of energy loss represents an appreciable part (7–9 %) of the gross energy of the food ingested. This loss seems to be quite constant under different dietary conditions and there is little scope for its reduction by selecting particular diets.

Regulating the loss of energy in the form of methane is a very attractive proposition: the two feasible directions are by the selective depression of methanogenic organisms or by diverting the reducing potential fixed as methane into some other route. It was attractive therefore to examine the possibility of using a supplement of unsaturated fatty acid (e.g. linolenic acid) in the expectation that it would be reduced and be subsequently made available to the animal at the expense of methane production (Blaxter, 1967). It was shown, however, that linolenic acid and saturated fatty acids in the series C_{10}–C_{18} all reduce methane production to less than 50 % of the normal level. There was some depression in cellulose digestion but it was not marked. Other compounds studied, notably sulphated fatty alcohols, have more striking effects in depressing methane production but these affect cellulose digestion appreciably. The limitation of methane wastage is clearly an area demanding attention but no fully satisfactory approach has yet been devised.

It has been observed, as can be readily predicted from knowledge of the metabolic processes, that the loss of energy as heat during the digestion process of ruminants is much greater than with other species. In the latter, loss as heat during digestion rarely exceeds 1 % of the heat of combustion of the material fed whereas the complex of activities in the rumen lead to a value of around 6 %. There would seem to be little scope to control this loss – it may, however, under certain circumstances spare oxidative metabolism to maintain body heat. The summation of the loss as methane and as heat represents an inevitable penalty of some 15 % carried by the ruminant in terms of its energetic economy, before beginning to consider the efficiency of utilization of the products of digestion.

It is also necessary to consider the extent to which the VFA's which are absorbed are available as a source of free energy for the animal. An evaluation of the efficiency of utilization of fatty acids (see Blaxter, 1967) can in effect be carried out by recognizing the magnitude of the heat increment, the portion of the total potential energy supply which becomes effectively wasted during utilization. It is known that in the case of the ruminant the heat increment is generally some two or three times its magnitude for other classes of animals. The heat increment is also greater for VFA's than glucose. Though results for individual fatty acids are available it is only those for the type of mixture that might be encountered in the rumen which have any applied meaning. There seems to be a fairly constant heat increment of about 15 % (i.e. efficiency of utilization of 85 %) for maintenance circumstances as opposed to a value of 6 % for glucose. This stage of utilization seems again to add a further penalty of 10 % to the ruminant for the possession of the microbiota in the rumen.

A great deal of attention has been given to the importance of rumen VFA ratios in relation to the efficiency of dietary utilization and the importance of having a minimal proportion of propionic acid has already been emphasized. There is also some reason to believe from feeding trials that an increasing proportion of propionic acid is advantageous in terms of efficiency of dietary utilization. Thus in experiments with cattle and sheep fed rations comprising different proportions of hay to flaked maize, the efficiency of utilization of metabolizable energy increased from 30 % to 60 % as the flaked maize replaced the hay. The molar percentage of acetate in rumen contents declined at the same time from about 70 % to 40 %. There are several examples that can be cited to substantiate this type of conclusion but there are also others where no difference has been found. It is difficult to eliminate other variables; digestibility, composition of body gain, effects upon food intake, flow through the alimentary tract, etc. It is perhaps more profitable to consider first principles.

A six-carbon hexose is first degraded to a triose which can either be oxidatively decarboxylated to acetate or reduced to propionate. In the formation of acetate an equivalent amount of carbon dioxide is released: this is a fully oxidized material and therefore does not represent any residual free energy that can be rendered available. But the reduction of carbon dioxide to methane does in effect constitute a loss. Whether to ascribe that to methane itself or to the actual acetate formation becomes merely an exercise in tautology. In the same way the formation of propionate has involved a reductive phase but this again can either be allocated to propionate or even acetate since the reducing potential is probably only available in parallel with the oxidative production of acetate. It may therefore be concluded that the advantage of propionate rests in the fact that its production might well have diverted a reducing potential from the wasteful route of carbon dioxide conversion to methane. One would therefore expect some sort of inverse relationship between propionate proportion and methane formation – though there is no real evidence to support this concept, there does not seem to have been a programme designed specifically to examine the problem in these terms. It can, however, be seen that in a predominantly flaked maize diet the methane production represents some 3 % of the feed energy (Blaxter & Wainman, 1964) whereas for a ration based on a concentrate added to a fibrous basal component loss as methane is within a range of 7–9 % (Blaxter, 1967).

It therefore becomes of interest to identify those factors which regulate rumen fatty acid proportion. This is essentially a function of the diet. As the proportion of what might be regarded as structural carbohydrate (i.e. dry matter minus ash, crude protein, ether extract and soluble carbohydrate) increases, the proportion of acetic acid rises. In the same way diets rich in starch or sugars favour propionic acid production and in general those foodstuffs that are rapidly fermented give rise to less acetic acid.

It is clear that the activities of the rumen impose upon the animal a penalty in energetic terms. Maintaining these energy losses at a minimum may be possible by restricting methane production either by encouraging a particular type of fermentation (propionic) or by directing the reducing potential in another direction. Advantages also can be expected with increasing proportions of propionic acid within rumen VFA's: this is essentially a function of the diet employed. The addition of products that can influence redox potential, pH or the actual organism (e.g. sulphite; Alhassan, Krabill & Satter, 1969) might also be advantageous.

Proteins

It has been recognized for some two decades that there is a fairly complex form of a nitrogen cycle in the ruminant. This can be represented as in Fig. 1. Protein that enters the rumen is hydrolysed by the microorganisms to yield peptides, amino acids and ammonia. The non-protein nitrogen (NPN) consumed can also produce amino acids and ammonia whilst any urea that re-enters the rumen from the blood system also produces ammonia. Simultaneously with these breakdown reactions there is a synthesis of microbial protein from the NPN compounds and ammonia in the rumen. The relative rapidity of the

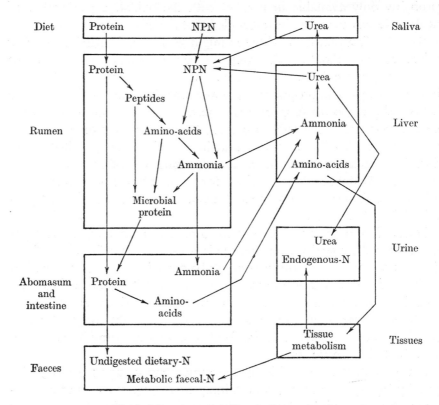

Fig. 1. Nitrogen metabolism in the ruminant.

breakdown and synthetic reaction exerts a controlling effect on the nutritive value of the protein fed. Two of the major factors that regulate this balance are the solubility of the ingested protein and the amount and type of carbohydrate present.

There is a continual passage along the alimentary tract of some unchanged protein, a proportion of microbial protein and some NPN. It is usually assumed that from this point the digestive processes are essentially the same as in non-ruminants. The significance of the rumen action therefore lies partly in the extent to which food protein is converted into microbial protein. The magnitude of this effect is dependent upon the relative nutritive values of the ingested protein and that synthesized in the rumen.

It is also important in evaluating the protein nutrition of the ruminant to recognize the importance of NPN that is absorbed from the rumen. Significant amounts of ammonia enter the ruminal veins but is not found in peripheral blood because of its quantitative removal in the liver, presumably by conversion to urea. In effect, a proportion of the nitrogen that follows such a route returns to the rumen either directly from the blood stream or via saliva as urea. It is extremely difficult to assess the quantitative significance of the proportion of the urea that is not so returned to the rumen but is lost in urine. Under circumstances of very low rumen ammonia there is a substantial movement of urea from blood to the rumen by direct transfer across the rumen wall (Egan, 1965). An estimate by Blaxter (1964) suggests that about 20 % of the nitrogen absorbed as ammonia is recycled but this was presumably for sheep on normal nitrogen intakes. By similar estimations it can be suggested that the nitrogen loss to the animal as urinary urea derived from ammonia absorbed from the rumen may approximate 20 % of the intake. This loss is, however, probably a direct function of mean rumen ammonia levels.

It was shown by Lewis (1957) that there was a close correlation between the concentration of ammonia in the rumen and of urea in blood. Though this has been subsequently confirmed in several instances there are also cases when anomalies in the relation between rumen ammonia and blood urea levels have been recorded. It is probably, however, still an attractive proposition to use blood urea as a monitor to establish whether an unfortunately large proportion of dietary nitrogen is being wasted as urinary urea.

Attempts to measure in quantitative terms the significance of the rumen to the protein nutrition of the host require an assessment of the extent of conversion of dietary protein to microbial protein and an evaluation of the nutritive value of the microbial protein. Such a quantitative approach is essential in order to record the productive significance of rumen metabolism.

It is now clearly recognized that the extent of degradation of a protein in the rumen is essentially a function of its own chemical and physical characteristics. It can be shown that a protein rapidly degraded leads to a high rumen ammonia peak achieved at an early time after the administration of the protein. Obvious extremes in this respect are the very rapidly degraded soluble protein casein and the relatively inert zein. It has recently been calculated (Chalupa, 1970) that glutamate

6

dehydrogenase in bacterial cells obtained from the rumen contents of steers fed a synthetic diet in which urea was the sole nitrogen source, functioning at half maximum velocity for 5 h. could potentially assimilate all the ammonia released from dietary urea in a 24 h. period. Since enzymes other than glutamate dehydrogenase are also involved in ammonia assimilation however, NPN utilization is probably limited more by cell permeability or insufficient amounts of energy, carbon skeletons and cofactors than by enzyme potential.

Rumen ammonia concentration may also be regulated by the carbohydrate component of the diet and an inert product such as cellulose exerts a far less marked effect upon rumen ammonia than does a soluble sugar. This is not a direct influence upon the degradation of protein but upon the utilization of the products of breakdown.

The degradative reactions that have been described provide a continual supply of peptides, amino acids and ammonia for the growth of rumen microorganisms, i.e. for the synthesis of microbial protein. The NPN products that are available include the fermentation products of ingested proteins, the nitrogenous compounds released by the autolysis of microorganisms and the urea entering the rumen from saliva and blood. There is still little quantitative information available with mixed cultures to indicate the relative importance of ammonia, amino acids and peptides for microbial protein synthesis.

A major objective for many decades has been the development of conditions under which the protein component of the ruminant's diet can be replaced by NPN to economic advantage. Considerable evidence has been empirically acquired to show for example that urea can replace almost all the protein of the diet by supplying nitrogen for rumen microbial protein synthesis. Most of the experimental work has been devoted to finding the optimum conditions for feeding urea and only indirect information on actual protein synthesis has resulted from feeding trials designed for other purposes. These include measurement of the changes in the proportion of ammonia and protein in rumen contents on feeding urea, measuring the synthesis of amino acids in the rumen on feeding urea, or by analysis of the total nitrogen of the microbial fraction after feeding.

A recent investigation by Smith (1969) has examined the problem of the extent of conversion of food to microbial nitrogen in more direct terms by comparing a mean value for the RNA/total N ratio in rumen microorganisms and whole rumen contents. It was found that for most diets some 70 % and 60 % of the total non-ammonia nitrogen was microbial in ruminal and duodenal digesta samples respectively.

Other methods of assessing the microbial contribution to the nitrogenous materials in rumen contents have included taking advantage of the solubility of zein in alcohol and the phosphorus content of casein, measuring the lysine or DNA content of diet and duodenal contents, following the rate of incorporation into protein of inorganic ^{35}S or determining the diamino-pimelic acid content of various fractions.

Comparisons of the results obtained by the various methods are difficult because of different approaches and especially of sites and conditions of study. There is, however, general agreement that with most normal diets about 60–70 % of the food nitrogen has been converted to microbial nitrogenous compounds on arrival at the small intestine. Decorticated groundnut meal, zein and certain maltreated proteins are more resistant to conversion and the value might be only of the order of 25–30 %. At the other extreme the soluble protein casein has generally been observed to be almost completely degraded before the duodenum: one must not, however, forget a significant loss as ammonia. As a general conclusion it may therefore be proposed that on average two-thirds of food protein nitrogen not lost as ammonia is converted to microbial protein during passage through the rumen.

It has already been emphasized that when assessing the value of conversion of food protein to microbial protein the extent of loss of the intermediate ammonia by absorption must be taken into account. One must also remember that of the total nitrogen entering the duodenum something of the order of 10 % is in the form of nucleic acids. These materials are probably well digested and absorbed but there is no reason to believe that they are of any value to the animal. The overall process of conversion therefore has to carry these two serious penalties before assessing its value to the host. A third penalty may be that of the digestibility of microbial protein. In general, ruminants show a greater relative loss of nitrogen in the faeces than do simple-stomached animals, especially with low-nitrogen diets. This viewpoint tends to be confirmed by the various assessments that have been recorded of the digestibility of rumen microorganisms (around 75% for bacteria and perhaps 85% for protozoa), usually with the rat as test animal (see Table 7).

The major determinant of the value of microbial protein is of course its amino acid composition. Though a considerable amount of attention has been given to the biological value of rumen microbial protein, the findings must be essentially regarded as having no significance, basically since the concept does not assist in the establishment of satisfactory dietary situations – values are not additive and not consistent in different physiological circumstances. Furthermore, since they are

Table 7. *Net protein utilization (NPU) values of rumen bacteria and protozoa determined in various laboratories*

Material	Test animal	True digestibility (%)	B.V.	NPU (B.V. × digestibility)	
Rumen bacteria	Rat	55	66	36	Johnson *et al.* (1944)
Rumen protozoa	Rat	86	68	58	Johnson *et al.* (1944)
Cultured bacteria	Rat	82	66	54	Johnson *et al.* (1944)
Bacteria	Rat	74	81	60	McNaught *et al.* (1954)
Protozoa	Rat	91	80	72	McNaught *et al.* (1954)
Bacteria	Rat	73	88	64	McNaught *et al.* (1950)
Bacteria (green fed)	Rat	62	80	50	Reed *et al.* (1949)
Bacteria (dry fed)	Rat	65	78	51	Reed *et al.* (1949)
Casein	Rat	101	80	81	Reed *et al.* (1949)
Casein	Sheep	86	73	63	Ellis *et al.* (1956)
Soyabean meal	Sheep	82	83	68	Ellis *et al.* (1956)
Urea	Sheep	80	54	43	Ellis *et al.* (1956)

related to the rat they are of limited application to the protein nutrition of the ruminant. The vital need to establish the dietary adequacy of rumen microbial protein is to relate the requirements of the ruminant to the amino acid composition of duodenal fluid.

The amino acid composition of rumen microorganisms in Table 8 are means of many analyses of rumen bacteria, rumen protozoa and mixed microorganisms: they are compared with the values for whole egg, a material usually considered to represent a good quality protein. It is often difficult from an appraisal of the amino acid composition of a diet to establish its adequacy especially when little information is available on amino acid requirements. It does, however, appear that the lysine, isoleucine and threonine levels are tolerably good whereas the methionine content is significantly lower in rumen microorganisms than in whole egg.

It is fairly clear from an examination of the information available and probably also *a priori* that the change in amino acid composition which results from the conversion of dietary protein to microbial protein in the rumen can be an advantage or a disadvantage depending on dietary circumstances. Provided the dietary protein is of reasonably good quality any apparent advantage is likely to be more than compensated by wastage of nitrogen as ammonia, conversion to nucleic acid nitrogen and relatively low digestibility. It has been observed on several occasions that protein materials are of better quality when administered beyond the rumen than by mouth. It is tempting to propose that every effort be made to limit rumen microbial activity

Table 8. *The amino acid composition of rumen microbial protein as compared with whole egg*

Values expressed g. amino acid per 100 g. protein.

	Rumen bacteria	Rumen protozoa	Mixed rumen organisms	Whole egg
Aspartic acid	10·6	12·0	7·6	10·5
Threonine	4·1	4·6	3·0	4·5
Serine	3·3	3·5	2·3	6·1
Glutamic acid	17·8	13·9	9·4	13·2
Proline	3·2	3·1	2·6	4·1
Glycine	3·9	4·4	3·0	2·3
Alanine	4·7	4·4	3·3	6·9
Valine	5·7	5·1	3·3	7·1
Methionine	2·1	2·0	1·5	3·0
Isoleucine	4·9	6·9	3·9	6·0
Leucine	6·4	8·1	5·1	9·1
Tyrosine	4·1	5·7	3·8	5·8
Phenylalanine	4·1	5·7	3·8	5·8
Lysine	6·0	9·6	7·5	7·9
Histidine	2·8	1·7	1·6	2·7
Tryptophan	1·5	1·3	1·8	1·9
Arginine	6·8	4·6	2·8	6·7

but it must be remembered that this must also severely adversely affect cellulose digestion, vitamin production and fatty-acid formation. Advantage can be more readily gained by directing rumen activity along desirable routes than by reducing it to a minimum.

AMINO ACID NEEDS OF RUMINANTS

It is as yet conventional not to pay any serious attention to the amino acid needs of the ruminant and to restrict dietary evaluation to that of the protein component as a whole. The reasons for this outlook are essentially twofold, each of which largely serves as the reason for not attempting to overcome the problems of the other. They are that no obvious method is apparent for determining the specific amino acid needs of the ruminant and that it is not possible to alter the dietary intake directly by changing the ration ingredients or supplementing the food with amino acids. In resolving this problem it is essential that the two difficulties be overcome simultaneously.

It has for some time become apparent that there may be a case for supplementing ruminant diets with amino acids. The assessments of the biological value of rumen bacterial protein usually arrive at a figure of around 75: this must mean (at least for the rat) that the supply of some amino acid is only 75 % of the ideal. Furthermore, there are several reports indicating responses from the supplementation of ruminant diets: advantages in terms of growth were reported by Loosli & Harris (1945) for methionine supplementation of urea-based diets for lambs but McLaren, Anderson & Borth (1965) were not able to show any response in terms of nitrogen retention to methionine or tryptophan supplementation under similar conditions. Though published information is somewhat confused concerning the possible advantages of amino acid supplementation of ruminant diets in terms of growth there has been some recent dramatic responses reported in wool growth. Reis (1967) demonstrated a marked improvement in wool growth in sheep following abomasal infusion of methionine or compounds which could give rise to methionine. The variable responses to orally administered amino acids are probably attributable to their degradation in the rumen. These observations emphasize the importance of the simultaneous resolution of the problem of defining amino acid needs and overcoming the hazard of rumen degradation of any supplements aimed at correcting an inadequacy in the diet.

The determination of the amino acid requirements of ruminants is complicated by the intervention of the rumen microbial population in its digestive processes. Conventional methods – for example, the construction of a growth-response curve – are not directly applicable. It is more profitable to seek some easily measured physiological response which allows identification of a requirement. The measurement of plasma amino acid levels in relation to the quantities of amino acids passing the duodenum of the sheep has been examined in this respect.

Plasma amino acid patterns are affected by many variables, most of which cannot be readily evaluated. The rate of entry of amino acids into the plasma pool is regulated by the quantity and composition of dietary and microbial protein reaching the duodenum, the patterns of amino acid release during digestion, and the rate of absorption and metabolism of the amino acids by the intestinal tissues. In addition to the rate of entry into plasma, the removal of amino acids from the plasma also regulates the level-removal by entry into cellular matter and by passage along both anabolic and catabolic routes. Since the actual amino acid level at any one time merely represents a small balance between major entries and withdrawals it is difficult to accord to a single value an absolute meaning. It is more appropriate to draw conclusions from relative changes.

For ruminant animals it is difficult to identify a limiting amino acid, but several authors have made use of plasma amino acid measurements in attempts to do so. In 1969 Virtanen suggested that since the plasma concentration of histidine in lactating cows fed purified diets with urea and ammonium salts as the sole source of nitrogen is very low, histidine may be limiting protein synthesis. Schelling, Hinds & Hatfield (1967) implicated methionine as the limiting amino acid from measurements of plasma amino acid levels in growing lambs fed diets containing 17–28 % soyabean meal. Under conditions of restricted protein intake abomasal administration of sulphur containing amino acids has significantly increased wool growth (Reis, 1967). It would seem likely that in view of the sheep's relatively large requirement for sulphur-containing amino acids for wool growth, methionine is most probably the first limiting amino acid, especially under conditions of low protein intake.

If protein synthesis is restricted by the supply of a limiting amino acid, increasing the quantity of the limiting amino acid available to the animal should increase protein synthesis, stimulating the demand for the other essential amino acids, and consequentially causing a lowering of their levels in the plasma amino acid pool. In a series of experiments conducted to test this hypothesis (Wakeling, Annison & Lewis, 1970) supplements of 5 g. of L-methionine or L-lysine or 5 g. of a 1:1 molar mixture of L-methionine and L-lysine were infused over a 30 min. period into the duodenum of a mature wether lamb. The lamb was continuously fed (i.e. once per hour) a restricted quantity of a low-protein diet consisting of barley, barley straw and molasses, and plasma amino acid levels were measured prior to each infusion and subsequently at 2, 4 and 8 h. after commencing the infusion. Large percentage increases in plasma methionine levels are a reflexion of the very low pre-infusion concentration of 0·1 mg./100 ml. The changes in concentration are somewhat inconsistent for individual amino acids, but the mean overall decrease of 22 % and 20 % for the methionine and methionine + lysine experiments respectively indicate methionine as the most likely first limiting amino acid. The mean overall change in concentration for the lysine experiments is plus 1 %, suggesting lysine was non-limiting. This technique may be used to identify the limiting amino acid but probably lacks the sensitivity necessary to evaluate a requirement.

The determination of a requirement by measurement of plasma amino acid levels is based upon the assumption that the plasma concentration of an amino acid will remain at a fairly low and constant value when the amino acid is supplied in quantities insufficient to satisfy the requirement, but in excess of requirement there will be a corresponding proportionate increase in the plasma concentration. The major difficulty in the experimental approach to the problem is in ensuring that the dietary regime will produce a deficiency in the supply of one or more of the essential amino acids, despite the activities of the rumen population. The passage of amino acids in the duodenum was measured

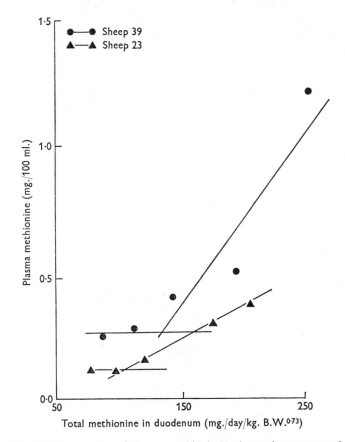

Fig. 2. The response of plasma methionine to increasing passage of methionine in the duodenum of the sheep.

for mature wether lambs continuously fed restricted quantities of a high-energy/low-protein diet, consisting mainly of cellulose, starch and glucose with the protein supplied by groundnut meal. Comparison of these values, with the quantities necessary to satisfy an approximate requirement calculated from pig requirements, indicated that, only in the case of methionine was there likely to be an appreciable deficiency.

In a series of experiments, graded levels of L-methionine were infused sequentially over 36 h. periods into the duodenum of two sheep. Plasma amino acid levels were determined in jugular blood sampled prior to infusion and at 4 h. intervals over the final 12 h. of each 36 h. period. The curve relating duodenal methionine flow and plasma methionine remained flat at lower infusion levels but deflected upwards

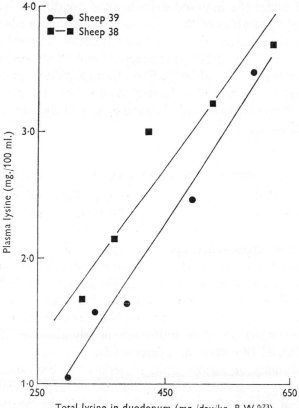

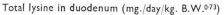

Fig. 3. The response of plasma lysine to increasing passage of lysine in the duodenum of the sheep.

at higher levels (Fig. 2). It is proposed that the inflection point of the curve identifies a requirement in terms of methionine passing the duodenum: this is equivalent to approximately 2·1 g./day for a 50 kg. sheep.

The changes in plasma concentration of other essential amino acids were somewhat inconsistent, both in the same animal and between different animals at the various infusion levels. The greatest decreases in plasma concentration were recorded at the higher infusion levels. Threonine showed the largest decrease followed in decreasing order of response by leucine, isoleucine, valine and lysine respectively, indicating that threonine was possibly the second limiting amino acid.

During a similar series of experiments with L-lysine the plasma lysine concentration increased linearly, with increasing passage of lysine in the duodenum over the whole range studied (Fig. 3). This is in agreement with the assumption that lysine was not the first limiting amino acid under the imposed experimental conditions. The changes in plasma concentration of the other essential amino acids were again rather inconsistent. However, the mean decrease in methionine concentration was 27 % and 21 % for sheep 39 and 38 respectively and for plasma threonine 28 % and 36 %. Even though lysine was non-limiting supplemental lysine caused a further depression in the plasma concentration of methionine and threonine, the likely first and second limiting amino acids.

FAT-FIBRE INTERACTIONS IN THE RUMEN

It is recognized for non-ruminant species of farm animals that lipid materials are better digested as neutral fat than as free fatty acids, that unsaturated fatty acids are more readily assimilated than saturated ones, and that digestibility falls off rapidly above chain lengths of 14 carbon atoms. There is little information available in the ruminant, however, to define the extent to which different types of fatty materials might be utilized. Since there is considerable hydrolysis and hydrogenation of lipids by microorganisms within the rumen, it seems possible that lipid materials would be uniformly poorly absorbed if composed mainly of 16 and 18 carbon unit fatty acids.

The digestibility of various lipid materials has been evaluated with adult sheep in metabolism cages (Andrews, 1967; Devendra, 1969). The output of fat not of immediate dietary origin was measured whilst a relatively fat-free basal ration was fed. Supplements of fat were included at the level of 8 % to this basal ration and both apparent and corrected digestibilities determined. Remarkably constant values at quite a high level were obtained for an animal fat, a vegetable fat, one of marine origin and an industrial by-product. These materials differed greatly in degree of esterification, unsaturation and also in the chain length of the predominant acids. By means of gas–liquid chromatographic techniques the digestibilities of the constituent fatty acids were also determined. There was only a relatively small drop in digestibility even in the case of stearic acid. When species differences are considered in the case of natural fats there seems to be something of a levelling out

Table 9. *Species differences in fat and fatty-acid digestibility*

	Sheep	Pig	Chick	Rat
Beef tallow	85	86	84	87
HEF	78	71	87	—
Herring oil	83	93	94	—
Soyabean oil	84	93	96	98
Lauric acid	90	85	69	81
Myristic acid	85	78	39	82
Palmitic acid	87	38	12	30
Stearic acid	77	23	4	15

of the values in the case of ruminants (Table 9). The marked drop in digestibility of the fatty acids encountered in non-ruminant species as the chain length of the fatty acids increases is not found in the sheep. Another set of observations that has recently become available (Steele & Moore, 1968) is not in agreement with these findings in that their values are 75 % for myristic acid, 64 % for palmitic acid and 46 % for stearic acid. It must be pointed out, however, that these estimates are apparent digestibility coefficients.

Dietary lipids exercise a considerable influence on the utilization of other components in the diet: this is especially the case in the ruminant. There is a substantial depression in fibre digestibility when a fat supplement is included in the diet of the ruminant. There is, however, considerable evidence to suggest that the depression resulting from the addition of an unsaturated fat supplement is much greater than with a saturated fat. These findings coincide with the magnitude of the effect in lactating animals on the depression of milk fat and with the extent of change imposed upon the acetate:propionate ratio in rumen contents. It is also found in some cases that the nature of the diet influences the extent to which lipid material is utilized. It has been found that there is a marked improvement in the utilization of beef tallow from 57 % with 10 % hay to 76 % as the fibre content increased to 70 %: the remainder was barley.

The exact mechanisms for the interaction between lipids and fibre leading to a reduced digestibility of the latter are not clear but several theories have been advanced as an explanation. One theory is that the effect is due to a coating of the fibrous portion of the diet with lipids, thereby preventing attack by the microorganisms. The second theory stems from studies of the effects of prolonged supplementation of 5 % maize oil in the diet on crude fibre digestibility. Cellulose digestibility was progressively depressed over a period of 40 days and recovery was not complete until 17 days after the oil was withdrawn from the diet.

These results suggested that lipid supplementation either modified the rumen population concerned with cellulose digestion or decreased the metabolic activity of certain microorganisms.

A third theory is that of inhibition of the activity of rumen microorganisms brought about by lipid supplementation. A review of the effect of fatty acids on the growth of microorganisms suggests that the fatty acids may inhibit or promote growth and where growth is interfered with, it is due to changes in cell permeability caused by adsorption of the fatty acids on the cell wall. This feature has recently been demonstrated using labelled fatty acids which were found to be weakly absorbed on the bacteria and was also associated with inhibition. The effect of saturated long-chain fatty acids with 12–20 carbon atoms on lactic acid bacteria also showed that inhibition was due to an antimetabolite effect. The theory of inhibition is not entirely supported by the work of Czerkawski, Blaxter & Wainman (1966), who showed that while Gram-positive methanogenic bacteria were inhibited and this led to a depression in methane production, the growth of cellulolytic bacteria was not affected.

On the other hand, the theory of inhibition or reduced microbial activity is supported by the work of Roberts & McKirdy (1965) with an *in vitro* method to measure total gas production as an index of microbial activity in beef cattle fed two levels of animal tallow (5 % and 10 %) and two levels of calcium (0·5 % and 0·7 %). They found that the 10 % level of lipid supplementation had a depressing effect on ruminal activity. Since there was a reduced feed intake in these studies it is not clear if this led to a reduced microbial activity. It was suggested, however, that reduced food consumption is caused by an adverse effect upon microbial activity. Calcium level appeared to have no effect, perhaps because the two levels studied were not sufficiently different to allow a critical appraisal.

A fourth theory is concerned with the observation that frequently, following dietary lipid supplementation, there is a simultaneous marked depression in fibre digestibility and a reduced retention of calcium due to excessive excretion of soaps in the faeces. It has also been shown (White *et al.* 1958) that the depression in fibre digestibility is alleviated by calcium supplements.

RECAPITULATION

The nature of the relationship between a mammalian organism and its alimentary microflora is essentially a function of the intrinsic physiological structure. It is amenable, however, to minor adjustment in diet and it is possible to exploit such adjustment to advantage in animal production. In the case of the ruminant the rumen microorganisms are in many respects an advantage to the host, but there are also drawbacks. Restricting the penalties of rumen action by an overall diminution of rumen activity is probably an objective doomed to failure – the effects upon overall physiological function will probably inevitably outweigh any minor advantages that result from restricting pathways of wastage and relative inefficiency.

REFERENCES

ALHASSAN, W. S., KRABILL, L. F. & SATTER, L. D. (1969). Manipulation of the ruminal fermentation. 1. Effect of sodium sulfite on bovine ruminal fatty acid concentration and milk composition. *J. Dairy Sci.* **52**, 376.

ANDREWS, R. J. (1967). *Fat digestibility in the ruminant*. Ph.D thesis, University of Nottingham.

ANNISON, E. F. (1954). Some observations on volatile fatty acids in the sheep's rumen. *Biochem. J.* **57**, 400.

ANNISON, E. F. (1954). Studies on the volatile fatty acids of sheep blood with special reference to formic acid. *Biochem. J.* **58**, 670.

ANNISON, E. F., BROWN, R. E., LENG, R. A., LINDSAY, D. B. & WEST, C. E. (1967). Rates of entry and oxidation of acetate, glucose, $D(-)$-β-hydroxybutyrate, palmitate, oleate and stearate, and rates of production and oxidation of propionate and butyrate in fed and starved sheep. *Biochem. J.* **104**, 135.

ANNISON, E. F., HILL, K. J. & KENWORTHY, R. (1968). Volatile fatty acids in the digestive tract of the fowl. *Brit. J. Nutr.* **22**, 207.

ARMSTRONG, D. G. & BEEVER, D. E. (1969). Post-abomasal digestion of carbohydrate in the adult ruminant. *Proc. Nutr. Soc.* **28**, 121.

BARNES, E. M. & SHRIMPTON, D. H. (1957). Causes of greening of uneviscerated poultry carcases during storage. *J. appl. Bact.* **20**, 273.

BLAXTER, K. L. (1964). In *The Role of the Gastro-intestinal Tract in Protein Metabolism*. Ed. H. N. Munro. Oxford: Blackwell.

BLAXTER, K. L. (1967). *The Energy Metabolism of Ruminants*, 2nd edition. London: Hutchinson.

BLAXTER, K. L. & WAINMAN, F. W. (1964). The utilization of the energy of different rations by sheep and cattle for maintenance and for fattening. *J. agric. Sci., Camb.* **63**, 113.

CHALUPA, W. (1970). Assimilation of ruminal ammonia by glutamate dehydrogenase. *J. Dairy Sci.* **53**, 667 (abstr.).

CRANWELL, P. D. (1968). Microbial fermentation in the alimentary tract of the pig. *Nut. Abs. & Rev.* **28**, 721.

CZERKAWSKI, J. W., BLAXTER, K. L. & WAINMAN, F. W. (1966). The effect of linseed oil fatty acids incorporated in the diet on the metabolism of sheep. *Brit. J. Nutr.* **20**, 485.

DEVENDRA, C. (1969). *The interaction between dietary lipids and fibre in ruminant nutrition*. Ph.D. thesis, University of Nottingham.

EGAN, A. R. (1965). The fate and effects of duodenally infused casein and urea nitrogen in sheep fed on a low-protein roughage. *Aust. J. agric. Res.* **16**, 169.

ELLIS, W. C., GARNER, G. B., MUHRER, M. E. & PFANDER, W. H. (1956). Nitrogen utilization by lambs fed purified rations containing urea, gelatin, casein, blood fibrin and soybean protein. *J. Nutr.* **60,** 413.

FRIEND, D. W., NICHOLSON, J. W. G. & CUNNINGHAM, H. M. (1964). Volatile fatty acid and lactic acid content of pig blood. *Can. J. Anim. Sci.* **44,** 303.

HABEL, R. E. (1965). In *Physiology of Digestion in the Ruminant.* Ed. R. W. Dougherty *et al.* Washington: Butterworth.

JAYNE-WILLIAMS, D. J. & COATES, M. E. (1969). In *Nutrition of Farm Animals of Agricultural Importance,* Part 1. Ed. Sir David Cuthbertson. London: Pergamon Press.

JOHNSON, C. B., HAMILTON, T. S., ROBINSON, W. B. & GAREY, J. C. (1944). On the mechanism of non-protein nitrogen utilization by ruminants. *J. Anim. Sci.* **3,** 287.

KEYS, J. E., VAN SOEST, P. J. & YOUNG, E. P. (1969). Comparative study of the digestibility of forage cellulose and hemicellulose in ruminants and non-ruminants. *J. Anim. Sci.* **29,** 11.

LENG, R. A., STEELE, J. W. & LUICK, J. R. (1967). Contribution of propionate to glucose synthesis in sheep. *Biochem. J.* **103,** 785.

LEWIS, D. (1957). Blood-urea concentration in relation to protein utilization in the ruminant. *J. agric. Sci., Camb.* **48,** 438.

LEWIS, D. & MORGAN, J. T. (1963). Fats and amino acids in broiler rations. 2. Metabolisable energy determination and energy-protein balance. *Br. Poult. Sci.* **4,** 3.

LOOSLI, J. K. & HARRIS, L. E. (1945). Methinoine increases the value of urea for lambs. *J. Anim. Sci.* **4,** 435.

McLAREN, G. A., ANDERSON, G. C. & BORTH, K. M. (1965). Influence of methionine and tryptophan on nitrogen utilization by lambs fed high levels of non-protein nitrogen. *J. Anim. Sci.* **24,** 231.

McNAUGHT, M. L., OWEN, E. C., HENRY, K. M. & KON, S. K. (1954). The utilization of non-protein nitrogen in the bovine rumen. 8. The nutritive value of the proteins of preparations of dried rumen bacteria, rumen protozoa and brewer's yeast for rats. *Biochem. J.* **56,** 151.

McNAUGHT, M. L., SMITH, J. A. B., HENRY, K. M. & KON, S. K. (1950). The utilization of non-protein nitrogen in the bovine rumen. 5. The isolation and nutritive value of a preparation of dried rumen bacteria. *Biochem. J.* **46,** 32.

MICHEL, M. C. (1962). Activité metabolique de la flore totale isolée de l'intestin du porc. *Amino Acides* **5,** 157.

NRC–NAS (1969). *US–Canadian Table of Feed Composition,* NRC–NAS Publ. 1684. Washington D.C.: National Academy of Science.

REED, F. M., MOIR, R. J. & UNDERWOOD, E. J. (1949). Rumenal flora studies in the sheep. 1. The nutritive value of rumen bacterial protein. *Aust. J. scient. Res.* B **2,** 304.

REIS, P. J. (1967). The growth and composition of wool. IV. The differential response of growth and of sulphur content of wool to the level of sulphur containing amino acids given per abomasum. *Aust. J. Biol. Sci.* **20,** 809.

ROBERTS, W. K. & McKIRDY, J. A. (1965). Report 14. Livestock Reserach Department of Animal Science, University of Manitoba.

ROBINSON, D. W., PRESCOTT, J. H. D. & LEWIS, D. (1965). The protein and energy nutrition of the bacon pig. IV. Digestible energy values of cereals in pig diets. *J. Agric. Sci., Camb.* **64,** 59.

SCHELLING, G. T., HINDS, F. C. & HATFIELD, E. E. (1967). Effect of dietary protein levels, amino acid supplementation and nitrogen source upon the plasma free amino acid concentrations in growing lambs. *J. Nutr.* **92,** 339.

SMITH, R. H. (1969). Reviews of the progress of dairy science. Section G, General. Nitrogen metabolism and the rumen. *J. Dairy Res.* **36**, 131.

STEELE, W. & MOORE, J. H. (1968). The digestibility coefficient of myristic, palmitic and stearic acids in the diet of sheep. *J. Dairy Res.* **35**, 371.

VIRTANEN, A. I. (1969). On nitrogen metabolism in milking cows. *Fedn Proc.* **28**, 232.

WAKELING, A. E., ANNISON, E. F. & LEWIS, D. (1970). The amino acid requirements of ruminants. *Proc. Nutr. Soc.* (In the Press.)

WARD, J. K., RICHARDSON, D. & TSIEN, W. S. (1961). Volatile fatty acid concentrations and proportions in the gastrointestinal tract of full-fed beef heifers. *J. Anim. Sci.* **20**, 830.

WHITE, T. W., GRAINGER, R. B., BAKER, F. H. & STROUND, J. W. (1958). Effect of supplemental fat on digestion and the ruminal calcium requirement of sheep. *J. Anim. Sci.* **17**, 797.

ECOLOGICAL ESSENTIALS OF ANTIMICROBIAL FOOD PRESERVATION

D. A. A. MOSSEL

Central Institute for Nutrition and Food Research TNO,
Zeist, The Netherlands
and the Catholic University, Louvain, Belgium

INTRODUCTION

Instances of the involvement of microorganisms with food production are numerous. They are of great economic significance for the food-manufacturing industries in general, but particularly pressing for developing areas of the World. Hence it seems justified to concentrate first of all on the latter.

Table 1. *Alleged, occasional or regular, annual losses in various areas of the World, due to food spoilage of biological origin (F.A.O. 1969)*

Country or continent	Commodity	Size of annual losses	
		Unit	Quantity
Africa	Postharvest (total)	%	30
Argentina	Bread	10^6 kg.	400
	Total	10^6 \$	350
Canada	Wheat	10^6 tons	13·5
Chile	Vegetables	%	50
	Fruits	%	40
India	Rice	10^6 tons	3
Latin America	Total crop	%	40

Lack of sufficient locally prepared food will be contributory to malnutrition and even starvation. The importance of proper preservation of raw materials, semi-prepared foods and manufactured products can therefore hardly be overrated. The losses actually occurring from microbial food spoilage seem hard to assess accurately. Some rough estimates for total biological losses of particularly vulnerable products, or in technologically not sufficiently advanced areas, are presented in Table 1. At first sight these figures seem quite out of proportion. However, the source from which these data were taken makes it likely that heavy losses of this order of magnitude might indeed occur as a result of microbial spoilage of foods.

The population in undernourished areas is generally suffering from an increased susceptibility for food-borne diseases of microbial and parasitic origin (Scrimshaw, Taylor & Grodon, 1968). Hence, keeping foods to be consumed in such areas safe is a second, urgent responsibility for food microbiologists; successful attempts of this kind will not only lead directly to considerably decreased morbidity data, but in addition systematically sanitize the environment (water, soil, vegetables, fish, slaughtered animals – and man) and hence result in lower infectious disease rates by a secondary mechanism as well. Safe food being generally, rather than exceptionally, available in these areas will also have an indirect economic benefit in that it will favour tourism, so often hampered by food- and environment-borne diarrhoea, rightly named 'tourists' disease' (Ringertz & Mentzing, 1968; Kean, 1969).

Finally, economists seem to agree on the paramount importance of developing areas exporting manufactured food products rather than agricultural commodities, including whole carcase meats. This sort of vital exportation of food products makes it imperative that microbiological standards for foods, as they exist in many of the economically advanced countries, are met by the local producers. The author's personal experience with attempts to manufacture food products complying with such standards in Latin American countries shows that this is not always an easy task, particularly where environmental sanitation has not yet greatly advanced.

Solutions to all these sorts of microbiological problems have also to be economically acceptable. This means that measures which are to be introduced have to be carefully studied in two respects: (i) being effective for the purpose of use; (ii) being absolutely required, i.e. possessing no element of unnecessary perfectionism. It is obvious that this goal cannot be hoped to be reached without very advanced and specific knowledge of the occurrence and fate of microorganisms in foods, or, in more academic terms: the ecology of safeguarding microbial wholesomeness and quality of foods.

ESSENTIALS OF FOOD ECOLOGY
General principles

Most fresh foods are initially contaminated internally, or from external sources. Even well protected materials such as the intact avian egg (Board, 1970) or the internal tissues of plants (Samish, Etinger-Tulczynska & Bick, 1963) are not necessarily sterile and, in fact, very little bruising or spontaneous breakdown of protective barriers is required to convert such products into very perishable commodities.

All foods are, in one of these ways, contaminated from the same ubiquitous foci: soil, dust, surface water, manure, decaying material and spoiled foods. However, this rather non-specific primary contamination does not as a whole contribute to the ultimate microbial spoilage or loss of wholesomeness of foods. Rather is a sharp selection occurring so that different foods deteriorate due to the development of fully different food-specific microbial populations. This is well known in practice. Chilled fresh meats spoil due to slime formation by bacteria of the *Pseudomonas/Acinetobacter/Alcaligenes* group; cured meats, on the contrary, turn sour because of extensive proliferation of mainly species of the genera *Micrococcus*, *Microbacterium* and *Lactobacillus*; fresh vegetables show 'soft' and other types of 'rots' caused by Gram-negative rod-shaped bacteria; whereas fruit juices tend to fermentation by yeasts and/or acidification by *Acetobacter* bacteria and dried staples, such as cereals, pulses and fruits turn mouldy (Mossel, 1971).

Such specific groups of organisms, predominating the spoilage flora of foods, are called, in accordance with the nomenclature followed in general ecology (Kershaw, 1964), 'the spoilage association of foods' (Westerdijk, 1949). Their specificity results from the selective effect of four groups of so-called ecological parameters.

The parameters of food spoilage

The following ecological parameters determine the spoilage association of foods.

Intrinsic parameters

These determinants of microbial proliferation reside in the physico-chemical composition of the food itself.

Very essential is the water-activity (symbol a_w) of the food, defined as the aqueous vapour pressure of a food divided by the vapour pressure of pure water at the same temperature (Scott, 1957). The value for a_w is inversely proportional to, and therefore a measure for, a food's osmotic pressure (Bone, 1969). Water activity values over 0·98 will offer almost equal potentialities of development to most groups of microorganisms. At a_w values below 0·95 only the more osmoduric bacteria, particularly catalase-positive and catalase-negative cocci as contrasted to Gram-negative rod-shaped bacteria, will grow well, and in addition virtually all yeasts and moulds. Below a_w values of approximately 0·88 most bacteria and yeasts will be inhibited, so that the spoilage association will be mostly fungal. With decreasing a_w values fewer and fewer fungi will be able to grow at a significant rate, and at a_w values below 0·60 even the most xerophilic moulds and some extremely osmophilic yeasts will cease to proliferate (Mossel & van Kuijk, 1955; Scott, 1957; Mossel, 1971).

Acidity of a food, expressed in pH and buffering power, is an equally important intrinsic selective factor. The pH range from neutrality down to approximately pH 6 will not exert any selective influence. However, below pH 6 the less aciduric, generally strongly proteolytic bacteria will start to be inhibited, their proliferation coming to a complete standstill at pH values somewhere between 5·0 and 4·5. Lactobacillaceae, yeasts and moulds will henceforth predominate in spoilage associations, some very acid-tolerant rod-shaped homofermentative *Lactobacillus* species being often the winners in the lowest pH bracket (Hays & Riester, 1952). At pH values well over 7, which only rarely occur in foods, *Pseudomonas* and *Vibrio* species will mostly predominate.

Much less selective is the influence of the nutrient composition of a food. No doubt, proteinaceous staple foods (meat, fish, eggs and dairy products) offer the best supply of nutrients to most microorganisms and indeed cause most of the food-borne disease outbreaks. However, the nutrient requirements of various groups of microorganisms do not vary so greatly as their attributes, dealt with earlier, do. The best example is perhaps the colonization of certain vegetables by rather exacting lactic acid bacteria which results in the transformation of the fresh vegetables to fermented products. The plant substrates seem to contain sufficient nutrients to support the growth of the fastidious Lactobacillaceae. That these organisms do not, normally, form the spoilage association of vegetables is explained by the fact that more rapidly growing bacteria will normally dominate unless prevented to do so by the selective environmental conditions which are essential for vegetable fermentation, namely a high degree of anaerobiosis and a slightly decreased water activity (Pederson, 1960).

The redox potential of a food is also exerting an influence on the organisms that will invade it. Meat carcases form an interesting example of the selective effect of the redox potential. At the same time when the surface has become rather oxidized due to exposure to air and is hence colonized by aerobic proteolytic Gram-negative rod-shaped bacteria, the deeper layers will have almost maintained their low initial redox potential and hence be attacked only, for example, by the clostridia that have contaminated these areas during an agonal stage of bacteriaemia (Ingram & Dainty, 1971).

Finally, components of foods that possess antimicrobial properties, such as tannins, essential oils, glycosides and various proteinaceous inhibitors, may influence the spoilage association, particularly because many of these exert a selective microbiostatic effect (Busta & Speck, 1968; Neeman, Lifshitz & Kashman, 1970; Al-Delaimy & Ali, 1970).

Processing factors

Heat processing of foods in hermetically sealed containers, which avoids any subsequent recontamination, obviously greatly influences spoilage associations. With increasing heat dissipation the relatively less heat-resistant types of microorganisms will gradually be eliminated (Mossel, 1971). Most proteolytic aerobic Gram-negative rods and yeasts will disappear first, followed by micrococci and thence corynebacteria. These broad groups of organisms having quite different metabolic attributes, their disappearance will be reflected in the physiology of spoilage of foods thus treated. Amongst the spores of *Bacillus* and *Clostridium* species that, with increasing heat treatment, are sequentially eliminated, such differences in metabolic pathways (glycolytic, proteolytic, lipolytic, pectinolytic) are at least as pronounced as in the other groups just dealt with, so that in so-called appertized foods similar selective influences may be observed.

Heat processing may also sufficiently change the chemical composition of foods to influence the character of the association ultimately spoiling the food. Smoking (Handford & Gibbs, 1964) is the best-studied example of this sort of phenomenon; browning reaction compounds, such as hydroxymethylfurfural, may also exert microbiostatic influences (Ingram, Mossel & Lange, 1955; Jemmali, 1969) and so do certain products of fat oxidation (Ramsey & Kemp, 1963). In the latter instance again the antimicrobial activity is rather selective, spores of Bacillaceae being virtually the only organisms that are inhibited by rancid fats (Alford & Smith, 1971).

The influence of processes such as full or partial drying, candying (= sugar preserving), acidulation or addition of antimicrobial preservatives on the microflora is obvious, since it will result in the changes of spoilage patterns discussed above under the headings a_w value, pH value and inhibitors. Less generally realized is the contaminating effect of some modes of food processing. A last-minute addition of, for example, some grated cheese to a thoroughly precooked dinner just before freezing, or the incorporation of certain rather heavily contaminated spices in sausage, may greatly affect the type of microflora ultimately predominating; and ill-conceived modes of deboning of fried meats and barbecued poultry have often conferred to such products the same microflora as encountered in the raw commodities, pathogens included.

Extrinsic parameters

The temperature range of stored foods is the predominant factor of this group. Chilled foods will only be colonized by so-called psychrotrophic organisms (Eddy, 1960). These include the Gram-negative non-fermentative rods referred to earlier, but also many micrococci, yeasts and moulds. Dependent on other intrinsic and extrinsic influences, one or more of these groups will ultimately dominate and determine the spoilage biochemistry accordingly (cf. Table 2). In the mesophilic temperature range (20–35°) all organisms, with the exception of psychrophiles and stenothermophiles, can flourish. From 35° to 45° obviously the more thermotrophic types of bacteria obtain increasing importance, and at temperatures over approximately 55° solely thermophilic organisms will take part in spoilage.

Table 2. *A review of psychrotrophic food spoilage organisms with reference to their susceptibility to a_w, pH and P_{O_2} values*

Taxonomic genus	Limit of a_w value for growth	Growth at pH < 4·5	Sensitivity to decreased P_{O_2}
Bacteria			
Pseudomonas, Acinetobacter, Alcaligenes, Flavobacter	0·98–0·95	−	+
Aeromonas	0·98–0·95	−	−
Enterobacter, Hafnia, Citrobacter, Klebsiella, Serratia, Proteus	0·98–0·95	−	−
Micrococcus	0·95–0·91	−	+
Group D – streptococci	0·93	−	−
Lactobacilli	0·91	+	−
Corynbacterium, Arthrobacter, Listeria, Microbacterium	0·98–0·95	−	±
Some *Clostridium* and *Bacillus* species	*c.* 0·95	−	±
Yeasts			
Candida, Debaryomyces, Monilia, Torulopsis	0·91–0·87	+	−
Osmophilic *Saccharomyces* species	*c.* 0·70	+	−
Halotolerant *Debaryomyces* species	*c.* 0·80	...	...
Moulds			
Alternaria, Aspergillus, Botrytis, Cladosporium, Fusarium, Margarinomyces, Mucor, Penicillium, Rhizopus, Sporotrichum, Thamnidium, and a few others	0·88–0·80	+	±

− = no growth, or insensitive. + = growth, or sensitive. ± = variable response.
... = insufficient data available.

The oxygen partial pressure (P_{O_2}) value in the gaseous environment of a food is an additional important parameter, although many groups of organisms generally thought of as obligate aerobes (moulds, bacilli) will grow at considerably lower oxygen partial pressures than expected (Tabak & Cooke, 1968; Mol & Timmers, 1970). Where P_{O_2} values are very low, i.e. much less than 1 mm., only organisms with anaerobic pathways will flourish and of course clostridia are the organisms of paramount public health significance in this context. Part of the gaseous atmosphere of foods often consists of carbon dioxide; this compound exerts a specific antimicrobial effect which tends to give rise to organisms such as Lactobacillaceae which tolerate both low P_{O_2} and high P_{CO_2} values (Gardner & Carson, 1967; Baran, Kraft & Walker, 1970).

Finally, the a_w value of a food, and in this way its association, can be modified by extrinsic influences. Moisture absorption from, or desorption to, the atmosphere is an obvious example. Moisture migration within closed packages, resulting from temperature changes during storage, is another often overlooked but yet quite important instance.

Implicit parameters

From intrinsic, processing and extrinsic selective influences a primary selection of the initial microflora of foods results. This does not, as a rule, proliferate as an entity, because these organisms may exert mutual influences leading to predominance of some at the expense of others. Because the size of these effects depends on the initial selection, which in turn results from intrinsic and extrinsic selective influences, it has been suggested that such factors be called 'implicit parameters' by analogy with the corresponding term used in mathematics (Mossel & Westerdijk, 1949).

The simplest example of an implicit parameter is the rate of development, a rather loose expression customarily used to indicate the three criteria of the Hinshelwood curve (Dean & Hinshelwood, 1966). As already indicated, the significant difference of growth rate between, for example, *Lactobacillus plantarum* and *Pseudomonas fluorescens* means that normally the latter will have developed 'fully', i.e. to viable counts of about $10^8/g$. almost before the former has come out of the lag phase; and only when pseudomonads are inhibited by, for example, anaerobic conditions will the Lactobacillaceae obtain a footing.

More complicated instances of implicit influences are based on synergism or antagonism. One group of organisms – for example streptococci – may favour the development of others – for example, lactobacilli or propionibacteria – by the biosynthesis of amino acids or vitamins which stimulate growth of the latter group. Or, certain yeasts may metabolize organic acids occurring in fruits or vinegar-preserved salads (Dakin & Stolk, 1968), and in this way increase the pH and clear the way for less aciduric bacteria. Similar metabolic processes may exert opposite effects and the resulting occurrences are then called antagon-ism. Rapidly growing non-fastidious bacteria often exhaust the supply of nutrients in a food, before more exacting organisms have proliferated to any extent (Dubos & Ducluzeau, 1969b); and bacteria that very quickly form acid from carbohydrates may inhibit slightly slower growing types of high pH sensitivity (Winter, Weiser & Lewis, 1953; Saleh & Ordal, 1955). Also, facultatively anaerobic bacteria sometimes decrease the redox potential of a food so rapidly that the proliferation of obligately aerobic organisms is soon checked (Dubos & Ducluzeau, 1969a).

This approach allows one to define in more precise terms what in food microbiology are loosely called 'robust' organisms. These will be rapidly growing types, with few nutritional requirements for growth and survival, and not too sensitive to: (i) moderate changes in intrinsic, extrinsic and implicit parameters (Pietkiewicz et al. 1969) influencing a_w value, pH value and P_{O_2} value; and (ii) changes in temperature (Mossel, 1963).

PRACTICAL IMPLICATIONS
Type of problems to be solved

From what has been said in the previous section, it is obvious that anti-microbial food preservation is bound to lead to disappointing results if it is not based on a thorough knowledge of food ecology. Two examples from daily practice may illustrate this further.

Salmonellosis

Although a great many Salmonella serotypes occur in foods and feeds of animal origin to which man is exposed, only five or so dominate in human epidemiology, notably Salmonella typhimurium, S. panama, S. stanley, S. enteritidis and S. newport (Guinée, Kampelmacher & Valkenburg, 1967; Le Minor & Le Minor, 1967; Pietkiewicz & Buczow-ski, 1969; Martin & Ewing, 1969). This has resulted in great errors in the prevention of salmonellosis, because it led to the assumption that foods, contaminated with other serotypes, would not play an aetio-logical role in human disease and hence did not have to be banned per se.

However, the examination of foods for Salmonellae is fraught with analytical uncertainties. In the (in the ecological sense) 'best' case, all Sal-monellae are missed because the enrichment method used for their detec-tion is unsuccessful due to strong antagonistic influences exerted by other food bacteria (Gundstrup, Grunnet & Bonde, 1969; Splittstoesser & Segen, 1970). Much more serious are the consequences of the occurrence of more than one Salmonella serotype in a contaminated food or feed, a situation which is the rule rather than the exception (Jacobs *et al.* 1963; Harvey & Price, 1967*b*; Wesselinoff & Toneff, 1968). Which type will ultimately be 'isolated' from such materials depends on: (i) the numbers of the various serotypes present; (ii) their viability (Corry, Kitchell & Roberts, 1969; Speck, 1970); (iii) how intrinsically 'robust' each of them is, particularly in terms of resisting the high degree of selectivity applied in current enrichment techniques; (iv) probability (Harvey & Price, 1967*a*). Clearly such results are of very low, if any, significance then for epidemiological purposes.

It seems that the reason for the predominance of serotypes such as *S. typhimurium* in human pathology lies in the frequency with which these organisms contaminate surface waters (Grunnet & Nielsen, 1969). This may be a consequence of their being voided in such large numbers with human stools and/or their particular adaptation to the aquatic environment. In addition, this results in frequent possibilities for contamination of slaughter animals and hence our meat supply. The exact causal relationships have yet to be assessed; however, the facts established at present show, beyond any doubt, how the food scientist can be led astray if he does not pay sufficient attention to ecological finesse.

Failures in food preservation

It is a sound principle of food microbiology that the preservation of food should primarily be achieved by physical methods such as drying, canning or freezing, because this avoids any potential risks of toxicity resulting from the use of antimicrobial food additives, the alternative of such physical methods. However, in some instances, the organoleptic quality of a food suffers greatly from such modes of processing, while in quite a few areas freezing, canning or even drying are not technologically or economically feasible. Hence there is a fully legitimate, logistically determined place for the use of chemical inhibitors of microbial growth in food preservation, obviously provided such an addition to food will not in any respect harm the health of the consumer (Chichester & Tanner, 1968). Successful chemical preservation of foods requires the application of a few ecological principles (cf. Table 3).

Firstly, a preservative should be used whose antimicrobial spectrum is in accordance with the spoilage association of the food to be preserved. No effect of the use of an inhibitor can be expected when it is, in essence, a bacteriostatic agent, while the spoilage association of the food in which the preservative is to be used is almost purely fungal, and *vice versa*. In more detail: a bacterial inhibitor which, at the concentrations used, exerts its action virtually only against Gram-positive bacteria,

Table 3. *Efficacy of antimicrobial chemical preservatives that are acceptable from a toxicological point of view*

	Antimicrobial activity		Acceptable unconditional max. daily intake (mg./kg. body wt.)
Compound	pH Dependence	Main spectrum	
Benzoic acid	+	*m, y*	5
Diacetates	±	*b*	15
Diphenyl	−	*m*	0·05
Esters of gallic acid	−	Rather broad antimicrobial	0·2
Esters of *p*-hydroxybenzoic acid	−	spectrum	2
Nitrites	+	*b*	0·4
Propionic acid	+	*m, b*	10
Sorbic acid	±	*m, y, c, b, gn*	12·5
Sulphites	±	*gn, y*	0·35

+ = very dependent; ± = moderately dependent; − = independent; *b* = spores of Bacillaceae; *c* = catalase-positive cocci; *gn* = Gram-negative rod-shaped bacteria; *m* = moulds; *y* = yeasts.

should not be considered for use on fish, for example, whose spoilage association is entirely Gram-negative.

Moreover, the spectrum of activity should not be too narrow. When the normal association of a food is successfully inhibited, quite often, as we have seen earlier, a group of organisms will come to the fore that is capable of invading and spoiling the food in principle, but which is generally suppressed by the more robust groups of organisms which dominate the normal spoilage association.

Neither should a preservative compound be too labile. Spontaneous chemical decomposition in a food is an obvious drawback. But slow inactivation by certain microbial groups (Laxa, 1930; Ingram, 1960a; Rehm, Wallnöfer & Lukas, 1963; Marth, 1966) or induction of resistance (Kneteman, 1953; Vaughn et al. 1969) also leads to unsuccessful preservation.

Also, similar to the situation in, for example, preservation of foods by heating (Stumbo, 1965), the efficiency of chemical preservation depends greatly on the initial concentration of potential spoilage agents in the food; the higher these initial loads the higher the concentration of the preservative required to attain a desired degree of keeping quality (Christian, 1963). Therefore chemical preservation of foods has to be supported by the highest attainable degree of food-plant sanitation.

There is a second, ecologically determined reason for using the latter measures. When food-processing equipment is not, or cannot be, fully cleaned, foci will be formed in which the more robust organisms will be enriched. Amongst these the types with the highest resistance, particularly against decreased a_w values and increased concentrations of preservatives, will eventually dominate. Such organisms are later voided into the food passing through the lines; they will thus lead to a complete failure of the attempted preservation since they will proliferate virtually unchecked despite the presence of the preservative.

Available methodology

Control of microorganisms of all sorts rests on the application of a few simple principles. The first is constant surveillance of the microbiological quality of all materials coming into the food factory: fresh foods and minor ingredients, as well as packaging material, drinking water and the air supply. The next step is sanitary supervision of the premises and production lines: the control of foci as well as guarding against vermin and bad sanitary practices of the staff. Highly essential also is a constant supervision of processing steps, such as those carried out for safety, as in the pasteurization of milk and egg products, or for improving the keeping quality of acid products such as fruit juices and salads, or dry goods (for example, marzipan) that do not present health risks but spoil rapidly unless preserved properly. Finally, the packaging, storage, transport and holding before service or culinary preparation require constant attention.

It is generally accepted that food protection should have this preventative character (Semple, 1969; Mossel, 1970). Nevertheless the effect of such measures has to be checked regularly and this requires laboratory examination of, for example, line samples, water, air, apparatus, and hands of staff. Also, laboratory examination of fully manufactured foods is in place, particularly to verify whether these will meet the requirements of importing countries, so essential for the food industry in developing areas, as we have seen earlier.

Examinations of this type present a few logistical as well as operational problems which are worth mentioning since they pertain to food ecology and their solution is not too difficult, when once appreciated.

Sampling

The importance of adequate sampling cannot be overrated. Where raw materials or manufactured foods are examined, an adequate number of samples has to be taken and, in addition, at such a degree of randomization that the data obtained on the samples have a known correlation with the condition of the consignment; this obviously requires thorough knowledge of the mode of distribution of pertinent organisms over the food material (Cowell & Morisetti, 1969). In the examination of food-manufacturing machinery, the predilection sites for contamination have to be known and these have to be sampled preferentially. When water and air are to be examined their generally low contamination rate should be taken into account by the examination of portions of sufficiently large size. When staff is to be examined at all for the excretion of pathogenic organisms, such examinations have to be repeated in a frequency known to guarantee the detection of a certain carrier rate.

All of these procedures result in rather large numbers of samples being submitted for laboratory examination. In order to avoid exceeding the capacity of the available analytical services, or making the testing financially unacceptable, a very rigorous limitation of the numbers of tests carried out with all samples is absolutely required (Wilson, 1955).

Choice of tests

Justified limitation of the numbers of examinations of each sample requires sufficient knowledge of the ecology of the materials under examination. It would, for example, be senseless to examine fresh meats for yeasts or fruit *purées* for *Pseudomonas aeruginosa*. Examination of materials has hence to be for those organisms only that have been found experimentally to be either a health risk or part of the spoilage association. The former are mostly carried out as presence-or-absence tests in aliquots of 10–20 g., the latter as plate counts on suitably selective media.

Nevertheless some examinations are useful in many instances. These are particularly enumerations of so-called index organisms, bacteria whose quantified presence indicates a potential health-or-spoilage risk. The Enterobacteriaceae group, for example, is a most useful group of index organisms for the evaluation of heat-treated foods of pH values greater than 4·5. Since all Enterobacteriaceae are eliminated by adequate heat-processing, aiming at safety, the occurrence of these bacteria in foods processed in this way indicates either inadequate heat-treatment or post-process recontamination, both unacceptable deficiencies. Because of the low frequency and very irregular spread of pathogenic bacteria in such foods, a negative result of a presence-or-absence test for the latter is of no significance whatsoever, despite the time and effort spent, which is considerably more than to conduct the much more meaningful enumeration of Enterobacteriaceae, which is exactly as simple as a 'coliform' test (Mossel, 1967). Similarly a sensitive presence-or-absence test for all yeasts in fruit juices indicates whether adequate preservation has been achieved and hence allows an estimation of shelf life (Ingram, 1960 b). Likewise, testing of dried foods for the development of mould colonies upon moistening with an aqueous solution of an antibacterial antibiotic will allow a very approximate estimation of the mould-spore count, thus allowing one to anticipate to what extent proliferation of moulds in the food will occur when local rehydration might take place (Krugers Dagneaux & Mossel, 1959). The latter test may merit increased interest in attempts to avoid foods and feeds becoming sources of mycotoxins (Moreau, 1968).

Where deficiencies are detected in this way, more detailed investigations are required to establish their cause. Such examinations may include enumerations of more specific organisms, or tests for residual enzyme activity, indicating insufficient heat-processing (Ingram & Roberts, 1966; Hobbs & Gilbert, 1970). Hopefully, deficiencies of this sort and the analytical follow-up which they require are exceptions in factories, where the microbiological quality of food is controlled systematically; hence the laboratory can as a rule cope with the work load which they involve.

Standards

The microbiological examination of line and end-product samples, machinery, staff, water and air presupposes that, when the results become available, they can be subsequently evaluated and particularly roughly translated for engineers and management in terms such as: 'reassuring', 'doubtful', or 'imminent'. Hence, the drafting of such standards is an urgent task for the food microbiologist (Hobbs & Gilbert, 1970; Mossel, 1970; Cox, 1970).

In order to assess what is attainable in practice, a few test runs have to be carried out under extremely rigorous supervision. A sufficient number of line and finished samples is then taken and examined and the standard defined as the levels of counts not exceeded by 90–95 % of the samples.

An essential element in the application of such standards are the tolerances (Peterson, 1970). Customarily a maximum of two out of ten samples tested may show elevated counts, without the consignment or the plant condition being considered substandard. Obviously the maximum value of the excesses allowed is limited. Dependent on the coefficient of variation of the enumeration, as determined by the distribution of the organisms, and factors such as their tendency to clumping, this maximal permissible excess varies from 1–4 times the value of the count set as the standard (Mossel, 1970).

It is clear that no tolerances are ever applied to any standards for pathogenic or toxinogenic organisms.

Streamlining of methodology

The steadily increasing numbers of examinations entrusted to the microbiological laboratory, despite all rationalizations and limitations already applied, makes streamlining, and if possible automation, of microbiological examination of foods a pressing necessity. Two modes of approach are at present available.

The first is called the replication principle (Lederberg & Lederberg, 1952). Food dilutions are first spread on to a rich infusion agar and the colonies thus obtained replicated on to a set of selective media which allows their tentative grouping (Wiseman & Sarles, 1956; Corlett, Lee & Sinnhuber, 1965; Mossel, 1970). The same procedure can be applied to anaerobic master plates, if obligately anaerobic organisms are to be included in ecological or other studies. A set of media which we have found useful for the tentative grouping of the aerobic part of the microflora of foods has been brought together in Table 4.

The second streamlined methodology, in principle lending itself to automation, is the agar-dip technique. Glass or plastic slides of convenient size are covered with thin layers of general-purpose agar and selective agar media. Sets of these are dipped in beakers containing dilutions of the food to be examined and absorb in this way a certain amount of inoculum. The colonies obtained after incubation can be enumerated in the usual way. So far only slides of approximately 2×7 cm., and covered solely with total-count and McConkey agar, are available in commerce and currently used in the bacteriological examination of urine (Cohen & Kass, 1967; Naylor & Guttmann, 1967; Arneil, McAllister & Kay, 1970). However, we have found this principle also very convenient for more general application in the differential microbiological examination of foods (Mossel, 1970).

Conditions of replication subculture

... aerobic master plates of rich infusion agar

Description of medium	Incubation			Taxonomic group predominantly selected by culture conditions	Colony type to be looked for	Further tests, primarily to be carried out with such colonies
	Temp.	Time (h.)	Po$_2$			
Sugar-free agar with 0·5% NaCl and 1 μg. crystal violet/ml.	21–24°	30–48	Aerobic	All Gram-negative, rod-shaped bacteria	Non pin-point colonies	Gram stain
Violet-red bile glucose agar	35–37°	18–20	Anaerobic	Enterobacteriaceae and Aeromonadaceae	Colonies surrounded by purple halo of precipitate	Oxidase reaction. Mode of attack on glucose
Chapman's mannitol 7·5% NaCl agar	30–32°	48	Aerobic	Catalase-positive cocci	Regular, smooth, creamy colonies	Gram stain. Catalase reaction
Baird-Parker's tellurite egg-yolk pyruvate agar	35–37°	28–30	Aerobic	*Staphylococcus aureus*	Anthracite black, smooth colonies, mostly but not invariably surrounded by a clear zone	Coagulase reaction
Azide crystal-violet kanamycin agar	35–37°	48	Anaerobic	Lancefield group-D streptococci	Small, regular, smooth colonies, bluish or white	Gram stain (broth culture). Catalase reaction. Growth at 45± 0·1°
Acetate agar of Rogosa *et al.* (pH = 5·8)	30–32°	48–72	Anaerobic	Lactobacillaceae	Rather small, regular colonies	Gram stain. Catalase reaction
Infusion agar with 10 μg. polymyxin B sulphate/ml.	30–32°	18–20	Aerobic	Bacillaceae	Irregular, dull, flat colonies	Gram stain
Blood infusion agar with 40 μg. nalidixic acid/ml.	30–32°	28–30	Aerobic	*Corynebacterium, Arthrobacter* and *Brevibacterium* spp.	Mostly pigmented, creamy and regular	Gram stain
Glycerol mannitol acetamide cetrimide phenol red agar	42±0·1°	18–20	Aerobic	*Pseudomonas aeruginosa*	Colonies surrounded by red zones	Oxidase reaction. Mode of attack on glucose, arginine and starch
Sucrose 5% NaCl bromothymol blue agar with 5 μg. tylosin/ml. and 2 μg. polymyxin B sulphate/ml.	35–37°	28–30	Anaerobic	*Vibrio parahaemolyticus*	Rather small, regular colonies surrounded by blue zones	Fermentative attack on starch agar containing 5% NaCl
Oxytetracycline, or gentamicine, glucose-yeast extract-agar	21–24°	48–72	Aerobic	Moulds and yeasts	Mycelial, resp. regular and glistening colonies	Alkaline methylene blue stain

REFERENCES

AJL, S. J., KADIS, S. & MONTIE, T. C. (1970). *Microbial Toxins*. New York: Academic Press.

AL-DELAIMY, K. S. & ALI, S. H. (1970). Antibacterial action of vegetable extracts on the growth of pathogenic bacteria. *J. Sci. Fd Agric.* **21**, 110.

ALFORD, J. A. & SMITH, J. L. (1971). Roles of microbial lipolysis in desired and spoilage changes in foods. *J. appl. Bact.* **34**. (In the Press.)

ARNEIL, G. C., McALLISTER, T. A. & KAY, P. (1970). Detection of bacteriuria at room temperature. *Lancet* i, 119.

AYLWARD, F. & HAISMAN, D. R. (1969). Oxidation systems in fruits and vegetables – their relation to the quality of preserved products. *Adv. Fd Res.* **17**, 1.

BARAN, W. L., KRAFT, A. A. & WALKER, H. W. (1970). Effects of carbon dioxide and vacuum packaging on color and bacterial count of meat. *J. Milk Fd Technol.* **33**, 77.

BEIJERINCK, M. W. (1908). Fermentation lactique dans le lait. *Archs néerl. Sci.* II, **13**, 356.

BOARD, R. G. (1970). The microbiology of the hen's egg. *Adv. appl. Microbiol.* **11**, 245.

BONE, D. P. (1969). Water activity – its chemistry and applications. *Fd Prod. Dv.* **3**, no. 5, 81.

BUSTA, F. F. & SPECK, M. L. (1968). Antimicrobial effect of cocoa on Salmonellae. *Appl. Microbiol.* **16**, 424.

CHICHESTER, D. F. & TANNER, F. W. (1968). Antimicrobial food additives. In *Handbook of Additives*, p. 137. Ed. Th. E. Furia. Cleveland: Chemical Rubber Co.

CHRISTIAN, J. H. B. (1963). Preservation of minced meat with sulphur dioxide. *Fd Preserv. Q.* **23**, 30.

COHEN, S. N. & KASS, E. H. (1967). A simple method for quantitative urine culture. *New England J. Med.* **277**, 176.

CORLETT, D. A., LEE, J. S. & SINNHUBER, R. O. (1965). Application of replica plating and computer analysis for rapid identification of bacteria in some foods. II. Analysis of microbial flora in irradiated Dover sole (*Microstomus pacificus*). *Appl. Microbiol.* **13**, 818.

CORRY, J. E. L., KITCHELL, A. G. & ROBERTS, T. A. (1969). Interactions in the recovery of *Salmonella typhimurium* damaged by heat or gamma radiation. *J. appl. Bact.* **32**, 415.

COWELL, N. D. & MORISETTI, M. D. (1969). Microbiological techniques – some statistical aspects. *J. Sci. Fd Agric.* **20**, 573.

COX, W. A. (1970). Microbiological standards for dairy products. *Chemy Ind.* **7**, 223.

DABBAH, R., EDWARDS, V. M. & MOATS, W. A. (1970). Antimicrobial action of some citrus fruit oils on selected food-borne bacteria. *Appl. Microbiol.* **19**, 27.

DAKIN, J. C. & STOLK, A. C. (1968). *Moniliella acetoabutans*: some further characteristics and industrial significance. *J. Fd Technol.* **3**, 49.

DEAN, A. C. R. & HINSHELWOOD, C. (1966). *Growth, Function and Regulation in Bacterial Cells*. Oxford: Clarendon Press.

DUBOS, F. & DUCLUZEAU, R. (1969 a). Étude du mécanisme de l'inhibition *in vitro* d'une souche de *Micrococcus* sp. par une souche de *Staphylococcus pyogenes*. *Annls Inst. Pasteur, Paris* **117**, 76.

DUBOS, F. & DUCLUZEAU, R. (1969 b). Étude *in vitro* du mécanisme de l'inhibition d'une souche de *Staphylococcus pyogenes* par une souche de *Micrococcus* sp. *Annls Inst. Pasteur, Paris* **117**, 86.

EDDY, B. P. (1960). The use and meaning of the term 'psychrophilic'. *J. appl. Bact.* **23**, 189.

FOOD AND AGRICULTURAL ORGANIZATION (1969). *Food Losses: The Tragedy and Some Solutions.*

GARDNER, G. A. & CARSON, A. W. (1967). Relationship between carbon dioxide production and growth of pure strains of bacteria on porcine muscle. *J. appl. Bact.* **30**, 500.

GRUNNET, K. & NIELSEN, B. B. (1969). Salmonella types isolated from the Gulf of Aarhus compared with types from infected human beings, animals and food products in Denmark. *Appl. Microbiol.* **18**, 985.

GUINÉE, P. A. M., KAMPELMACHER, E. H. & VALKENBURG, J. J. (1967). Salmonella isolations in The Netherlands, 1961–1965. *Zentbl. Bakt. ParasitKde*, Abt. I, Originale, **204**, 476.

GUNDSTRUP, A. S. P., GRUNNET, K. & BONDE, G. J. (1969). Salmonella antagonists and Salmonella enrichment media. *Health Lab. Sci.* **6**, 221.

HANDFORD, P. M. & GIBBS, B. M. (1964). Antibacterial effect of smoke constituents on bacteria isolated from bacon. In *Microbial Inhibitors in Food*, p. 333. Ed. N. Molin and A. Erichsen. Stockholm: Almqvist and Wiksell.

HARVEY, R. W. S. & PRICE, T. H. (1967*a*). The isolation of salmonellas from animal feedingstuffs. *J. Hyg., Camb.* **65**, 237.

HARVEY, R. W. S. & PRICE, T. H. (1967*b*). The examination of samples infected with multiple Salmonella serotypes. *J. Hyg., Camb.* **65**, 423.

HAYS, G. L. & RIESTER, D. W. (1952). The control of 'off-odor' spoilage in frozen concentrated orange juice. *Fd Technol.* **6**, 386.

HOBBS, B. C. & GILBERT, R. J. (1970). Microbiological standards for food: public health aspects. *Chemy Ind.* no. 7, p. 215.

INGRAM, M. (1960*a*). An influence of carbon source on the resistance of a yeast to benzoic acid. *Annls Inst. Pasteur, Lille* **11**, 167.

INGRAM, M. (1960*b*). Fermentation tests to detect yeasts in fruit juices and similar products. *Annls Inst. Pasteur, Lille* **11**, 203.

INGRAM, M. & DAINTY, R. H. (1971). Meat spoilage. *J. appl. Bact.* **34**. (In the Press.)

INGRAM, M., MOSSEL, D. A. A. & LANGE, P. DE (1955). Factors, produced in sugar-acid browning reactions, which inhibit fermentation. *Chemy Ind.* no. 3, p. 63.

INGRAM, M. & ROBERTS, T. A. (1966). Microbiological principles. In *Proc. Symp. Food Irradiation*, Karlsruhe, p. 267. Wien: International Atomic Energy Agency.

JACOBS, J., GUINÉE, P. A. M., KAMPELMACHER, E. H. & KEULEN, A. VAN (1963). Studies on the incidence of Salmonella in imported fish meal. *Zentblt. VetMed.* **10**, 542.

JEMMALI, M. (1969). Influence of the Maillard reaction products on some bacteria of the intestinal flora. *J. appl. Bact.* **32**, 151.

KEAN, B. H. (1969). Turista in Teheran. *Lancet* ii, 583.

KERSHAW, K. A. (1964). *Quantitative and Dynamic Ecology*, p. 137. London: Arnold.

KNETEMAN, A. (1953). Some forms of adaptation of micro-organisms to extreme conditions in food industries. *VI Int. Congr. Microbiol.*, Rome **7**, 336.

KRUGERS DAGNEAUX, E. L. & MOSSEL, D. A. A. (1959). The applicability of Emmerling's principle ('mould test') in food microbiology. *Antonie van Leeuwenhoek* **25**, 152.

LAXA, O. (1930). *Margarinomyces bubaki* – ein Schädling der Margarine. *Zentbl. Bakt. ParasitKde*, Abt. II, **81**, 392.

LEDERBERG, J. & LEDERBERG, E. M. (1952). Replica plating and indirect selection of bacterial mutants. *J. Bact.* **63**, 399.

LE MINOR, L. & LE MINOR, S. (1967). Activités du Centre français des Salmonella de l'Institut Pasteur. 5e Rapport 1964–1966. *Revue Hyg. Méd. Soc.* **15**, 221.

MARTH, E. H. (1966). Degradation of potassium sorbate by *Penicillium* species. *J. Dairy Res.* **49**, 1197.

MARTIN, W. J. & EWING, W. H. (1969). Prevalence of serotypes of Salmonella. *Appl. Microbiol.* **17**, 111.

MEEDENIYA, K. (1969). Investigations into the contamination of Ceylon desiccated coconut. *J. Hgy.* **67**, 719.

MOL, J. H. H. & TIMMERS, C. A. (1970). Assessment of the stability of pasteurized comminuted meat products. *J. appl. Bact.* **33**, 233.

MOREAU, C. (1968). *Moisissures toxiques dans l'alimentation.* Paris: Lechevallier.

MOSSEL, D. A. A. (1963). La survie des Salmonellae dans les différents produits alimentaires. *Annls Inst. Pasteur, Paris* **104**, 551.

MOSSEL, D. A. A. (1967). Ecological principles and methodological aspects of the examination of foods and feeds for indicator microorganisms. *J. Ass. offic. Analyt. Chem.* **50**, 91.

MOSSEL, D. A. A. (1970). Mikrobiologische Qualitätsbeherrschung in der Lebensmittelindustrie. *Alimenta* **9**, Sondernummer, 47.

MOSSEL, D. A. A. (1971). Physiological and metabolic attributes of microbial groups associated with foods. *J. appl. Bact.* **34**. (In the Press.)

MOSSEL, D. A. A. & KUIJK, H. J. L. VAN (1955). A new and simple technique for the direct determination of the equilibrium relative humidity of food. *Fd Res.* **20**, 415.

MOSSEL, D. A. A. & WESTERDIJK, J. (1949). The physiology of microbial spoilage in foods. *Antonie van Leeuwenhoek* **15**, 190.

NAYLOR, G. R. E. & GUTTMANN, D. (1967). The dip-slide: a modified dip-inoculum transport medium for the laboratory diagnosis of infections of the urinary tract. *J. Hyg., Camb.* **65**, 367.

NEEMAN, I., LIFSHITZ, A. & KASHMAN, Y. (1970). New antibacterial agent isolated from the Avocado pear. *Appl. Microbiol.* **19**, 470.

PEDERSON, C. S. (1960). Sauerkraut. *Adv. Fd Res.* **10**, 233.

PETERSON, A. C. (1970). Development of microbiological standards for ready to eat foods. *Q. Bull. Ass. Fd Drug Off. U.S.* **34**, 114.

PIETKIEWICZ, K. & BUCZOWSKI, A. (1969). Salmonellosis in man in Poland, 1957–1966. *Publ. Hlth Rep., Wash.* **84**, 712.

PIETKIEWICZ, K., HAMON, Y., LE MINOR, L. & CHABERT, Y. A. (1969). Bactériocines et facteurs de résistance de *Salmonella panama. Annls Inst. Pasteur, Paris* **117**, 645.

RAMSEY, C. B. & KEMP, J. D. (1963). Inhibiting action of oxidized pork fat on the germination of spores of *Bacillus subtilis. J. Fd Sci.* **28**, 562.

RAMSEY, G. B., FRIEDMAN, B. A. & SMITH, M. A. (1959). *Market Diseases of Beets.* United States Department of Agriculture Handbook, no. 155.

RAMSEY, G. B. & SMITH, M. A. (1961). *Market Diseases of Cabbage.* United States Department of Agriculture Handbook, no. 184.

REHM, H. J., WALLNÖFER, P. & LUKAS, E. M. (1963). Zur Kenntnis des Abbaus der Sorbinsäure durch Mikroorganismen im natürlichen Substrat. *Dt. LebensmittRdsch.* **59**, 197.

RINGERTZ, O. & MENTZING, L. (1968). Salmonella infection in tourists. 1. An epidemiological study. *Acta path. microbiol. scand.* **74**, 397.

SALEH, M. A. & ORDAL, Z. J. (1955). Studies on growth and toxin production of *Clostridium botulinum* in a precooked frozen food. II. Inhibition by lactic acid bacteria. *Fd Res.* **20**, 340.

SAMISH, Z., ETINGER-TULCZYNSKA, R. & BICK, M. (1963). The microflora within the tissue of fruits and vegetables. *J. Fd Sci.* **28,** 259.

SCOTT, W. J. (1957). Water relations of food spoilage microorganisms. *Adv. Fd Res.* **7,** 83.

SCRIMSHAW, N. S., TAYLOR, C. E. & GRODON, J. E. (1968). *Interactions of Nutrition and Infection.* W.H.O., monograph series, no. 57.

SEMPLE, A. B. (1969). Imported foods. *R. Soc. Hlth J.* **89,** 21.

SPECK, M. L. (1970). Selective culture of spoilage and indicator organisms. *J. Milk Fd Technol.* **33,** 163.

SPLITTSTOESSER, D. F. & SEGEN, B. (1970). Examination of frozen vegetables for Salmonellae. *J. Milk Fd Technol.* **33,** 111.

STUMBO, C. R. (1965). *Thermobacteriology in Food Processing.* New York: Academic Press.

TABAK, H. H. & COOKE, W. B. (1968). The effects of gaseous environments on the growth and metabolism of fungi. *Bot. Rev.* **34,** 124.

VAUGHN, R. H., MARTIN, M. H., STEVENSON, K. E., JOHNSON, M. G. & CRAMPTON, V. M. (1969). Salt-free storage of olives and other produce for future processing. *Fd Technol.* **22,** 832.

WESSELINOFF, W. & TONEFF, M. (1968). Salmonellen in eingeführtem Tierkörpermehl. *Berl. Münch. tierärztl. Wschr.* **81,** 426.

WESTERDIJK, J. (1949). The concept 'association' in mycology. *Antonie van Leeuwenhoek* **15,** 187.

WILSON, G. S. (1955). Symposium on food microbiology and public health: general conclusion. *J. appl. Bact.* **18,** 629.

WINTER, A. R., WEISER, H. H. & LEWIS, M. (1953). The control of bacteria in chicken salad. II. Salmonella. *Appl. Microbiol.* **1,** 278.

WISEMAN, R. F. & SARLES, W. B. (1956). A plating technique for screening intestinal coliform bacteria. *J. Bact.* **71,** 480.

MICROBES AS BIOLOGICAL CONTROL AGENTS

J. R. NORRIS

Shell Research Limited, Sittingbourne, Kent, England

INTRODUCTION

The rapid development of interest in the use of microorganisms to control pest insects has been a feature of the biological control field during the past ten years, and it will form much of the substance of this review, but the existence of infectious disease in insect populations has been known for a long time and the use of disease-producing microorganisms to control field pests is not by any means a recent concept. Aristotle described diseases of the honeybee in his *Historia Animalium* and, in 1834, Agostino Bassi showed experimentally that the fungus *Beauveria bassiana* caused an infection in the silkworm and suggested that microorganisms might be used for control purposes. Today we know that insects are susceptible to a wide variety of infectious agents, including bacteria, viruses, fungi, protozoa, rickettsia and nematodes. Representatives of each of these groups have been tested for the inability to control field infestations but relatively few have yet reached a stage where their commercial exploitation seems justified.

One of the problems to be overcome in the use of a disease-causing organism to control a field pest is the necessity to produce the infective agent in a form in which it will remain viable during storage and, following application in the field, for long enough to ensure its effective distribution and use. It is perhaps natural therefore that most success has so far attended the use of microorganisms which have resistant–dormant phases in their life-cycles, and the major part of this review will be concerned with the spore-forming insect pathogens and the polyhedrosis viruses which are today the most successful biological control agents.

The successful use of microorganisms to control natural populations of pest species has been restricted almost entirely to insects and mites but, even so, the subject is an enormous one and the literature formidable. Rather than attempt a superficial survey of the whole field, I shall concentrate on the agents which are successful, or almost successful, control agents in a commercial sense, and these are concerned entirely with insect pests. For an excellent treatment of the whole subject I would refer the reader to the recently published book *Microbial Control of Insects and Mites*, edited by Burges & Hussey (1971). My aim will be to extract principles from the work which is now available, to identify problem areas in the development of microbial control agents, and to attempt to forecast the future development of the subject.

THE MEANING OF 'BIOLOGICAL CONTROL'

One generally accepted definition of biological control is the simple 'the use of biota to control biota'. This is meaningful in relation to simple situations involving the use of pathogens and predators to control native pest populations, but it is open to criticism when some of the insect pathogenic microorganisms are concerned. Clearly, the use of *Bacillus popilliae*, or of some of the viruses which spread rapidly in an insect population, to build up a persisting endemic condition fall within this definition. So too do some of the viruses which cause spreading but short-lived infections in insect populations and must be repeatedly used during a growing season to protect a crop. But what of an organism like *Bacillus thuringiensis*, where it is the toxins produced by the bacterium which are the effective insecticidal agents and the living organism plays a secondary, though possibly important, role in population control? In this instance, persisting infections in insect populations are not set up following dissemination of the organism in the field, but the toxins separated from living cells show little commercial potential when used alone in the field and the use of spore-toxin mixtures of *Bacillus thuringiensis* can be considered as a valid example of a biological control agent. Clearly, it would be nonsense to consider as a biological control agent any toxic product which was produced by a living organism. Such an approach would imply that an antibiotic was a biological control agent. For the purpose of this review, I shall define a biological control agent as any living organism used, either alone or together with its toxic products, to control another species where the introduction of living material into the host population is an essential part of the control procedure. This definition involves no consideration of the spread or persistence of infection in a natural community, and is a definition of convenience, justified because it brings together a group of microbial agents which exhibit underlying scientific similarities and present similar problems to the field worker, and particularly to government regulatory authorities.

MYXOMATOSIS

Although the major part of this review is concerned with the microbial control of insects, the lessons to be learned from the myxomatosis episode are fundamental to the subject and merit at least brief attention. The subject is discussed in detail by Fenner & Ratcliffe (1965).

In 1895, Professor Guiseppi Sanarelli, the Director of the Hygiene Institute in the University of Sienna, was invited by the Government of Uruguay to set up a Hygiene Institute at Montevideo. In the process of setting up this Institute, Sanarelli introduced domestic European rabbits (*Oryctolagus caniculus*) into Uruguay for the production of immune sera. In 1896, these rabbits showed the symptoms of a devastating disease which was quite unlike anything known in Europe. The disease was infectious and highly lethal, and characteristically produced mucinous tumours on the skin of the infected animals. Sanarelli named the disease 'infectious myxomatosis' of rabbits and suggested that it was caused by a virus. Aragão (1943) showed that the common wild rabbit of Brazil (*Sylvilagus brasiliensis*) is the native host of the virus in Brazil. Infection of this host is manifest only by the production of a single localized tumour of the skin. Aragão (1943) also showed that the virus was transferred mechanically from these tumours to other rabbits (domestic or native) by mosquito bites, thus providing an explanation for the apparently spontaneous appearance of the disease in domestic rabbit colonies as well as for its seasonal incidence. In the European rabbit, myxomatosis results in the discharge of highly infectious material in the conjunctival secretions and the production of a viraemia lasting for several days, as well as to extensive skin lesions which could serve as a virus source for biting arthropods. Infection can be transferred between European rabbits by contact, but this is of minor importance in nature. In epizootics in Europe, bites of arthropods are much the most important mode of transfer. In Australia, and also in parts of Europe, mosquitoes are of outstanding importance in the transmission of infection. In Britain the rabbit flea is the major vector.

The European wild rabbit is a major animal pest in Australia, and efforts to introduce the virus for control purposes were made as long ago as 1926. A series of trials up to the Second World War were unsuccessful and the disease was not successfully introduced until 1950 as a result of liberations of the virus made between May and November in the Murray Valley, Australia. The weather in the summers 1950–1 and 1951–2 was abnormal, the first providing favourable conditions for mosquito transmission in northern New South Wales and Queensland, while the moderately dry season of the second favoured the spread of the mosquito *Anopheles annulipes* in Victoria and southern New South Wales. The season 1952–3 was also exceptionally favourable for the breeding of mosquitoes and other insect vectors, and millions of rabbits were killed by myxomatosis in south eastern Australia. Four-fifths of the rabbits in this area were estimated to have died.

In June 1952 the disease was introduced into France. This was a different virus strain from the one used in Australia. Again mosquitoes and other biting insects acted as vectors, and the disease developed rapidly, reaching most of France by the end of 1953 and spreading into Belgium, Luxembourg, Germany, The Netherlands and Spain. The first outbreaks in England occurred towards the end of 1953 in Kent and spread to most areas of the country so effectively that, by the end of 1955, well over nine-tenths of the wild rabbits in Britain had been killed by the disease. In Britain it is the rabbit flea (*Spilopsyllus cuniculi*) which is the most effective vector.

In Australia and Europe, the disease has persisted until the present time but there has been a dramatic change in the mortality rates. The highly virulent viruses first introduced had a mortality rate usually exceeding 99 %. Within a year of the introduction of the virus into Australia, strains with a 90 % mortality rate had appeared in the field. These have remained dominant ever since, but still more attenuated strains (causing mortality sometimes as low as 20 %) have been re-covered since 1955. In Europe the fully virulent strains have persisted longer and on a wider scale, but here also attenuated strains have become common. In Australia the involvement of the mosquito in the summer spread of the virulent virus, which results in the death of the host animal in a few days, militates against the over-winter survival of the virus in the absence of the vector. Long persistence of infectious lesions is an essential condition of successful survival of the virus through the winter. It is probably this factor which has led to the rapid emergence of attenuated strains of virus in Australia. In Britain the spread of virulent virus was slower than in Australia, and was effected by man and the rabbit flea which continue to be the major vectors of myxomatosis. Virulent strains persisted much longer as significant features of the population, and the virulence of attenuated strains has remained somewhat higher than in Australia. Infected fleas would much more frequently leave a dead animal than a live one, and it has been suggested that flea transmission would lead to selection for lethality of viruses. These findings suggest that the mode of transmission is of overriding importance in determining the relative frequency of emergence of attentuated strains of virus.

Changes in resistance of the rabbit host have also been noted. Under the rigorous selective stress of regular exposure to strains of virus killing in excess of 90 % of the population, a rapid increase in resistance was seen in Australian rabbits such that, in a period of 7 years of exposure to successive epizootics, resistance to a standard virus which originally killed 90 % of wild rabbits had developed to such an extent that it would kill only 30 % of newly caught young ones. From the point of view of effective control in the field, resistance on the part of the rabbit population is likely to be of far greater significance than the emergence of attenuated virus strains.

The net result of these changes in the nature of the virus and the host has led to a sharp decrease in the effectiveness of biological control of the rabbit, such that today myxomatosis is only a minor factor in controlling Australian rabbits despite the fact that there is usually an epizootic each summer. In Britain the last few years have seen the re-emergence of the rabbit population, and damage by the rabbit is again a major factor in agricultural economics.

The myxomatosis episode has taught us a great deal about the use of disease processes for control of pest species in the field. Three lessons are particularly important. First, successful introduction of a micro-organism into a host population is dependent on a variety of environmental factors, and failure to achieve success in initial experiments should not be regarded as unduly ominous for the success of the programme. Secondly, natural selection of the host and of the infective microorganism may occur in a fairly short time in such a way as to lead to a fall off in control. This, however, will depend on a number of environmental factors and, as we shall see later, decreased effectiveness is not necessarily a feature of long-established infections in nature. Thirdly, the myxomatosis episode underlines the need for proper studies of the factors controlling the spread and persistence of an infectious disease in a natural population, areas in which our knowledge is often sadly deficient but which are, nevertheless, fundamental to the full exploitation of the potential of biological control methods.

MICROBIAL CONTROL OF INSECTS
Bacteria

Over 90 species and varieties of pathogenic bacteria have been isolated from insects (Falcon, 1971), and many more of doubtful pathogenicity have also been reported (Steinhaus, 1946). Two different groups of aerobic spore-forming bacteria, the *Bacillus popilliae/B. lentimorbus* group and *Bacillus thuringiensis*, have achieved commercial application in controlling insects in the field. Attempts have been made to use other species for control purposes but these have been relatively unsuccessful. The reader is referred to Heimpel (1961) for an account of work with *Bacillus cereus* and to a review article by Angus (1965) for a general discussion concerning the use of entomopathogenic bacteria for microbial control purposes. More general review articles concerning microbial control are Heimpel (1965), Grison (1967), Rivers (1967) and Tanada (1967). Bucher (1960) suggested classifying bacterial insect pathogens into four categories, namely obligate pathogens, crystal-forming sporeformers, facultative pathogens and potential pathogens. Table 1 is modified from Falcon (1971) and lists the more important microorganisms classified according to Bucher's proposal.

Table 1. *Classification of some bacterial insect pathogens: modified from Falcon (1971)*

Sporeformers		Non-sporeformers	
Obligate	Facultative	Potential	Facultative
	Crystalliferous		
Bacillus	*Bacillus thuringiensis*	*Pseudomonas*	*Serratia*
popilliae	group of	*aeruginosa*	*marcescens*
B. *lentimorbus*	organisms	P. *chlororaphis*	
B. *lentimorbus*		P. *fluorescens*	
var. *australis*	Non-crystalliferous	P. *reptilivora*	
B. *fribourgensis*	B. *cereus*	P. *septica*	
Clostridium brevifaciens		P. *putida*	
C. *malacosomae*		*Aerobacter* spp.	
		Cloaca spp.	
		Proteus vulgaris	
		P. *mirabilis*	
		P. *rettgeri*	

Bacillus popilliae/Bacillus lentimorbus

The obligate pathogens of the genus *Bacillus* have a narrow host range and infect only certain closely related beetles of the family Scarabaeidae. Dutky (1963) provides an extensive list showing the susceptibilities of different species to these bacteria. The organisms cause disease conditions in scarabaeid grubs, which are known as the milky diseases because of the characteristic opacity of the moribund larvae. Best known is the disease of the Japanese beetle (*Popillia japonica*). The discovery and practical application of the milky-disease bacteria for the control of this pest in the U.S.A. was the first major successful attempt at biological control of an insect pest species, and the success of the programme has become one of the classic examples of microbial control resulting from colonization by a microorganism. The Japanese beetle was introduced into the United States, and became a pest of lawns, pastures, shrubberies and other plants through a large part of the New England states and into Canada. The various larval instars feed actively on the roots of grasses and other plants during the late summer months. Pupation takes place in the soil during May and adult beetles, emerging about the second week in June, live for some 40 days during which time the female will lay up to 60 eggs. Larvae suffering from milky disease are occasionally seen in native populations. Microscopic examination shows the haemocoele to be filled with oval spore-forming rods which are responsible for the characteristic opacity of the body fluid. Each sporangium contains, in addition to the spore, a second refractile body known as the parasporal body. The milky diseases are caused principally by two closely similar bacteria, *B. popilliae* and *B. lentimorbus*. Both have been developed for microbial control purposes, but it is *B. popilliae* which has been studied most extensively because it has shown most potential in the field.

Larvae ingest viable spores of the bacterium in nature and these germinate in the gut, the resulting vegetative cells penetrating the wall of the gut to multiply in the body cavity. Infected larvae will usually live for a considerable time, and extensive growth of microorganisms occurs in the haemolymph. The pathogenicity of the bacteria is not fully understood. Much of the effect of the bacterium on the host can be attributed to the removal from the blood of nutrients and essential growth factors which become locked away in the growing microbial cells. Toxins are also involved since cell-free filtrates of cultures are lethal when injected in small amounts into larvae. Whether the refractile parasporal body plays any role in the disease is not known. The refractile body of *Bacillus fribourgensis*, a closely related organism which alone has been examined in detail, contains nucleic acids and there is no evidence that it dissolves or is in any way activated when taken into the larval gut (Lüthy & Ettlinger, 1967).

The presence of viable spores is essential for infection of larvae in the field. Unfortunately, attempts to produce spores by cultivation of the microorganism *in vitro* have been largely unsuccessful and material for use in the field has been produced by infection of larvae collected in the field by a method originally developed by the United States Department of Agriculture, Bureau of Entomology in 1939. Spore suspensions, formulated with a powder carrier, produce a dust containing 10^6 spores/g. for field use. This spore powder is applied to soil in 2 g. amounts at intervals of a few feet, and the disease spreads in the insect population by various means, including the wandering of infected larvae, dispersal by water and wind, passive transfer by birds and animals and by human activity. Such a treatment programme enables the disease to spread throughout the Japanese beetle population in the treated area within three seasons. Reduction in the numbers of beetles over the treated area has been spectacular and there is no evidence of increased resistance on the part of the pest to the microorganism. For many years spore dusts containing no less than 10^8 viable spores/g. of either or both *B. popilliae* and *B. lentimorbus* have been produced by Fairfax Biological Laboratories in the U.S.A. and marketed under the trade name 'Doom'.

Although control of the pest has been encouraging, application of the material has been limited by the difficulty of producing sufficient amounts of viable spore dust for dissemination. Considerable efforts have been made to develop cultural conditions to enable *in vitro* production of viable spores to be achieved but so far with little success. Vegetative cells of *B. popilliae* are readily produced in shake flasks and fermenters (Sharpe, 1966) and spores have appeared in liquid cultures containing activated carbon, but yields have been low (Haynes & Rhodes, 1966). A recent report (Sharpe, St Julian & Crowell, 1970) describes a new strain of *B. popilliae* which produces reasonable numbers of spores when grown on agar *in vitro*. The resulting cells were of considerably impaired pathogenicity, but progress is clearly being made and the next few years may well see the development of successful *in vitro* cultivation methods for the production of infective material.

The milky-disease bacteria are in general a promising group of bacterial pathogens whose use extends far beyond the control of the Japanese beetle. Several similar bacteria have been isolated from scarabaeid larvae in other parts of the world (Falcon, 1971). Promising results have been reported from Australia where again the major obstacle is the development of artificial culture techniques for the bacteria.

Bacillus thuringiensis was first discovered in 1902 by the Japanese bacteriologist Ishiwata, who isolated an aerobic spore-forming bacterium from diseased silkworms and showed it to be the cause of the infection. The early history and the development of control preparations have been well reviewed by a number of authors (Norris, 1963; Krieg, 1967; Martouret, 1967; Angus, 1968 a; Norris 1970). *B. thuringiensis* is, in effect, a name for a group of closely related bacteria (Krieg, 1967). They form a closely knit group of aerobic spore-forming bacteria characterized by the production of a protein crystal inclusion body produced at the same time as the spore within the developing sporangium. This protein crystal is a toxin responsible for many of the symptoms of infection in caterpillars (Norris, 1969 a). *B. thuringiensis* is a naturally occurring pathogen of lepidopterous larvae, and it has been shown that some 130 species of caterpillar are susceptible to infection with the organism. In nature, epizootics of *B. thuringiensis* infection occur (Norris, 1969 b). *B. thuringiensis* is an easily grown organism which sporulates to produce fully pathogenic spore/crystal mixtures when grown on simple laboratory, or cheap industrial, media. The introduction of spore/crystal mixtures into native populations of pest species does not result in the establishment of a persisting infection in the population, although high initial kills are often achieved. The situation is therefore quite different from that found with the obligate pathogens. Adequate amounts of material for field use are readily produced but they must be used more as a direct-action chemical insecticide than as a persisting biological control agent.

Toxic protein crystals are produced by all types of *B. thuringiensis*. Indeed, the ability to produce the protein crystal is the main diagnostic feature of these organisms. Although the crystals vary from strain to strain in shape, size and antigenic composition, they appear to play similar roles in pathogenicity. The end-product of the growth of *B. thuringiensis* on culture media is a mixture of spores and crystals released by the breakdown of the residual wall of the sporangium. The crystal is soluble only with difficulty, dissolving at high pH values under reducing conditions. Solutions of protein are highly toxic when fed to caterpillars such as the silkworm (*Bombyx mori*), causing gut paralysis followed by general body paralysis in as little as 60 min. Mixtures of spores and crystals are pathogenic for a wide range of lepidopterous larvae and for a few sawflies but are without effect for other kinds of insects and are apparently non-toxic to other life forms. The protein crystal on ingestion dissolves in the alkaline reducing conditions encountered in the mid-gut of most lepidopterous larvae, and the protein is digested by the complex of proteolytic enzymes in the gut with the release of one or more toxic fragments. The details of this activation procedure are far from completely understood and are extremely complex. The reader is referred to an article by Cooksey (1971) for a full discussion of this work and of the relationships between the observations of various authors.

Although the effects of the protein toxin on lepidopterous larvae have been extensively studied, there are still aspects of the mode of action that are poorly understood. Suggested modes of action include changes in gut-wall permeability (Fast & Angus, 1965), interference with ion permeability of the gut wall (Angus, 1968b; Ramakrishnan, 1968), nerve blockade (Cooksey et al. 1969) and complex formation with proteins (Faust, 1968).

The only conclusion that one can draw from these various publications is that many of the observed effects of toxin are probably secondary in nature and that the primary effects will only be understood when the nature and properties of the components present in protein solutions are fully understood and the action of gut enzymes in activating protoxin have been fully characterized.

Considerable attention has been paid to the structure of the protein crystal and the nature of the sub-unit within. Norris (1969a, 1971) has provided direct electron-microscope evidence that the basic sub-unit is a rod-shaped structure of average length 11·8 nm. and width 4·7 nm. in support of X-ray diffraction data reported by Holmes & Monro (1965).

Crystal synthesis and spore formation proceed at the same time in *B. thuringiensis* cells and the two processes appear to be intimately associated with one another. Asporogenic mutants of *B. thuringiensis* usually fail to produce crystals. But this is not invariably the case, and mutants producing apparently normal crystals but failing to complete the normal sporulation cycle have been isolated. Acrystalliferous mutants which produce fully mature spores are much more readily isolated (Fitz-James & Young, 1959; H. J. Somerville & J. R. Norris, unpublished observations). Smirnoff (1963) has shown that crystal formation can occur in the absence of spore production when cultures are grown at low temperatures. The protein of the crystal is synthesized from amino acids that result from the breakdown of cellular protein during a massive protein turnover in the cell early in sporulation. Antigens typical of the protein crystal can be detected in the cell at an early stage in spore formation coincident with the appearance of the forespore membrane and exosporium. At this time, the beginnings of crystal formation can be seen in the electron microscope, and Somerville & James (1970) have shown that the crystal begins to develop on the exosporial membrane of the developing spore. Somerville, Delafield & Rittenberg (1968, 1970) have suggested that the crystal is a result of over-production of a spore-coat protein, and they have accumulated evidence in support of this idea. They have shown immunological and biochemical similarities between crystal protein and spore-coat protein and we are presented with the interesting concept of the crystal protein arising as a result of a malfunction in the normal control mechanisms involved in spore production. Why the protein should crystallize on the exosporial membrane is far from clear. The membrane is known to possess a regular structure of dimensions rather similar to the lattice spacings of the protein crystal (Somerville & James, 1970).

The protein toxin of *B. thuringiensis* is a highly toxic material when tested against caterpillars. Angus & Norris (1968) reported the results of testing protein extracted from several *B. thuringiensis* biotypes against *Bombyx mori*. The results show wide variations from type to type and this is typical of work in which toxins of different origins have been compared for their effects on different insect species. Some preparations give LD_{50} values as low as 0.09 μg./g. larvae, a value which is comparable with some of the more widely used chemical insecticides.

The presence of a heat-stable toxin in the supernatants of cultures of certain strains of *B. thuringiensis* was first demonstrated by McConnell & Richards (1959) who showed that autoclaved culture filtrates killed larvae of the Wax moth (*Galleria mellonella*) when injected into the body cavity. Further studies have shown that the toxin is active against a wide range of species, including lepidoptera, diptera, hymenoptera, coleoptera and orthoptera, and these have been summarized by Norris (1969a) and Rogoff (1966). The effects of feeding small doses of exotoxin to larvae are only seen at moulting or during metamorphosis. Purified exotoxin kills housefly larvae when fed in doses as low as 0.5 μg./ml. of food and the LD_{50} for *Galleria mellonella* by injection is of the order of 0.005 μg./larva (C. B. C. Boyce, personal communication).

Exotoxin is a high molecular-weight (800–900) adenine nucleotide containing one phosphate group per adenine molecule. In addition, an unusual sugar, allomucic acid, is present in the molecule (Bond, 1969; Farkas *et al.* 1969). Farkas *et al.* (1969) have published a suggested structural formula for exotoxin (Fig. 1) but it is uncertain whether this is correct in all details since there are significant differences between the findings of the main groups working in this area (Bond, 1969; Šebesta, Horská & Vankova, 1969; Farkas *et al.* 1969; de Barjac & Dedonder, 1968). It could well be that the different groups are working with similar, closely related but not identical microbial products.

Exotoxin inhibits the DNA-dependent RNA polymerase of *Escherichia coli* (Šebesta & Horská, 1968; Šebesta *et al.* 1969). Inhibition is competitive with ATP in the sense that it is partly reversed by the addition of ATP. Exotoxin apparently inhibits the polymerization step. Whether the same mode of action is responsible for intoxication *in vivo* is not known, but the indications are that it probably is. Insect larvae are more sensitive to effects of toxin at moulting and pupation when RNA synthesis is maximal.

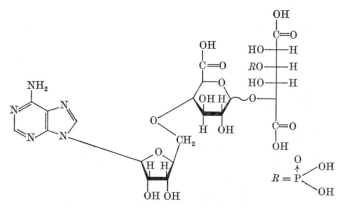

Fig. 1. Suggested structure for the exotoxin produced by *Bacillus thuringiensis*. From Farkas *et al.* (1969).

Smirnoff & Berlinguet (1966) demonstrated the presence of an additional toxin in certain batches of commercially available *Bacillus thuringiensis* preparations. When the wettable powder was extracted with water, the sterile filtered extract contained a toxin which was rapidly inactivated by heat, oxidation and ultraviolet irradiation. The toxin was lethal on feeding for 19 species of sawflies but its lability will probably prevent its use in the field for control purposes. The nature of the toxin is not known, although it is probably a polypeptide.

In addition to these clearly defined toxins several exocellular enzymes, including phospholipases and hyaluronidases, have been implicated in the disease process but it seems unlikely that they play a major part in the intoxication process under field conditions (Rogoff, 1966).

Field application of Bacillus thuringiensis

The early commercial development of *B. thuringiensis* was extensively reviewed by Hall (1963). Commercial preparations containing *B. thuringiensis* have been produced by at least 12 manufacturers in five countries in recent years. In the U.S.A. hundreds of tons are manufactured each year and production is rising annually. Commerical prepara-

Table 2. *Some registered uses of Bacillus thuringiensis in the U.S.A.: from Falcon (1971)*

Vegetable and field crops

Alfalfa caterpillar	*Colias eurytheme*	Alfalfa
Artichoke plume moth	*Platyptilia carduidactyla*	Artichokes
Bollworm	*Heliothis zea*	Cotton
Cabbage looper	*Trichoplusia ni*	Beans, broccoli, cabbage, cauliflower, celery, collards, cotton, cucumber, kale, lettuce, melons, potatoes, spinach, tobacco
Diamondback moth	*Plutella maculipennis*	Cabbage
European corn-borer	*Ostrinia nubilalis*	Sweet corn
Imported cabbageworm	*Pieris rapae*	Broccoli, cabbage, cauliflower, collards, kale
Tobacco budworm	*Heliothis virescens*	Tobacco
Tobacco hornworm	*Manduca sexta*	Tobacco
Tomato hornworm	*M. quinquemaculata*	Tomatoes

Fruit crops

Fruit-tree leaf roller	*Archips argyrospilus*	Oranges
Orange dog	*Papilio cresphontes*	Oranges
Grape-leaf folder	*Desmia funeralis*	Grapes

Forests, shade trees, ornamentals

California oakworm	*Phryganidia californica*
Fall webworm	*Hyphantria cunea*
Fall cankerworm	*Alsophila pometaria*
Great Basin tent caterpillar	*Malacosoma fragile*
Gypsy moth	*Porthetria dispar*
Linden looper	*Erannis tiliaria*
Salt-marsh caterpillar	*Estigmene acrea*
Spring cankerworm	*Paleacrita vernata*
Winter moth	*Operophtera brumata*

tions of *B. thuringiensis* are currently registered in the United States for use on more than 20 agricultural crops, in addition to trees, ornamental shrubs and forests for the control of at least 23 insect species (Table 2). Registration of a product in the United States requires that the United States Department of Agriculture Registration Authority is satisfied not only that the product is safe but also that field testing has shown that it gives effective control of a registered pest species.

Insecticides based on *B. thuringiensis* are used for short-term control. They neither persist from year to year nor spread extensively in the population, and so they must be applied very efficiently to give thorough coverage of the treated crop. Many observations have shown that the pathogenicity for a particular insect varies extensively from one strain of *B. thuringiensis* to another, and that a particular strain of the bacterium can have a markedly different effect on a range of host species (Herfs, 1963; Burgerjon & Biache, 1967; Angus & Norris, 1968; Rogoff *et al.* 1969). The careful selection of a bacterial strain for a particular control purpose is an important feature of modern use of *B. thuringiensis*, and the marketing of specific varieties of bacterial preparation for use against different pest species may well become realistic in the near future.

Fungi

Beauveria bassiana

Attempts to use the fungi *Beauvaria bassiana* and *Metarrhizium anisopliae* to control insect pests have been made since the end of the last century. Field trials have been numerous and present a variable picture ranging from impressive success to total failure. The work is well summarized by Müller-Kögler (1965). Unlike the bacteria or viruses, the fungi invade the integument of the host and do not depend upon ingestion for infectivity. It seems probable that an invasive organism of this kind will depend on environmental factors, particularly temperature and humidity, for its effectiveness and this appears to be so from many reports in the literature. More recently attention has been directed to the importance of actual dosage rates of spores, and trials using high dose rates have shown promising results. The last few years have seen a renewed interest in the fungi for control purposes. Spores are readily produced by culture, and experimental materials are available from several potential manufacturers for field testing; thus Nutrilite Products Incorporated in the United States provide spore suspensions either as a wettable powder (Biotrol FBB 5 Wettable) or as a dust (Biotrol FBB Dust) containing 5×10^9 or 5×10^8 spores/g. respectively.

Fungi have advantages over the other microbial agents in that they act as contact insecticides and have a very broad spectrum of activity, attacking a wide range of pest species representing many widely differing genera. The main disadvantage of the agents is their slow action, since invasion of the insect body is a lengthy process. The importance of this consideration, however, can be minimized by careful choice of time of spraying in relation to the life-cycle of the host. For a recent and detailed account of the use of fungi for microbial control of insects, the reader is referred to an article by Roberts & Yendol (1971).

Nematodes

Although nematodes can hardly be considered as microorganisms, they merit attention at this point since they present tangible possibilities for biological control of insects. With the most interesting organism,

Nematode DD 136, invasion of the host is intimately connected with the transfer of a bacterium from the nematode to the insect.

Nematodes have been found associated with representatives of most insect orders as well as with mites. A recent account of the use of nematodes for microbial control of insects is given by Poinar (1971). Entomogenous nematodes invariably have a free-living soil stage and actively search out the host insect. The relationship between micro-organism and nematode has been described for a strain of *Neoaplectania carpocapsae* by Poinar (1967). This nematode was originally known as the DD 136 nematode and its relationship with an associated bacterium was first reported by Dutky (1959).

The bacterium *Achromobacter nematophilus* is carried in the ventricular portion of the intestinal lumen of the infective stage nematodes and released through the anus when the nematode enters the body cavity of a host (Poinar, 1966). The neoaplectanid nematodes are fairly easy to grow on artificial media and this, together with their wide host range, has led to their introduction as biological control agents in many localities. This has sometimes been on a large scale, as, for instance, the trials against Japanese beetle conducted in New Jersey and reported by Glaser, McCoy & Girth (1940). Since 1959 most attention has been paid to the DD 136 nematode/bacterium complex, particularly promising results being obtained against larvae of the Codling moth (*Carpocapsa pomonella*), the Colorado potato beetle (Welch, 1958) and the Tobacco budworm (Chamberlain & Dutky, 1958). As with the fungi, climatic conditions appear to play a significant role in determining the success or failure of field trials with the DD 136 nematode/bacterium complex (Poinar, 1971).

One of the prerequisites for the use of entomogenous nematodes as biological control agents is the production of large numbers at relatively low cost. The DD 136 strain of *N. carpocapsa* was initially propagated in waxmoth larvae which yielded up to 160,000 infective stage nematodes/insect (1·5 million/g.). House, Welch & Cleugh (1965) described a method for rearing *N. carpocapsa* on a dog-food medium, and Nutrilite Products Incorporated have established a mass-breeding programme of this strain in pure culture containing only *A. nematophilus* using a dog-food medium. Since 1967 more than 100 million infective stage worms have been sent to various agencies throughout the United States for field tests against a number of insect pests, particularly the Corn rootworm (*Diabrotica* spp.) and other soil pests. Cost of production has been quoted at about U.S. $1 per 1 million nematodes (Poinar, 1971).

The nematode will multiply under field conditions in the presence of a host, and it is possible that a single application may give continued control throughout an entire growth season. If this proves to be the case, the cost of control using the nematode would be considerably lower than that of any soil insecticide now available. Not all field trials reported in the literature have been successful by any means, however, and it is clear that considerably more work is required before the full potential of preparation of this kind can be properly assessed.

Viruses

The most rapidly expanding field of microbial control at the present time concerns the exploitation of a wide range of virus diseases of insects. Over 300 insect viruses are known to exist, and many of these cause rapidly spreading epizootics among insect populations in nature. Many insect viruses are unlike animal and plant viruses in that the individual virus particles are encased, either singly or in large numbers, in protein crystals which are insoluble in water. These crystals, which are produced in large numbers in infected larvae and released when the cadaver disintegrates, protect the virus particles and ensure that they remain infective even if stored for years outside living tissues.

There are two main groups of insect viruses: the polyhedroses, in which many hundreds of virus particles are contained in each polyhedral crystal, and the granuloses, in which each crystal contains only one virus particle. Both types spread rapidly in insect populations and are potentially valuable as insecticides. Most attention has been paid to the nuclear polyhedrosis viruses which develop in the nuclei of host cells following ingestion and affect the skin, fat body and haemolymph of larvae. These tissues become packed with many-sided crystals. Millions of polyhedra are liberated when the skin of the insect finally ruptures and they spread among the population by the effects of wind, rain and other insects.

The fact that a virus disease is capable of controlling an animal population over very large areas was amply demonstrated by the effects of myxomatosis on the rabbit population. Similar dramatic demonstrations of population control by virus action are known in the insect world. As long ago as 1915 a virus disease was found to control the caterpillars of the Gypsy moth in the forests of southern Europe, and we are probably still seeing the effects of an epidemic of granulosis virus disease which spread through the populations of the Large Cabbage White caterpillar (*Pieris brassicae*) in the United Kingdom in 1954–5. Populations of the Large Cabbage White butterfly were substantially decreased in size in a short period of time, and since then there have been no reports of large numbers of this species.

The first clear example of the control of a pest insect species by a virus occurred during the 1930s. The European spruce-sawfly became a serious pest in Canada and parts of the U.S.A. where it was free from its natural enemies, and spread rapidly, destroying thousands of acres of spruce trees. A chance introduction of a nuclear polyhedrosis virus from Europe checked the sawfly population and the species gradually declined until now it is no longer a serious pest. In 1949 the Canadians imported another sawfly virus from Europe and this has been so effective that it has generally replaced chemical control for that particular insect. The period since the Second World War has seen a tremendous development in the subject of insect virology both at the fundamental level and also from the point of view of biological control. The period falls naturally into two sections. The first covers the period 1944–60 and saw numerous attempts to apply insect viruses for biological control against foliage pests in nature. This period was characterized by the utilization of virus prepared from diseased insects collected in the field, a process which was, necessarily, expensive and inefficient. Nevertheless, progress was made since in some applications, particularly in forestry, the actual amount of virus which must be disseminated is quite small. Thus (Rivers, 1964), working with a nuclear polyhedrosis virus of the Pine sawfly, found that the virus from only 15 diseased larvae mixed with 1·5 gal. of water could be used to treat an acre of young pine trees effectively by the use of powered knap-sack spraying machines. A persisting virus epizootic was initiated in the population and similar results followed the spot treatment of individual trees and subsequent natural spread of disease from these localized infection sites.

A second method of producing virus was to rear larvae under laboratory conditions and purposely infect them with the disease agent. Most leaf-eating caterpillars live only on a diet of foliage, and frequently their requirements are highly specific. Attempts to rear large numbers of larvae for virus production involve the collection of considerable amounts of foliage of particular plant species and the process was again costly and time-consuming. Relatively little progress was made until the development of artificial rearing diets about 1960 (Getzin, 1962; Briggs, 1963; Berger, 1963; Ignoffo, 1963). It was this development by workers in the U.S. Department of Agriculture that issued in the present vigorous phase in the development and exploitation of viruses for biological control. Several laboratories and two industrial organizations in the United States have developed the use of synthetic rearing diets to a point where it is possible to mass-rear sufficient susceptible larvae to justify commercial production of the large amounts of virus required for use against crop-pest insects.

Table 3. *Susceptibility of insect tissue explants or cell lines to alien insect pathogenic viruses: from Ignoffo (1968)*

Host tissue	Virus species*
Antheraea pernyi	*Bombyx mori* NPV, CPV
A. eucalypti	*Sericesthis* NIV
	Tipula NIV
	Chilo NIV
Bombyx mori	*Galleria* DNV
Galleria mellonella	*Sericesthis* NIV
Laodelphax striatella	*Chilo* NIV
Lymantria dispar	*Bombyx mori* NPV, CPV
	Antheraea pernyi NPV
	Pieris brassicae GIV
Nephotettix cincticeps	*Chilo* NIV

* NPV, Nuclear polyhedrosis virus; CPV, cytoplasmic polyhedrosis virus; GIV, granulosis inclusion virus; NIV, non-inclusion virus; DNV, densonucleous virus.

Table 4. *Specificity of the Heliothis zea nuclear polyhedrosis virus to plants, other insects, invertebrates and vertebrates: from Ignoffo (1968)*

PLANTS: Cotton, Corn, Sorghum, Bean, Soybean, Snapbean, Kidney Bean, Tomato, Tobacco, Radish	External application	Negative: no apparent phytotoxicity or pathogenicity
INVERTEBRATES: Grass Shrimp, Brown Shrimp, Oyster	*Per os*, topical	Negative
INSECTS: *Heliothis*, five-species; Tobacco hornworm, Tomato hornworm, Wax moth, Cabbage looper, Beet armyworm, Fall armyworm, Southern armyworm, Lucerne moth, Honeybee, Housefly	*Per os*, intrahaemocoelic	Positive only for *Heliothis*. Negative for all other listed species
VERTEBRATES: Man, Monkey, Dog, Rabbit, Guinea Pig, White Rat, White Mice, Chicken, Chicken egg, Quail, Sparrow, Mallard, Killifish, Spotfish, Rainbow Trout, Bluegill, Black Bullhead, White Sucker, Sheepshead Minnow	General, *per os*, inhalation, topical intradermal, intracerebral, intravenous, intra-muscular, intracellular, intraperitoneal, tissue cells	Negative: clinical and laboratory; cutaneous or respiratory sensitivity; toxicity, pathogenicity, teratogenicity, carcinogenicity

Insect viruses are, typically, very narrow in their host range, often infecting only one insect species in nature. Active attempts in the laboratory to infect insects with viruses from other species meet with widely differing results for different viruses. For example, the nuclear polyhedrosis virus of the silkworm, *Bombyx mori*, has been successfully transferred to other species, including the waxmoth, *Galleria mellonella*. The nuclear polyhedrosis virus of *Heliothis* spp., however, is narrowly specific within the genus *Heliothis*. Ignoffo (1968) presents a detailed list of successful and unsuccessful attempts to cross viruses between different species. Interestingly enough, invertebrate tissue cultures prove to be less specific in the viruses they will support than do the intact insects. Table 3 is taken from Ignoffo (1968) and lists the sensitivity of insect tissue explants or cell lines to heterologous insect pathogenic viruses. Irrespective of whether the species specificity of an insect virus is broad or narrow within the insect world, attempts to infect life forms other than insects or tissue cultures of mammalian origin have not met with success. Table 4, also taken from Ignoffo (1968), lists attempts to infect a range of other species with the important *Heliothis zea* nuclear polyhedrosis virus, which is likely to be the first crop virus to achieve market potential.

The infective particle of a polyhedrosis virus is the small viral unit embedded in the protein polyhedron. The protein has a marked protective effect and the viability of virus when stored outside the infected animal is satisfactory for insecticidal purposes (see, for example, Ignoffo, 1964). With the development around 1963 of synthetic rearing diets for use on a large scale, industrial organizations, particularly in the United States, began the development of virus insecticide materials for extensive field testing and ultimate marketing. Although several different viruses have been developed, most attention has been concentrated on the nuclear polyhedrosis virus of Cotton bollworm (*Heliothis zea*), which has shown considerable promise in field trials, and the nuclear polyhedrosis viruses of the Cabbage worm (*Trichoplusia ni*) and of the Armyworm (*Prodenia* spp.).

Formulation presents little difficulty, although viruses tend to lose activity when spread in the field. This is probably largely due to damage by ultraviolet irradiation, and ultraviolet absorbents are incorporated in the formulations. One material available for experimental field trials is formulated as a wettable powder or as a dust and is said to be stable on storage in cool temperatures for two years. Disease takes some time to develop, so that timing of the application in the field is an important matter. Preparations of the nuclear polyhedrosis virus of *Heliothis* must be used when the larvae are small. First instar larvae will die very rapidly, often within a day or two, but it may take from 4 days to a week to kill a third instar larva. Viruses must be ingested by the insect, and really effective coverage of foliage is an important feature of field application. Repeated spraying during the growing season is required since persisting infections are not usually established.

SPECIFICITY OF BIOLOGICAL CONTROL AGENTS

The microorganisms discussed in this review differ widely in the extent of their specificity for insect hosts. *Bacillus popilliae* and *B. lentimorbus* are specific for larvae of the scarabaeid beetles and in effect are used only for one species, *Popillia japonica*, the Japanese beetle. *Bacillus thuringiensis* is known to attack a wide range of lepidopterous larvae including many pest species, but there is an underlying pattern of adaptation of particular biotypes of the microorganism to different insect species and it seems likely that *B. thuringiensis* will be used most effectively in the field when we know sufficient about the underlying mechanisms involved to select particular bacterial strains for particular field applications. The fungus *Beauveria bassiana* and the nematode insecticides have a low order of specificity, being active in the laboratory and in the field against a very wide range of insect species. The viruses are highly specific, often affecting only one species.

Traditionally the grower has favoured the broad-spectrum insecticides capable of killing each of the different pest species growing on his crops, and in this sense a high degree of specificity is a disadvantage for a microorganism. On the other hand, specificity usually implies safety, and the acceptability of microbial insecticides for wide dissemination in the environment may well prove to be an important factor in their further development.

SAFETY OF INSECT PATHOGENS

The pattern of registration of pesticidal chemicals differs in different countries. Since the majority of commercial activity concerning microbiological control agents occurs in the United States, I will restrict my discussion to the situation in that country. The United States was among the first countries to enact laws requiring the registration of materials used to treat food crops. The Federal Insecticide, Fungicide and Rodenticide Act was established in 1947. The Pesticide Regulation Division (P.R.D.) of the United States Department of Agriculture (U.S.D.A.) was established to carry out the dictates of that legislation. On receipt of a petition for tolerance (for a pesticide), the P.R.D. is requested and responsible for certifying that the product is useful for the purpose defined. In doing so, they review the efficiency data, and the label presented which describes directions for use and precautions to be taken in order that it can be safely applied and will not adversely affect the environment, in particular birds, fish and other wildlife. They also review the residue data and provide judgement as to whether or not it reflects what residues, if any, are likely to occur when the product is used as directed.

On certification by the U.S.D.A., the Food and Drug Administration (F.D.A.) of the United States Department of Health, Education and Welfare officially files the petition and begins a review of the toxicological, metabolic and residue data. The level of tolerance of pesticides permitted on crops at harvest (the raw agricultural commodity) is based on the residue data. Such residues must of course not lead to any human health hazard.

Only after these extensive reviews have been completed will the product receive a tolerance or an exemption from a tolerance. Concurring in this decision are the U.S.D.A., F.D.A. and in addition the Public Health Service and the U.S. Department of the Interior.

There is no *a priori* reason why a microbial agent should be treated in any way other than as a chemical agent for insect control. Since the agent being used is a viable one and apt to multiply in the environment, it must be proved safe for all other forms of life. In the case of microbial agents, the regulatory agencies have made it clear that they require proof of safety sufficient to allow the granting of an exemption of tolerance. The petitioner must supply the following information:

(1) A method of production of a microbial agent and the demonstration of an adequate technique for production of a uniform product and for standardization of that product.

(2) Evidence that extraneous biotypes have been eliminated or controlled to an acceptable level.

(3) The efficacy of the product must be demonstrated by data from thorough field testing in various parts of the country, on all of the insect host plants that are proposed for treatment in the petition.

(4) The safety of the product for vertebrates, plants and beneficial insects must be established by applying a protocol of tests that have the approval of the P.R.D.

(5) There must be an exemption from tolerance requirement issued by the F.D.A. This exemption must cover all food or feed crops that are to be treated or that may be contaminated.

Bacillus popilliae

The early work on *B. popilliae* was carried out before the establishment of protective legislation in the United States. Although there have not been any harmful effects following the broadcast use of the milky-disease organisms, extensive toxicity tests, of the kind which would be required for registration today, have never been carried out although the matter is now said to be under consideration by the Agricultural Research Services of the U.S.D.A. (Heimpel, 1971).

Bacillus thuringiensis

Initial field-work and safety studies including volunteer tests were necessary for the commercial development of *B. thuringiensis* in the United States, and work of this kind occupied the period 1956–68 when the F.D.A. granted a temporary exemption from tolerance and the P.R.D. issued an experimental permit allowing the use of *B. thuringiensis* on vegetable crops. In 1960 the F.D.A. granted full exemption from tolerance, allowing commercial application of preparations based on *B. thuringiensis* to food and forage crops. Most of the experimental work on safety carried out with *B. thuringiensis* is given in a paper by Fisher & Rosner (1959). The tests were extensive and detailed, involving intraperitoneal inoculations, inhalation toxicity in mice, allerginicity tests of formulated material in guinea pigs, inhalation and ingestion of formulated material by human volunteers, and acute oral toxicity measurements for formulated material on rats.

As mentioned earlier, some strains of *B. thuringiensis* produce an exotoxin which can be present in commercially available preparations. It is now well established that the exotoxin is profoundly toxic to mice when administered intraperitoneally and subcutaneously (Šebesta *et al.* 1969; Bond *et al.* 1969; DeBarjac & Riou, 1969). Associated with this toxicity is pathological damage to liver and kidney. Few data are available concerning the oral toxicity of the exotoxin or the pathological changes associated with oral administration. It is too early at the moment to define accurately the toxic hazards of exotoxin when considered as an insecticide. Consideration of the nature and probable mode of action of the exotoxin must raise the question of the possible teratagenic effects of the material. More work is required before decisions can be made concerning the widespread use of exotoxin for insecticidal purposes.

Fungi

In several cases *Beauvaria* spp. have been isolated from the tissues of vertebrates and have been implicated as the causal agent of pulmonary infection in other animals, including the giant tortoise. Toxicity and pathogenicity tests have been carried out on a fairly substantial scale. Schaerffenberg (1968) reported on animal tests with *Beauvaria bassiana* and *Metarrhizium anisopliae*. Subcutaneous and intravenous injections of spore suspensions showed no pathogenic effects, but feeding experiments suggested that physically stressed animals may be susceptible to some action by these fungi. There have been several reports from scientists working with *B. bassiana* of moderate to severe allergic reactions to spore preparations, and it is clear that regulatory agencies will require extensive experimental data on toxicity and pathogenicity as well as data on the mutability of these fungi before extensive field use can be approved.

Nematodes

Little information has been published concerning the host spectrum of nematode DD 136. However, manufacturing organizations have for some time been carrying out safety tests in animals. The nematode appears to be only a parasite of insects. According to present knowledge it does not attack any form of plant. It also appears that it is unable to survive at the body temperature found in warm-blooded animals. Naturally, with an agent like this nematode which is capable of continuing to multiply in the insect and hence of initiating a persisting infection for at least a few months, it will be necessary for extensive safety tests to be carried out before the material can be used commercially.

Viruses

At the present time the immediate future of biological control is considered by many workers in the field to be tied up with the use of insect viruses, and it is these viruses which are attracting most attention at the present time. The safety of viruses for use in this way has been the subject of a series of contacts and conference between the P.R.D. and the F.D.A. and representatives of industry and of the Agricultural Research Services of the U.S.D.A. since 1962. Tests by Ignoffo & Heimpel (1965) using the nuclear polyhedrosis virus of *Heliothis zea* showed this material to be non-toxic for mice and guinea pigs when administered by inhalation, or intraperitoneal or intracerebral injection, or when fed *per os*. Purified polyhedra had no allergenic effect on guinea pigs. Similar tests using the nuclear polyhedrosis virus of

Trichoplusia ni were carried out by Heimpel (1966) with similar results. Subsequently, preparations based on several different viruses have been tested extensively for toxicity under widely differing conditions of administration to mammals, invertebrates, fish, birds and humans. Heimpel & Buchanan (1967) fed 5·8 billion polyhedra over the course of 5 days to 20 volunteers. No significant changes were detected in the condition of these individuals, and the authors concluded that the virus had no harmful effects. At the present time, two commercial preparations of *Heliothis zea* nuclear polyhedrosis virus are lodged with the United States authorities for approval for field use and exemption from tolerance requirement. The concept of spreading a virus widely in nature demands close attention from regulatory authorities, and the United States authorities are being understandably and commendably cautious in demanding very extensive safety testing of these first preparations. With the approval of the first virus preparation a precedent will have been set, and Heimpel (1971) has expressed the opinion that subsequent testing might well be of a less extensive nature in view of the high level of specificity of the insect viruses.

DEVELOPMENT OF RESISTANCE TO MICROBIAL CONTROL AGENTS IN INSECT POPULATIONS

Burges (1971) presents a detailed examination of the possibilities of pest resistance to microbial control agents. He comments that he has found no records of indisputable resistance to microbial agents in field trials or control programmes and makes the point that few such agents have been used for very long and that it is doubtful whether anything less than a high level of resistance could have been detected by the bioassay techniques employed. The evidence sometimes presented in support of development of resistance in nature can often be challenged on a basis of the assay technique used and the interpretation of the statistical data obtained.

Resistant stocks of insects have certainly been produced in the laboratory. Pasteur reared a stock of silkworms resistant to the protozoan responsible for pebrine, a discovery which as much as any other saved the silkworm industry of France (Steinhaus, 1949). Selective methods have also produced a stock of silkworms more resistant by a factor of 16-fold to a cytoplasmic polyhedrosis virus (Watanabe, 1967). Well-documented work by David & Gardiner (1965) reported the rearing of cultures of the Cabbage-White caterpillar (*Pieris brassicae*) with susceptibility to granulosis disease varying by as much as a

1,000-fold. Thus a degree of resistance as high as some of those obtained against chemicals has been recorded against insect pathogens.

Most of the cases of resistance reported have concerned virus diseases. There has been no report of increased resistance to spores and crystals of *B. thuringiensis*, although the housefly (*Musca domestica*) has been reported to have developed a 14-fold resistance to the exotoxin of *B. thuringiensis* in 50 generations (Harvey & Howell, 1965).

Biological control agents have one advantage over chemicals with regard to resistance. They are alive and are themselves subject to variation which may lead to the natural development of more effective strains or even to a capacity to infect new species. This feature of microorganisms can sometimes be used to commercial advantage. For instance, the manufacturers of insecticides based on *B. thuringiensis* have, by a process of mutation and strain selection, boosted the toxicity of their preparations very considerably over the past five years. The improved field effectiveness of present-day formulations compared with the earlier materials marketed is due in no small part to this strain selection and improvement, and there is no reason to think that the process has yet ended.

There is scant evidence on which to predict whether resistance will become of widespread importance in practice. It shows no signs of doing so at present in such well-known microbial control successes as the use of *B. popilliae* against the Japanese beetle and of the virus of the Larch sawfly for the control of that forest pest. However, microbial control is generally still in its infancy and it is not possible to predict the effects of increased selection pressure on host resistance. The myxomatosis episode (see p. 199) has underlined the importance of a proper knowledge of population dynamics of the host species and of the way in which a host and a pathogen can react to one another when they come into contact on a very large scale. One possibility open to the biological control world which is denied to the manufacturer of chemical insecticides is that of genetic manipulation of microbial agents in the laboratory, so supplementing any natural tendency for the pathogen to defeat developing resistance by the evolution of more effective strains. The future success of biological control programmes will depend to a large extent on the ability and ingenuity of the insect pathologist in the laboratory.

FUTURE PROSPECTS FOR MICROBIAL
CONTROL OF INSECTS

In order to be effective as insecticides, microbial preparations must meet certain basic requirements. They must be effective in the field, economically competitive with existing chemical insecticides, and raise no environmental pollution problems. Undoubtedly the biggest attraction of microbial control agents is their safety. As government regulations on the use of chemical insecticides increase, and as insects develop resistance to chemicals on the market, it can be expected that the need for biological control agents will increase (Dulmage, 1971).

The main problem in the development of microbial agents so far has been variable results in field trials. However, there has been a marked general improvement in performance during the last few years, particularly with *B. thuringiensis* preparations. The early commercial preparations of *B. thuringiensis* which became available about 1958 were not uniform in quality, tended to settle quickly in spray containers, and varied in potency. Strain selection and improvement, research with spreaders, stickers, anti-weathering agents, emulsifiers and stabilizing agents and improved production methods have resulted in better and more reliable commercial products. Information on how and when to use preparations most effectively in the field has also developed. This has enabled the manufacturers to increase the actual toxicity of their marketed material without increasing the cost, and Falcon (1971) forecasts that the use of *B. thuringiensis* as one of the few available broad-spectrum, yet selective and ecologically safe, agents in insect control will accelerate in the 1970s. It has been argued that the growth of microbial control agents in the insecticide industry has been slow, but it should be recognized that the problems faced by the industry in developing these new and very different kind of control agents are considerable. Dulmage (1971) estimates that the world-wide investment in industrial investigations of insect control with microbial pathogens probably totals no more than U.S. $600,000/year. This should be contrasted with the U.S. $100 million/year spent by the industry in investigating chemical pesticides. Seen in terms of research and development investment, the growth of microbial control agents during the past ten years has been sufficiently impressive to justify some optimism concerning its future.

Some pathogens have the power of persisting and spreading in the environment. With such agents, even though the production and application costs may be high, the treatment may well be commercially sound. For example, the cost of introducing the virus of the European spruce sawfly (*Diprion hercyniae*) into Canada probably did not exceed Canadian $50,000 but the introduction of this virus over fifteen years ago appears to have contributed to the control of the pest ever since. There are other examples of spectacular return on investment for agents that give long-term control, and the economics of this kind of preparation are such as to indicate that most of the future research and development work will be conducted not by industry but by public agencies.

Relatively few of the microbial pathogens under investigation at the present time do in fact persist and spread in the population. *Bacillus thuringiensis* and many of the viruses under development must be applied in relatively high doses, repeatedly during the growing season of a crop. In these cases the economics are simple. If the use of the pathogen results in satisfactory control and helps to avoid restrictions on insecticidal residues, and if it is cheap enough in comparison with available (and acceptable) chemical insecticides, it is a potentially economically viable agent. At the present time, *B. thuringiensis* is marketed at a price which is comparable with that of many chemical insecticides employed for the control of lepidopterous larvae. Present indications are that viruses when marketed will be slightly more expensive than chemical control agents, but the manufacturers believe that this will be compensated by the increased safety and acceptability of the agents in use. Interesting possibilities for the further development of virus insecticides stem from the dramatic developments which have occurred in insect tissue culture techniques over the past five years (Vago, 1967). The technical problems of producing infective virus by growth in insect cell lines are formidable, but progress is rapid and some workers in this field are enthusiastic about the possibility of tissue culture being used for virus production purposes on an economic basis (Vasiljević, 1970).

The effectiveness of *B. popilliae/lentimorbus* is well demonstrated and, if current progress in producing infective organisms by growth on *in vitro* media is maintained, the large-scale economic production of infective spores will be possible in the near future, and the rapid extension of the present control programme with dramatic effects on Japanese beetle infestations in the United States and Canada can be expected.

The effectiveness of *B. thuringiensis* as an insecticide is also well demonstrated for certain types of crop, particularly leaf vegetable crops such as cabbage and lettuce and for forestry applications (Falcon, 1971). Significant pointers for future developments are to be seen in the use of *B. thuringiensis* preparations against the Cabbage looper in Arizona and California, which has long been resistant to many chemical insecticides and where chemicals that are active against it may leave toxic residues on the edible plants. By using a carefully selected chemical insecticide which controls other pests, such as the thrips and aphids, in combination with *B. thuringiensis*, an acceptable control of the pest complex can be achieved without danger of build-up of toxic residues. In these areas a common practice is to withdraw the chemical insecticides some 30 days before harvest and apply *B. thuringiensis* from then until harvest. For tobacco insect control, commercial preparations of *B. thuringiensis* are as effective as chemical insecticides for control of Tobacco hornworm (*Manduca sexta*) (Rabb & Guthrie, 1964) and for Tobacco budworm (*Heliothis virescens*) (Creighton, Kinard & Allen, 1961). In recent years commercial preparations have been used extensively for control of these pests on tobacco in the U.S.A., especially for the production of tobacco free from chemical insecticide. *Bacillus thuringiensis* also plays an important part in integrated control programmes being developed for tobacco (Gentry, Thomas & Stanley, 1969).

Bacillus thuringiensis is compatible with a wide range of chemical pesticides and supplements. Herfs (1965) lists 26 pesticidal agents which are compatible with the microorganisms, and 35 supplements including wetting agents, emulsifiers and adhesives. As a consequence, *B. thuringiensis* preparations can be used in combination with many pesticides and supplements and it seems probable that much of the commercial future of *B. thuringiensis* will be as a component in carefully designed integrated control programmes.

Several insect viruses are known to be persistent and effective control agents for forestry pests, and application in this field will undoubtedly continue to expand. The ability of a virus disease to spread and control an insect population is often a question of the host ecology. An interesting and characteristic example of the successful use of a virus disease in this way is the recently reported control of the Coconut Rhinoceros Beetle (*Orychtes rhinoceros*) on islands in Western Samoa (Marschall, 1970). It is more difficult to forecast the future potential for the non-persistent virus diseases. Nevertheless, it is these which are receiving most research and development attention at the present time, and the next few years will undoubtedly see the extensive experimental application of these materials.

Much of the future for microbial control must depend on the changing attitude to pest control. We see today a developing pattern of agriculture towards intensive, large-scale, single-crop operation, and this, coupled with the present interest in controlling environmental pollution and preventing interference with natural animal and plant populations, will almost certainly lead to more widespread development and adoption of integrated control programmes. If this happens, the narrow-spectrum microbial insecticides with their high, innate safety may be expected to play an increasing role in insect control. Certainly the past few years have seen a sharp increase in the amount of research and development effort being directed towards these agents, and we can expect that entirely new microorganisms will become available for study during the next few years. Application of integrated control methods will require a much more detailed understanding of the population dynamics of insect pests than we at present possess. Falcon (1971), discussing the future potential of bacterial control agents, comments that the future of these materials looks extremely bright in the dawning era of ecological pest control.

There is a further aspect of the subject. The effectiveness of many of the microbial pathogens depends on their remarkable specificity for the insect as opposed to other forms of life. A study of the scientific basis for this specificity could open up whole new areas in our understanding of the way microorganisms bring about disease and invade host bodies. When we understand the reasons for this specificity we may well find ourselves in a better position to exploit the knowledge by the design of highly specific insecticidal molecules.

REFERENCES

ANGUS, T. A. (1965). Bacterial pathogens as microbial insecticides. *Bact. Rev.* **29**, 364.

ANGUS, T. A. (1968*a*). The use of *Bacillus thuringiensis* as a microbial insecticide. *Wld Rev. Pest Control* **7**, 11.

ANGUS, T. A. (1968*b*). Similarity of effect of valinomycin and *Bacillus thuringiensis* parasporal protein in larvae of *Bombyx mori. J. Invert. Path.* **11**, 145.

ANGUS, T. A. & NORRIS, J. R. (1968). A comparison of the toxicity of some varieties of *Bacillus thuringiensis* Berliner for silkworm larvae. *J. Invert. Path.* **11**, 289.

ARAGÃO, H. DE B. (1943). O virus do maxoma no coelho do mato (*Sylvilagus minensis*), sua transmissão pelos *Aedes scapularis e aegypti. Mem. Inst. Osw. Cruz.* **38**, 93.

BERGER, R. S. (1963). Laboratory techniques for rearing *Heliothis* species on artificial medium. *U.S.D.A. A.R.S.*-33-84, 4 pp.

BOND, R. P. M. (1969). The natural occurrence of allomucic (allaric) acid: a contribution to the assignment of structure of an insecticidal exotoxin from *Bacillus thuringiensis* Berliner. *Chem. Comms.* **1**, 358.

BOND, R. P. M., BOYCE, C. B. C., BROWN, V. & TIPTON, J. D. (1969). Some chemical and biological studies on an exotoxin from *Bacillus thuringiensis* var. *thuringiensis* Berliner. *Proc. 494th Meeting Biochem. Soc.*

BRIGGS, J. D. (1963). Commercial production of insect pathogens. In *Insect Pathology: An Advanced Treatise*, vol. 2, p. 519. Ed. E. A. Steinhaus. New York: Academic Press Inc.

BUCHER, G. E. (1960). Potential bacterial pathogens of insects and their characteristics. *J. Insect. Path.* **2**, 172.

BURGERJON, A. & BIACHE, C. (1967). Contribution à l'étude du spectre d'activité de différentes souches de *Bacillus thuringiensis*. *Entomologia exp. appl.* **10**, 211.

BURGES, H. D. (1971). In *Microbial Control of Insects and Mites*. Ed. H. D. Burges & N. W. Hussey. London: Academic Press Inc.

CHAMBERLAIN, F. S. & DUTKY, S. R. (1958). Tests on pathogens for the control of tobacco insects. *J. Econ. Ent.* **51**, 560.

COOKSEY, K. E. (1971). In *Microbial Control of Insects and Mites*. Ed. H. D. Burges and N. W. Hussey. London: Academic Press Inc.

COOKSEY, K. E., DONNINGER, C., NORRIS, J. R. & SHANKLAND, D. (1969). Nerve blocking effect of *Bacillus thuringiensis* protein toxin. *J. Invert. Path.* **13**, 461.

CREIGHTON, C. S., KINARD, W. S. & ALLEN, N. (1961). Effectiveness of *Bacillus thuringiensis* and several chemical insecticides for control of budworms and hornworms on tobacco. *J. Econ. Ent.* **54**, 1112.

DAVID, W. A. L. & GARDINER, B. O. C. (1965). Resistance of *Pieris brassicae* (Linnaeus) to granulosis virus and the virulence of the virus from different host races. *J. Invert. Path.* **7**, 285.

DE BARJAC, H. & DEDONDER, R. (1968). Purification de la toxine thermostable de *Bacillus thuringiensis* var. *thuringiensis* et analyses complementaires. *Bull. Soc. chim. Biol.* **50**, 941.

DE BARJAC, H. & RIOU, J.-Y. (1969). Action de la toxine thermostable de *Bacillus thuringiensis* var. *thuringiensis* administrée à des souris. *Rev. Path. comp. med. exptl* **69**, 367.

DULMAGE, H. T. (1971). In *Microbial Control of Insects and Mites*. Ed. H. D. Burges and N. W. Hussey. London: Academic Press Inc.

DUTKY, S. R. (1959). Insect microbiology. *Adv. appl. Microbiol.* **1**, 175.

DUTKY, S. R. (1963). In *Insect Pathology: An Advanced Treatise*, vol. 2, p. 75. Ed. E. A. Steinhaus. New York: Academic Press Inc.

FALCON, L. A. (1971). In *Microbial Control of Insects and Mites*. Ed. H. D. Burges and N. W. Hussey. London: Academic Press Inc.

FARKAS, J., ŠEBESTA, K., HORSKÁ, K., SAMEK, Z., DOLIJS, L. & SORM, F. (1969) Structure of the exotoxin of *Bacillus thuringiensis* var. *gelechiae*. *Colln Czech. Chem. Commun.* (Engl. edn.), **34**, 1118.

FAST, P. G. & ANGUS, T. A. (1965). Effects of parasporal inclusions of *Bacillus thuringiensis* var. *sotto* Ishiwata on the permeability of the gut wall of *Bombyx mori* (Linnaeus) larvae. *J. Invert. Path.* **7**, 29.

FAUST, R. (1968). *In vitro* chemical reaction of the δ-endotoxin produced by *Bacillus thuringiensis* var. *dendrolimus* with other proteins. *J. Invert. Path.* **11**, 465.

FENNER, F. & RATCLIFFE, F. N. (1965). *Myxomatosis*. Cambridge University Press.

FISHER, R. & ROSNER, L. (1959). Insecticide safety: Toxicology of the microbial insecticide, Thuricide. *J. Agric. Fd Chem.* **7**, 686.

FITZ-JAMES, P. C. & YOUNG, J. E. (1959). Comparisons of species and varieties of the genus *Bacillus*. Structure and nucleic acid content of spores. *J. Bact.* **78**, 743.

GENTRY, C. R., THOMAS, W. W. & STANLEY, J. M. (1969). Integrated control as an improved means of reducing populations of tobacco pests. *J. Econ. Ent.* **62**, 1274.

GETZIN, L. (1962). Mass rearing of virus-free cabbage loopers on an artificial diet. *J. Insect. Path.* **4**, 486.

GLASER, R. W., McCoy, E. E. & GIRTH, H. B. (1940). The biology and economic importance of a nematode parasitic in insects. *J. Parasit.* **26**, 479.

GRISON, P. (1967). Réalisations et perspectives actuelles de la lutte microbiologique. *Phytiat. Phytopharm.* **16**, 62–74.

HALL, I. M. (1963). In *Insect Pathology: An Advanced Treatise*, vol. 2, p. 477. Ed. E. A. Steinhaus. New York: Academic Press Inc.

HARVEY, T. L. & HOWELL, D. E. (1965). Resistance of the housefly to *Bacillus thuringiensis* Berliner. *J. Invert. Path.* **7**, 92.

HAYNES, W. C. & RHODES, L. J. (1966). Spore formation by *Bacillus popilliae* in liquid medium containing activated carbon. *J. Bact.* **91**, 2270.

HEIMPEL, A. M. (1961). Pathogenicity of *Bacillus cereus* Frankland & Frankland and *Bacillus thuringiensis* Berliner varieties for several species of sawfly larvae. *J. Insect. Path.* **3**, 271.

HEIMPEL, A. M. (1965). Microbial control of insects. *Wld Rev. Pest Control* **4**, 150.

HEIMPEL, A. M. (1966). Exposure of white mice and guinea pigs to the nuclear-polyhedrosis virus of the cabbage looper, *Trichoplusia ni*. *J. Invert. Path.* **8**, 98.

HEIMPEL, A. M. (1971). In *Microbial Control of Insects and Mites*. Ed. H. D. Burges and N. W. Hussey. London: Academic Press Inc.

HEIMPEL, A. M. & BUCHANAN, L. K. (1967). Human feeding tests using a nuclear-polyhedrosis virus of *Heliothis zea*. *J. Invert. Path.* **9**, 55.

HERFS, W. (1963). Zur Technik der Wirksamkeitsbestimmung von *Bacillus thuringiensis* Präparaten (sporen-, endotoxin-komplex) an Raupen im Laboratorium. *Entomophaga* **8**, 163.

HERFS, W. (1965). Die Verträglichkeit von *Bacillus thuringiensis* Präparaten mit chemischen Pflanzenschutzmitteln und mit Beistoffen. *Z. PflKrankh. PflPath. PflSchutz* **72**, 584.

HOLMES, K. C. & MONRO, R. E. (1965). Studies on the structure of parasporal inclusions from *Bacillus thuringiensis*. *J. molec. Biol.* **14**, 572.

HOUSE, H. L., WELCH, H. E. & CLEUGH, T. R. (1965). Food medium of prepared dog biscuit for the mass-production of the nematode DD 136 (Nematoda; Steinernematide). *Nature, Lond.* **206**, 847.

IGNOFFO, C. M. (1963). A successful technique for mass rearing cabbage loopers on a semisynthetic diet. *Am. Ent. Soc. Am.* **56**, 178.

IGNOFFO, C. M. (1964). Production and virulence of a nuclear polyhedrosis virus from larvae of *Trichoplusia ni* (Hübner) reared on a semisynthetic diet. *J. Insect. Path.* **6**, 318.

IGNOFFO, C. M. (1968). Specificity of insect viruses. *Bull. Ent. Soc. Am.* **14**, 265.

IGNOFFO, C. M. & HEIMPEL, A. M. (1965). The nuclear polyhedrosis virus of *Heliothis zea* and *Heliothis virescens*. V. Toxicity pathogenicity of virus to white mice and guinea pigs. *J. Invert. Path.* **7**, 329.

KRIEG, A. (1967). Neues über *Bacillus thuringiensis* und seine Anwendung. *Mitt. biol. Bundanst. Ld- u. Forstw., Berlin-Dahlem* (125), 106 pp.

LÜTHY, P. & ETTLINGER, L. (1967). In *Insect Pathology and Microbial Control*, p. 252. Ed. by P. A. van der Laan. Amsterdam: North-Holland Publishing Co.

MARSCHALL, K. J. (1970). Introduction of a new virus disease of the Coconut Rhinoceros Beetle in Western Samoa. *Nature, Lond.* **225**, 288.

MARTOURET, D. (1967). L'utilisation des micro-organismes entomopathogènes en agriculture: potentialités actuelles et perspectives offertes pour *Bacillus thuringiensis* Berliner. *C. r. hebd. Séanc. Agric. Fr.* **53**, 154.

McCONNELL, E. & RICHARDS, A. G. (1959). The production by *Bacillus thuringiensis* Berliner of a heat-stable substance toxic for insects. *Can. J. Microbiol.* **5**, 161.

MÜLLER-KÖGLER, E. (1965). Pilzkrankheiten bei Insekten. *Anwending zur biologischen Schadlingsbekämpfung und Grundlagen der Insektenmykologie.* Berlin: Verlag Parey.

NORRIS, J. R. (1963). Bacterial insecticides. *Sci. Progr.* **L 1**, 202.

NORRIS, J. R. (1969a). Macromolecular synthesis during sporulation of *Bacillus thuringiensis.* In *Spores*, vol. IV. Ed. L. L. Campbell. Bethesda, Md.: American Soc. of Microbiologists.

NORRIS, J. R. (1969b). The ecology of serotype 4B of *Bacillus thuringiensis. J. appl. Bact.* **32**, 261.

NORRIS, J. R. (1970). Sporeformers as insecticides. *J. appl. Bact.* **33**, 192.

NORRIS, J. R. (1971). In *Microbial Control of Insects and Mites.* Ed. H. D. Burges and N. W. Hussey. London: Academic Press Inc.

POINAR, G. O. (1966). The presence of *Achromobacter nematophilus* in the infective stage of a *Noaplectana* sp. (Steinernematidae: nematoda). *Nematologica* **12**, 105.

POINAR, G. O. (1967). In *Insect Pathology and Microbial Control*, p. 197. Ed. P. A. van der Laan. Amsterdam: North-Holland Publishing Co.

POINAR, G. O. (1971). In *Microbial Control of Insects and Mites.* Ed. H. D. Burges and N. W. Hussey. London: Academic Press Inc.

RABB, R. L. & GUTHRIE, F. E. (1964). Resistance of tobacco hornworms to certain insecticides in North Carolina. *J. Econ. Ent.* **57**, 995.

RAMAKRISHNAN, N. (1968). Observations on the toxicity of *Bacillus thuringiensis* for the silkworm *Bombyx mori. J. Invert. Path.* **10**, 449.

RIVERS, C. F. (1964). Viral insecticides. *Discovery*, September 1964, p. 27

RIVERS, C. F. (1967). In *Insect Pathology and Microbial Control*, p. 252. Ed. P. A. van der Laan. Amsterdam: North-Holland Publishing Co.

ROBERTS, D. W. & YENDOL, W. G. (1971). In *Microbial Control of Insects and Mites*, Ed. H. D. Burges and N. W. Hussey. London: Academic Press Inc.

ROGOFF, M. H. (1966). Crystal-forming bacteria as insect pathogens. *Adv. appl. Microbiol.* **8**, 291.

ROGOFF, M. H., IGNOFFO, C. M., SINGER, S., GARD, I. & PRIETO, A. P. (1969). Insecticidal activity of thirty-one strains of *Bacillus* against five insect species. *J. Invert. Path.* **14**, 122.

SCHAERFFENBERG, B. (1968). Untersuchungen über Wirkung der insecktentötenden Pilze *Beauveria bassiana* (Bals) Vuill. und *Metarrhizium anisopliae* (Metsch.) Sorok. auf Warmblüter. *Entomophaga* **13**, 175.

ŠEBESTA, K. & HORSKÁ, K. (1968). Inhibition of DNA-dependent RNA polymerase by the exotoxin of *Bacillus thuringiensis* var. *Gelechiae. Biochim. biophys. Acta* **169**, 281.

ŠEBESTA, K., HORSKÁ, K. & VANKOVA, J. (1969). Isolation and properties of the insecticide exotoxin of *Bacillus thuringiensis* var. *Gelechiae. Colln Czech. Chem. Commun.* (Engl. edn.) **34**, 891.

SHARPE, E. S. (1966). Propagation of *Bacillus popilliae* in laboratory fermenters. *Biotechnol. Bioengng* **8**, 247.

SHARPE, E. S., ST. JULIAN, G. & CROWELL, C. (1970). Characteristics of a new strain of *Bacillus popilliae* sporogenic *in vitro. Appl. Microbiol.* **19**, 681.

SMIRNOFF, W. A. (1963). The formation of crystals of *Bacillus thuringiensis* var. *thuringiensis* Berliner before sporulation at low temperature incubation. *J. Insect Pathol.* **8**, 376.

SMIRNOFF, W. A. & BERLINGUET, L. (1966). A substance in some commercial preparations of *Bacillus thuringiensis* var. *thuringiensis* toxic to sawfly larvae. *J. Invert. Path.* **8**, 376.

SOMERVILLE, H. J., DELAFIELD, F. P. & RITTENBERG, S. C. (1968). Biochemical homology between crystal and spore protein of *Bacillus thuringiensis*. *J. Bact.* **96**, 721.

SOMERVILLE, H. J., DELAFIELD, F. P. & RITTENBERG, S. C. (1970). Urea-mercapto-ethanol-soluble protein from spores of *Bacillus thuringiensis* and other species. *J. Bact.* **101**, 551.

SOMERVILLE, H. J. & JAMES, C. R. (1970). Association of the crystalline inclusion of *Bacillus thuringiensis* with the exosporium. *J. Bact.* **102**, 580.

STEINHAUS, E. A. (1946). An orientation with respect to members of the genus *Bacillus* pathogenic for insects. *Bact. Rev.* **10**, 51.

STEINHAUS, E. A. (1949). *Principles of Insect Pathology*. New York: McGraw Hill.

TANADA, Y. (1967). In *Pest Control*, Ed. W. W. Kilgore and R. L. Doutt. New York: Academic Press Inc.

VAGO, C. (1967). In *Methods in Virology*, Ed. Maramorosch and Koprowsky. New York: Academic Press Inc.

VASILJEVIĆ, L. (1970). Conceptions et modalités pratiques d'utilisation des pré-parations entomopathogènes à base de virus en agriculture. Mimeographed paper presented to the Commision de Pathologie des Insectes et de Lutte Microbiologique Symposium, Amsterdam, 1970.

WATANABE, H. (1967). Development of resistance in the silkworm *Bombyx mori* to peroral infection of a cytoplasmic-polyhedrosis virus. *J. Invert. Path.* **9**, 474.

WELCH, H. E. (1958). Test of a nematode and its associated bacterium for control of the Colorado potato beetle, *Lepitinotarsa decemlineata*. *Am. Rep. Ent. Soc. Ontario* **88**, 53.

MICROBIAL ACTIVITY IN AQUATIC ENVIRONMENTS

JOHN D. H. STRICKLAND

Institute of Marine Resources, Scripps Institution of Oceanography, La Jolla, California 92037, U.S.A.

INTRODUCTION

I would like to introduce you to some problems. These arise out of studying the ecology of marine organisms in a rather specific habitat, that of the open ocean or a large deep lake, far from the interference of terrestrial flora or the relatively concentrated habitats of the water/air and water/sediment interfaces.

I am concerned with microbial activity but I would like to be wide in my definition of microbes, including not only bacteria, fungi and yeasts but extending it to include the photosynthetically active plant plankton and the zooplankton, the particle-feeding animal plankton that feed on the plants and do not themselves exceed a millimetre or so in length.

Before the last decade there had been surprisingly little study of the ecology of bacteria in open waters. Mention should be made of the pioneer work of Waksman from the United States (working in the sea) and Rodina from Russia (working in lakes). Pioneers in the marine field such as Claude Zobell at Scripps Institution and Ferguson Wood from Australia were mainly concerned with documenting the physiology and substrate specificity of marine organisms isolated and maintained at substrate levels far greater than those ever met with in nature. An excellent review of this work is found in Zobell's book (Zobell, 1946). The book by Kuznezow translated into German (Kuznezow, 1959) gives a good picture of the work done in lakes.

The plant and animal plankton have, of course, been studied much more extensively than the bacteria but even here work on truly ecological aspects of the physiology and nutrition of plants has started in earnest only in the last two decades. It is becoming clear that a surprising number of common factors seem to govern the growth kinetics of creatures as different as bacteria and diatoms and copepods.

In the following account I would like to outline the work done recently to find three things: the biomass of living microbial organisms in the sea or a lake; the *in situ* rates at which these organisms metabolize substrates (light is included as a 'substrate' when considering the phyto-

plankton); the substrates present in the water, their nature and concentrations.

The prime aim of the plankton ecologists is, of course, to understand
the quantitative nature of the flux of matter as it goes from photosynthetically fixed plant tissue to all other trophic levels. This includes
a study of the role of microheterotrophs in degrading and remineralizing
matter lost in the transfer of food from lower to higher trophic levels.
In the cold deep waters of the earth most of the metabolic activity of
living matter takes place in microorganisms so that a knowledge of their
kinetic processes becomes essential for a better understanding of the
ecology of the planet as a whole.

THE TOTAL BIOMASS OF MICROORGANISMS IN WATER BODIES

An assessment of the total biomass of microorganisms in a sample of
sea or lake water has proved to be surprisingly difficult. The small
animals and plant plankton can be settled in a glass-bottom cylinder and
viewed with an inverted microscope. Identification and volume estimates
can then be made for each different organism. One difficulty has been to
find suitable chemicals for preserving and fixing the plants and animals.
Many of the small protozoa in tropical waters disintegrate in the
presence of formaldehyde and other chemicals. The greatest problem,
however, is simply one of manpower to accomplish the impossibly
tedious feat of viewing and counting the hundreds of samples which can
accumulate from even one modest field study.

With the smallest organisms, such as bacteria, we have further problems because of the difficulties of recognizing living matter amongst the
mass of detritus present in all water samples. A very good example of
techniques of direct microscopy is given by Jannasch (1958) and
Jannasch & Jones (1959). Despite the protests of the few expert micro-
scopists who claim to be able to make a quantitative count of bacteria,
using suitable staining technology, I am far from convinced that any
really reliable routine technique can be applied and, even if this were so,
we are again faced with the insoluble problem of coping with the large
number of samples within a reasonable period of time.

Plating viable microbes on to nutrient agar is a possible alternative
method for determining biomass but there is considerable doubt as to
the relation of plate counts to the number of living microbes in the
original sample. Clumping occurs and many species simply do not
multiply on agar containing high concentrations of substrate. Plate

counts probably underestimate the true biomass of heterotrophs by a factor of from 10 to 1,000.

Using 'classical' techniques for assessing biomass, the values found for deep waters of the open ocean range from a hundredth to a few tenths of micrograms of bacterial carbon per litre, depending on location and depth (see summary by Kriss, 1963). Although these measures are underestimates they have been heavily criticized on the grounds that they are high due to contamination (Sorokin, 1964). Alternative methods are badly needed to settle the issue. Quantities of heterotrophs as high as 0.4 μg. carbon/l. could contribute significantly to the turnover of organic matter in deep water, but if much less than this were present the effect would be almost certainly unimportant. To my knowledge only two really new approaches have been applied to the problem recently.

In the euphotic zone, where the majority of living matter is in the form of small plants, an acceptable estimate of the biomass of plants can be found from a determination of chlorophyll a, for which very sensitive fluorometric methods are now available (see Yentsch & Menzel, 1963; Holm-Hansen et al. 1965). Historically this approach has been used extensively by marine workers even with very primitive methods for estimating chlorophyll. The problem is to find a suitable factor for converting estimates of plant chlorophyll to, say, plant carbon, as this factor is clearly a variable depending on the species of plant, its state of nutrition, degree of illumination, as well as other factors. We now have sufficient experience, however, to assign values for the ratio plant carbon:plant chlorophyll a with some success. Values given by Strickland (1965) for carbon and various other elements and metabolites still apply pretty well. Values for the carbon:chlorophyll a ratio vary from about 30 with well-nourished coastal phytoplankton crops to 90 for phytoplankton in the oligotrophic tropical oceans.

When we sample below 100 m. or more in seas or lakes, all of the living microorganisms are heterotrophs or phagotrophs with no characteristic pigments and one needs a characteristic metabolite that can be assayed for in extremely low concentrations.

Fortunately adenosine triphosphate (ATP) appears to be just that ideal metabolite. Following a suggestion from Levin et al. (1964), Holm-Hansen & Booth (1966) devised a technique for measuring ATP concentration in samples taken from any depth in sea water or lakes. The method is simple and extraordinarily sensitive. Based on the well-known firefly reaction, the light emitted when ATP activates a mixture of luciferin and luciferase, the light is measured by a sensitive photomultiplier. The limit of detection is about 10^{-5} μg. ATP/l. or better. The conversion of ATP to weight of living material or organic carbon presents us with the same set of problems as was found with chlorophyll. From the analysis of many phytoplankton species (O. Holm-Hansen, unpublished data) and marine bacteria (Hamilton and Holm-Hansen, 1967) we find a carbon:ATP ratio of 250 applying for most aquatic microflora. Of course, scatter is quite bad but a fairly good indication of the mass of living matter can be obtained. Adenosine triphosphate is lost very rapidly from cells on death, and its concentration drops in senescent cells so that, in deep waters, the ratio given above may be somewhat too low. Thus, the ATP method as now used *underestimates* the living organic carbon in samples taken from great depths in seas and lakes. Profiles analysed to date (Holm-Hansen & Booth, 1966; Holm-Hansen, 1969, and unpublished data) show typical ATP values ranging from 0·3 or more μg. ATP/l. in the plankton-rich surface waters to 0·05 μg. ATP/l. in subsurface maxima (at a few hundred metres) to less than 0·005 μg. ATP/l. in waters below 1,000 m. The latter value corresponds to a standing stock of living heterotrophs in deep waters of 0·15 μg. carbon/l. and less. These values agree with the suspected data of Kriss. Unpublished data taken from Lake Tahoe show that this deep fresh water body has a surprisingly high standing stock of microorganisms. Of course ATP does not necessarily measure bacteria but any protozoa or any living animals and plants present in less than the size of the screening net used for taking the sample (typically 75–150 μm.). The ATP method has scarcely been used at depths below 1,000 m., but what work has been done indicates that, at 4,000 m., the amount of living matter is present at 0·05 μg carbon/l. or less. However, for a good 2–3,000 m. of depth, the concentration may well exceed 0·1 μg. carbon/l.

Bernard (1963), Kimball, Corcoran & Wood (1963), Fournier (1966) and others have all reported large numbers of comparatively large cells in deep water but a visual search for particles in this category (Hamilton, Morgan & Strickland, 1966) failed to find anywhere near enough for this matter to account for the observed ATP until a depth of 3,500 m. when the visual counts and the ATP estimates gave the same results. The question whether deep-water heterotrophic activity takes place by bacteria or small ciliates and similar organisms is still wide open. Dr Wiebe, of the University of Georgia, has carefully examined deep-sea detritus and finds some of it singularly free from bacteria (private communication), confirming the findings of Riley, Wangersky & Van Hermet (1964). However, the total amount of heterotrophs is impressive considering the vast bulk of water involved, and a better knowledge of the *in situ* activity of these microorganisms is essential for a proper knowledge of planetary ecology.

DIRECT ESTIMATES OF GROSS METABOLIC ACTIVITY

The photosynthetic activity of the micro-plankton in surface waters can be measured satisfactorily by a variety of means. What we now need is a measure of respiratory activity (in the dark), especially in cold deep waters, where a direct approach by measurements of biochemical oxygen demand or oxygen uptake in the Warburg apparatus is far too insensitive.

A more promising attack is to determine *in situ* reproductive rates by the submerged-slide technique (ZoBell & Allen, 1933; Henrici, 1936), which has been modified extensively by Kriss and his coworkers in Russia. This technique has been heavily attacked because of the anomalous effects brought about by the presence of surfaces and the fact that only a relatively small fraction of the total population will adhere to glass. The submerged bottle method of Ivanoff (1955) is perhaps better, but all of these techniques hold the common problem in lack of adequate manpower for handling the number of samples involved.

Riley (1951) published a paper in which he attempted to estimate the respiration rates in deep Atlantic waters from a knowledge of the temperature and the availability of oxygen and phosphate over wide areas of the north Atlantic, and by making various hydrodynamical assumptions. He arrived at a figure for around 1 μl./l./year for deep-water oxygen consumption. Munk (1966) and Aron & Stommel (1967), with better data and theory to go on, estimated rates about an order of magnitude higher.

A question to be answered is: do microorganisms in nature behave as if they were in batch or continuous culture, or is the situation too complex to be simulated by any controlled environment? In certain upwelling conditions, mainly found near the west coasts of continental masses, the euphotic zone is periodically injected with massive doses of nutrients and the plant plankton develop explosively very much like a batch culture. Most of the time, however, the topmost parts of the ocean are only very slowly replenished from beneath and the plant material is recycled *in situ* many times. The plants use ammonia and other trivalent forms of nitrogen liberated by the zooplankton as they graze the plant crop. In deep water, where concentrations of both substrate and hetero-trophs are extremely low, we see the suspended state of the die-off phase of a batch culture, or it may be a great chemostat with the substrate entering the bottom deposits by sedimentation or adsorption, the whole working with a huge turnover time. It is not inconceivable that the main transfer from the sub-euphotic zone to the bottom may be predominately via great activity occurring only as events widely spaced in time or space as comparatively large animals (such as fish) fall under gravity.

Four new approaches have been suggested over the past few years which give some hope of mastering the problem of how to make direct determinations of respiratory activity in deep water. A lot of the most relevant work is yet unpublished but I would like to summarize it. In the first, respiration is measured on a concentrate of the particulate matter. The second approach concentrates particulate matter completely from the water by filtration and measures the electron-transport activity of the filtered material. The third method measures the ATP present in the water and assumes a connexion between temperature, ATP content and respiratory activity. The fourth method measures the uptake of ^{14}C-labelled carbonate by the microflora using counting techniques.

Respiration rates on concentrates

Dodson & Thomas (1964) described an ingenious way of concentrating microplankton. A fine net is fastened to the bottom of a plastic pipe which is placed inside a slightly larger pipe closed at the bottom with a plate. The sea water to be concentrated is piped into the anulus between the two pipes and upwells through the filter into the interior of the first pipe from which it is continually removed. The detritus and micro-organisms concentrate in the anulus. Pomeroy & Johannes (1968) adapted this approach to concentrate samples from the topmost 500 m. of the sea but using a 0·5 μm. membrane filter to replace the netting. Typically they obtained concentrates of up to a 1,000-fold on which direct respiration methods were possible in a few hours at *in situ* temperatures. The main question about the technique is how many organisms survive the concentration process alive and with their original metabolic activity? Comparisons of ATP and reductase power (see p. 237) by direct measurements before and after concentration indicate quite a serious loss (up to 80 %; Holm-Hansen, Pomeroy & Packard, 1970). This may be partially corrected by an empirical factor.

The work described by Pomeroy & Johannes (1968) points to an ocean system where much of the carbon fixed by the plants in the euphotic zone is reconsumed in the top 500 m., largely by micro-organisms who are responsible for much of the turnover of the surface organically fixed energy. The deeper layers of the ocean (below about 1,000 m. with variations according to locale) are fed by the residuum of this process which is too small to be measured satisfactorily by the concentration technique as at present developed.

In samples taken beneath the euphotic zone, respiration rates from microorganisms are variable but of significant magnitude. These rates seem positively related to the general productivity of the surface waters, being high, for example, beneath the Peru Current (100 μl. O_2/l./year). In waters below a few hundred m. in Antarctica, rates were surprisingly low (Pomeroy et al. 1969; Pomeroy & Wiebe, 1970).

Measurement of the electron-transport system (ETS)

Aleem (1955) made the novel suggestion that the population of zooplankton could be measured by its electron-transport activity as indicated by the oxidation of triphenyl tetrazolium chloride to give a deeply coloured dye. The method was not particularly successful, largely because plant plankton gave little reaction, but it was re-established by Curl & Sandberg (1961), who suggested that the respiratory activity of aquatic organisms could be predicted from their succinate dehydrogenase (SDH) activity as measured by dye production when various tetrazolium salts were reduced. Quite satisfactory results were obtained and these are supported by the work of Packard & Taylor (1968) working on respiration of Artemia salina and improved enzyme techniques.

Packard (1969) investigated the possibility of measuring the total respiratory activity of living cells by measuring both SDH activity and the activity of $NADH_2$ and $NADPH_2$ reductases. Total ETS activity is measured as dye release after incubation for 20 min. at 35° at pH 7·7 with 2-p-iodophenyl-3-0-nitrophenyl-5-phenyltetrazolium chloride. The dye produced on reduction is extracted with a tetrachloroethylene-acetone mixture and measured spectrophotometrically. From a knowledge of the specific absorption of the dye, the number of electrons and hence the amount of oxygen used per hour can be calculated. To calculate in situ activity, an activation energy approach was used with an average E value of 14 kcal. per mole being obtained by work on several marine organisms, including bacteria.

The ETS activity measures the respiratory capacity of the organism. Their actual activity may be only a fraction of this; Packard reasoned a fifth. Data from the California Current and the Eastern Tropical Pacific show an exponential decrease in depth from surface values of several hundred μl. O_2/l./year to about 7 μl. O_2/l./year at 500 m. Below 500 m. the values decreased much more slowly with depth to about 5 μl. O_2/l./year or less at 3,000 m.

Respiratory activity from ATP concentrations

This approach is at present only hypothetical but is being actively pursued at the present in my laboratory. It is based on the clearly speculative notion that most microorganisms will be found to have respiration rates that are tied approximately to the total amount of ATP present and the temperature–respiration relationship found for the ETS activity and temperature. We are measuring ATP profiles simultaneously with Pomeroy concentrate respiration profiles and with Packard ETS profiles. We are in the process of experimenting with laboratory cultures to find a relationship of respiration with ATP and temperature more exactly. Eventually it is hoped to produce a paper showing a comparison between all of these methods in the hope that they can be reconciled or shown to measure different facets of the same process.

The existing data are interesting. At a depth of 200 m. in the Atlantic (not often the same location) we find respiration rates of 100 μl. O_2/l./ year from Riley's calculations, 140 μl. O_2/l./year by the concentration–respiration method, 30 μl. O_2/l./year from the ETS method and 125 μl. O_2/l./year as a best guess using ATP data. At 500 m. the corresponding data are 200 μl. O_2/l./year (Riley), 200 μl. O_2/l./year (concentration–respiration), 15 μl. O_2/l./year ETS, and 35 μl. O_2/l./year (ATP) respectively, showing good agreement except for the high values by the concentration–respiration method. At several thousand metres, the concentration–respiration method cannot be used for lack of sensitivity. The ATP and ETS data at this depth so far show agreement at about 3–5 μl. O_2/l./year, which is appreciably more than Riley's estimate of 1 μl. O_2/l./year or less in the Atlantic at the same depth but more in accordance with the magnitudes suggested by Munk (1966) and Aron & Stommel (1967).

Estimates of metabolic activity from $^{14}CO_2$ uptake rates

This approach has been used extensively by Russian limnologists (ref. Romanenko, 1964 a, b). Uptake of carbonate in the dark is measured by adding radioactive carbonate (50 or 100 μCi) to a small bottle of sea water and incubating it at in situ temperatures for 1–2 days. The sample is filtered through a 0·45 μm. membrane filter and the washed filter dried and counted in a scintillation counter. Large amounts of carbonate in deep water limit the sensitivity of this method, but it should be adequate. Our Russian colleagues have produced a considerable bulk of evidence indicating quite convincing relationships between the uptake of carbonate and the growth and respiration of heterotrophs in anoxic waters. I have not yet seen any published data for deep ocean waters but Y. I. Sorokin (private communication) has data for waters from between 200 and 500 m. showing good agreement with the values found in these depth ranges by the ETS and ATP methods. The values published by Seki & Zobell (1967) seem much too high.

A GENERAL APPROACH TO
GROWTH KINETICS: GROWTH RATE AS A FUNCTION
OF FOOD CONCENTRATION

Evidence has accumulated over the past two decades to show that aquatic microorganisms grow at a rate limited by the concentration of one or at most two nutrient elements in the environment. Plant growth may also be limited by the intensity of illumination in addition to nutrients. In all cases the observed data fit an equation of the following form, using the nomenclature of Herbert, Elsworth & Telling (1956):

$$\mu = \mu_m \frac{s}{K_s + s}, \tag{1}$$

μ is the rate of growth in a medium with the limiting nutrient at concentration s; μ_m is the maximum value for μ when a very large concentration of s is in the medium; K_s is a constant with the units of concentration and is the concentration of s at which μ is half of μ_m; μ_m has the units of reciprocal time. The equation has been in common usage since its early applications to bacterial growth by Hinshelwood (1946). It resembles the Langmuir adsorption isotherm and the equation commonly used in Michaelis–Menten enzyme kinetics.

Based on his own thesis work and the experience of workers in my laboratories, and elsewhere, John Caperon (1967) made a very satisfactory synthesis to show that much microbial growth kinetics can be predicted by one equation:

$$B + C \underset{k_2}{\overset{k_1}{\rightleftharpoons}} [CB] \underset{k_4}{\overset{k_3}{\rightleftharpoons}} C + P, \tag{2}$$

where B is a unit of food, C an acquisition site, $[CB]$ is the concentration of occupied acquisition sites in the system with a fixed number of acquisition sites, and P a particle of ingested food. By making several reasonable assumptions, Caperon was able to derive equations for growth that ended in the same form as equation (1). Caperon showed that

$$\mu_m = \frac{k_3 c_0}{q}, \tag{3}$$

where q is the amount of P required to make a new individual, and c_0 is the total concentration of acquisition sites in the organism. The other constant, K_s, was given by

$$K_s = \frac{k_3}{k_1}. \tag{4}$$

Caperon (1967) was able to show that his equations predicted the common logistic equation. This is the first time that this equation has been given a theoretical basis:

$$\frac{dp}{dt} = p.K(1 - K^1.p),\qquad(5)$$

where p is population and K and K^1 are constants.

It can be shown, using Caperon's equations and assumptions, that if k_2 is kept to 10 % or less of k_1 and k_3, then a rather complicated equation for ingestion rate as a function of food concentration given by Caperon reduces to equation (1). With $\mu_m = k_1 c_0$ and

$$K_s = \frac{k_3}{k_1}.\qquad(6)$$

Both growth rate and ingestion can be expressed by the same form of equation (1). I would like to make a reference to a few examples of the use of this equation for calculating rates of growth or ingestion as a function of the supply of food.

Zooplankton feeding

It is interesting to speculate how large a creature must be before its feeding behaviour can no longer be expressed by equation (1). Marine bacteria would clearly be expected to conform but not so a single-mouthed herbivorous copepod several millimetres in length. Most of the animals in this size range filter water with a flapping motion of their appendages, and the amount of water filtered by this motion has been considered a constant for each species. The animal then behaves as a pump with the pumping rate being (typically) for *Calanus* around 100 ml./animal/24 h. Under these conditions, a graph portraying ingestion rate on the ordinate and food concentration on the abscissa would be initially a straight line. Doubling the concentration of the food in the water would lead to the pump filtering twice the amount of food. Clearly, at some point, the food concentration could get so high that filtration was affected so that the straight line no longer represented ingestion properly. Then the curve will roll over to a maximum or perhaps show a negative slope at sufficiently high food concentration.

Unfortunately, the relatively large number of experiments that have been made on a copepod filtration rates have too high a scatter of points for equation (1) to be rigorously tested. On the other hand, data of Ryther (1954), Rigler (1961), Mullin (1963), McMahon & Rigler (1965),

Conover (1966), Burns & Rigler (1967), Haq (1967), and Parsons, Le Brasseur & Fulton (1967) fit equation (1) as well as any other that has been suggested and used. What part of the animal is behaving like the [CB] in Caperon's equation is, of course, unknown, and a decrease of ingestion rate at very high food concentrations (which is by no means conclusively established) remains to be explained.

There is more to herbivorous feeding than this, as the animals show an extraordinary preference for selecting large particles over small. When doing this the animals hunt and seize their food. The kinetics of their behaviour may well fall better in the forms suggested by Cushing (1968) or Harris (1968) using analogies with insect feeding.

Micro-zooplankton feeding

A large number of small protozoa are not caught in the more commonly used plankton nets because the mesh of the nets is too large, but nevertheless these animals constitute a significant fraction of the total plankton (Beers, Stewart & Strickland, 1967). No feeding data have been collected for organisms in this size range using very low concentrations of substrate, to the best of my knowledge, except those by my colleague Dr Hamilton in these laboratories.

Hamilton & Preslan (1970) reared a *Uronema* sp. collected from 300 m. exclusively on resuspended marine bacteria in substrate-free sea water. The behaviour of this prorozoan in a chemostat was almost exactly in accordance with chemostat theory and hence ingestion obeyed equation (1). The value for μ_m was 0·15 h.$^{-1}$ and K_s was about 300 μg. carbon/l. The organisms made feeding vacuoles that were rapidly packed with bacteria which began dissolution in less than 30 sec. The vacuole finally closed and moved to the rear of the organism. Suspected faecal material was observed. At very low rates and very low concentrations of protozoa in the flask, chemostat kinetics broke down in a way resembling the behaviour noticed by Jannasch (p. 245), indicating a slowing down of growth when the *Uronema* concentrations were low. The organisms appear to excrete metabolites capable of enhancing their growth rates. Values for K_s were far too high for deep-ocean bacteria to be used as food if they occurred at their average concentrations. The protozoan clearly can only survive on very high concentration of bacteria such as to be suspected on the surfaces of solids. It is interesting to note that *Uronema* spp. have been reported from inside copepod carcasses by Russian workers.

The phytoplankton

Plankton plants and light

It is not immediately obvious, and it is, in fact, rather unexpected to find that plant growth is a function of light intensity as if the intensity of light could be treated as a substrate concentration in equation (1). Curves of production rate (P) *versus* light intensity (I) which abound in the literature are notoriously difficult to obtain without scatter, but the best examples fit equation (1) at least as well as any other (see, for example, Caperon, 1967, and his use of data from McAllister, Shah & Strickland, 1964). Tamyia *et al.* (1953) long ago found that equation (1) describes the growth of *Chlorella* cultures in light-limiting environments.

Eppley & Sloan (1966), in these laboratories, have attempted to rationalize the use of equation (1) by considering the light absorbed by chlorophyll *a* in concentrations per μ^3 of algal cell. They assumed that the light absorbed by chlorophyll in the unit-cell volume was utilized with an efficiency, Y, constant for all species, sufficient to produce a steady state of critical substrate for cell growth which could be treated as if from an external source, as in the more common usage of equation (1). Values of Y varied with light intensity and other constants were temperature-dependent, but Eppley & Sloan (1966) developed a final equation which predicted growth rate quite well for over a dozen species from a knowledge only of the content of cell chlorophyll *a*, cell volume, and temperature. The equation even seemed to take into account the drop in chlorophyll content brought about in chlorotic cells as a result of nitrogen starvation. However, this treatment was clearly of no use when plankton were exposed to light–dark cycles, as is generally the the case in nature. Much more work needs to be done on this subject.

Planktonic plants and dissolved nutrients

A particularly fruitful application of equation (1) has been found with the uptake of nitrate, ammonia, and vitamins by marine phytoplankton. Hyperbolas were found when uptake rates were plotted against concentration of ammonia or nitrate. Eppley & Coatsworth (1968), MacIsaac & Dugdale (1969) and Dugdale (1967) have discussed the approach very fully especially with respect to *in situ* applications to ecology.

Caperon's treatment predicts that the value for μ_m for uptake will not be the same as μ_m for growth (($k_1 c_0$) and ($k_3 c_0$)/q respectively) but that K_s is the same for both growth and uptake k_3/k_1. Thus K_s can be found by making uptake measurements, which are much easier to make than growth measurements, and can then be used with a direct evaluation of

μ_m to predict rates at all substrate concentrations. W. H. Thomas (unpublished observations) has had fair success in predicting *in situ* photosynthetic values at a given light intensity from a knowledge of only μ_m and K_s and the *in situ* ammonia concentrations. Values for μ_m and K_s will be found in Eppley & Thomas (1969) and Eppley, Rogers & McCarthy (1969). In general, open-ocean species have smaller K_s values for both nitrate and ammonia than coastal species which are often exposed to high concentrations of nitrate. Eppley *et al.* (1969) calculated curves for growth *v.* nutrient concentration at a high and low light intensity for four species, and were able to show that the growth rates varied in a manner that made sense with the observed facts in the field. Thomas & Dodson (1968) give some kinetic data for phosphate confirming the validity of equation (1) for the eastern tropical Pacific phytoplankton. Phosphate, however, rarely if ever is a limiting nutrient in sea water.

Water-soluble vitamins as growth factors for phytoplankton

Of all the marine phytoplankton so far grown in axenic cultures, more than half need one, or combinations of more than one, of the growth factors vitamin B_{12}, biotin and thiamin. The amounts needed are extremely small, in the order of 10^{-9} g./l. The two main questions being asked are the extent of plant growth which is possible from a given amount of vitamin in the sea, and the exact effect of a given concentration of vitamin on the kinetics of growth of the phytoplankton.

Carlucci & Silbernagel (1969) found that 3×10^5 cells of the oceanic diatom *Cyclotella nana* were formed from one pg. (10^{-12} g.) of vitamin B_{12}. Droop (1957) found the yield of the small flagellate *Monochrysis lutheri* to be about 8×10^5 cells/pg. of vitamin B_{12}.

Carlucci & Silbernagel (1969) found that corresponding values for biotin and the dinoflagellate *Amphidinium carteri*, and thiamin and the rock-pool flagellate *Monochrysis lutheri*, were both about 0.25×10^5 cells/pg.

From these data one can calculate the maximum yield of plant material to be expected from typical concentrations of free vitamin found in the ocean. In general, the concentrations of vitamins found in coastal waters are more than enough to yield even the highest concentration of plants that have been measured. In the open ocean, conditions may be limited, and Menzel & Spaeth (1962) have shown that the seasonal injection of vitamin B_{12} into the surface waters of the Sargasso Sea is accompanied by a slight blooming of diatoms that are known to require the vitamin.

The problem of the kinetic influence of vitamin concentration is much more complex. It is clear from the work of Carlucci & Silbernagel

(1969), who measured growth rate by the rate of uptake of [^{14}C]carbonate, that the growth rate of algae is directly related to the concentration of vitamin per unit of cell substance. Most of the vitamin in a solution enters the plant cell very rapidly, leaving the solution depleted. The kinetics from equation (1), however, using initial added concentrations of the vitamin, give reasonable values for μ_m and K_s. Values for K_s are 2·9 (vitamin B_{12} and C. nana), 4·0 (biotin and A. carterae), and 125 ng./l. (thiamin and M. lutheri). These K_s values are higher than those reported by workers using algal cell numbers to track growth.

The subject has been dealt with in great detail by Droop (1966, 1968) using the rock-pool flagellate Monochrysis lutheri and vitamin B_{12}. It is clear that the K_s values obtained from chemostat experiments with population densities much higher than those found in nature give much too high an estimate (2–3 ng./l.) compared with values found from low cell density batch-culture work (K_s values of 0·1 ng./l. or less). The yield coefficient (number of cells produced from a given amount of vitamin) was also much lower in a chemostat ($0·2 \times 10^6$ cells/pg.) at high algal cell density than when determined by a batch-culture at low cell densities ($0·8 \times 10^6$ cells/pg.). The problem seemed to be associated with complexing or deactivation of the vitamin in the water by substances produced by the plant. Droop (1968) showed directly, using ^{57}Co-labelled vitamin B_{12}, that there were not large concentrations of vitamin in solution deactivated by protein. He confirmed the fact that nearly all of the vitamin was rapidly taken into the cell and that kinetics of uptake depended on the amount of vitamin held in cells and the number of activated adsorption sites on the cell surface. The overall kinetics are composed of uptake and subsequent growth. The uptake was rapid and followed an isotherm equation with a K_s value of 2·5 ng./l. Inhibitors seemed to be excreted by the algae which inhibit adsorption of vitamin B_{12} by young cells by direct action on the cell surface rather than on the vitamin solution. Inhibitors are excreted from more than one species in culture, and we may be dealing with one substance acting similarly on many species. Effective K_s values are increased tenfold according to the preconditioning of the cells, and the application of exact mathematical treatment to all cells in all environments is becoming complex. Droop (1968) gives an excellent illustration.

Bacteria

As might be expected, work done to quantify the feeding characteristics of bacteria has been centred on equation (1). There is a massive literature affecting high bacterial concentration flow systems especially in anaerobic environments. Very little had been done with chemostats and the limiting concentrations of substrate and bacteria applicable to the deep waters of fresh and salt water areas of the World. Chemostat literature on this is limited almost exclusively to the research of Dr H. W. Jannasch working at Woods Hole Institute of Oceanography. Most other investigations with sea- and freshwater bacteria in the presence of low concentrations of substrate are based on a technique first suggested by Dr T. R. Parsons and myself (Parsons & Strickland, 1961). This involves the addition of known ^{14}C-labelled organic substrates in natural waters. I would like to discuss these two approaches more fully.

CHEMOSTAT WORK WITH LAKE AND SEA BACTERIA

The underlying interest in the chemostat as a tool in ecology is undoubtedly due to the possibility that natural populations might behave as if they were in gigantic chemostats. This is probably so during periods of eutrophication when there is an introduction of nutrients into the euphotic zone from upwelled sources or from sewage outfalls. Deep undisturbed water masses may well behave more as if they were low-density terminal batch-cultures with organisms in a senescent state. Dr Jannasch's work is summarized nicely in Jannasch (1964, 1967, 1969). All six strains of marine bacteria which he studied showed normal chemostat kinetics with three substrates (lactate, glycerol and glucose) when population densities were high, but when growth rates and population densities were low, growth efficiency was decreased. Values for K_s appeared to increase at low population densities compared with high population densities. Substrate concentrations increased at lowest growth rates. Thus natural sea water appears to be a sub-optimal medium unless preconditioned. This preconditioning can be partly brought about by adding reductants (ascorbic acid), by utilizing substances produced by the bacteria themselves, or by having a fairly high bacterial concentration in the water (much higher than any densities found in lakes and oceans). Thus, in practice, the existing bacteria in the sea are probably not scavenging substrate at as high an efficiency as is theoretically possible. Existing substrate concentrations are much higher than they would be if the bacteria could metabolize as predicted by their true K_s value.

THE USE OF ^{14}C-LABELLED SUBSTRATES
ADDED TO NATURAL WATERS

The Parsons–Strickland technique has been tested, expanded, and used extensively in lakes by Drs Hobbie and Wright, and in the oceans by Dr Hamilton of my laboratories and Drs Vaccaro and Jannasch at the Woods Hole Oceanographic Institute. A procedure has been used by European workers in lakes, but as yet I have seen no reports of its use in the ocean.

The procedure is to add a known number (D) of disintegrations per minute (typically, 5 or more microcuries) of ^{14}C-labelled substrate to about 100–200 ml. of water in a bottle together with variable amounts of unlabelled substrate. After a known time of incubation, typically 5 h., preferably at *in situ* temperatures, the water is filtered on a membrane filter (0·45 μm. pore size) and the activity of the natural flora on the filter counted by a Geiger or scintillation counter. This finds the number of disintegrations, d, taken up. If s is the substrate concentration of the natural substrate (e.g. glucose) and a is the amount of substrate added, then, if equation (1) applies,

$$\frac{d(s+a)}{Dt} = \mu_m \frac{s+a}{K_s+s+a},$$ (7)

where t is the time of incubation in hours. Any isotopic fractionation during uptake is ignored.

This form of the equation is difficult to use and it can be transferred into a linear form in at least three ways. The Linweaver–Burke form of the above equation reduces to:

$$\frac{Dt}{d} = \frac{K_s+s}{\mu_m} + \frac{a}{\mu_m}.$$ (8)

Plotting Dt/d against a gives $(K_s+s)/\mu_m$ when $a = 0$ and (K_s+s) at the point where Dt/d is 0. Hence values for (K_s+s) and μ_m can be found. A more rigorous way of transforming equation (1), the EADIE PLOT, was suggested by Dowd & Riggs (1964) and was used by Dr Hamilton when he was a member of my group.

Values for K_s must be assumed to be small compared with s if the value of s is to be determined. This is a poor approximation, however, as the value of K_s+s may vary widely; this value certainly sets a practical upper limit for s in any water body. However, it is better to use an alternative approach which enable both K_s and s to be determined separately. One approach is to use a bacteria with a known K_s value obtained by making kinetic plots as described above in organic-free (Norite-treated) water, where the value of s is zero. Once a bacterial culture of this type is found, the natural bacteria of flora are removed from the water samples by filtration and the filtrate assayed for substrate by a fixed supplement of the assaying bacterium with a known K_s value. If the natural flora present in the water samples give a good Linweaver–Burke plot, then an elegant way at getting at values for both K_s and s is to dilute the sample with an equal volume and double the volume of Norite-treated water. When Dt/d is 0, a then equals either $(K_s + s)$ or $\frac{1}{2}(K_s + s)$ or $\frac{1}{4}(K_s + s)$ when the respective plots are made. This enables values for both K_s and s to be determined separately (Allen, 1968; Wright & Hobbie, 1966). The total uptake of substrate is clearly not equal to d/Dt multiplied by $(s + a)$ as much substrate may be metabolized by the cell and rejected before counting is made.

Hamilton & Austin (1967) tested glucose labelled in different atoms with ^{14}C and found that the amount of substrate retained by bacteria was only a fraction of the amount metabolized. With uniformly labelled glucose, only about 30 % was retained by the cell and hence μ_m values as recorded in nature are at least only a third of the true rate of metabolism of the substrate. Observed uptake rates should be multiplied by a factor of at least three, or turnover times decreased to a third of their measured values. This has been confirmed by Hobbie & Crawford (1969) working in lakes, who suggested a correction factor of at least two.

Hamilton et al. (1966) found K_s and μ_m values for glucose and six marine isolates obtained from chemostats of low concentration. Values for μ_m varied from 12 to 650 multiplied by 10^{-12} $\mu g./cell/h.$, and K_s from 6 to 315 $\mu g.$ carbon/l. with no apparent interrelationship. Cells with low K_s values could have high or low μ_m values and vice versa. Values for both μ_m and K_s were found to be sensitive to temperature changes, with apparent activation energies of about 30,000 and 20,000 cal. respectively.

The fact that a mixture of species in a natural population should give a good Linweaver–Burke plot at all is interesting. This can only occur if, in fact, the observed kinetics are coming from one predominant species or if several predominant species have substantially the same uptake characteristics. Until more species are present in about equal concentrations, and have μ_m and K_s values very different from each other, bad plots must result. Such samples have been found off the coast of California very often, and Vaccaro & Jannasch (1966) found that, whereas nearly all of their samples from the Atlantic gave good plots, only about one out of 14 stations off Peru in upwelling waters could be treated kinetically. Filtering off the natural flora and adding a bacterium with known μ_m and K_s values gave excellent plots from which a value for s could be found. If one added, say, glucose, and incubated the population for a short time, then one predominant flora took over and one obtained quite good plots (see also Vaccaro, 1969).

Vaccaro & Jannasch (1966) gave the first really thorough testing of the Parsons–Strickland technique to the sea using a glucose-metaboliz-ing bacterium *Achromobacter aquarmarinus* in the form of washed-out cells from batch-cultures. The K_s value for this bacterium was assumed to be about 3 μg. carbon/l. Glucose concentrations in the Atlantic varied between 0 and about 25 μg. carbon/l. The K_s values for the populations in the Atlantic at 6° varied from about 3 to 106 μg. carbon/l. so that the measured values ($K_s + s$) could have been bad overestimates of s. *In situ* velocities of assimilating substrate carbon were about 0·001–0·18 μg. carbon/l./h. in the top 100 m.

Vaccaro *et al.* (1968) were able to measure values of s with a bacterium of known K_s value, and to compare these values with those obtained by an independent direct enzymic method (Hicks & Carey, 1968). Agree-ment was excellent, and natural concentrations of glucose in surface waters of the Atlantic were up to 60 μg. carbon/l. The glucose concen-tration decreased rapidly with depth.

Hamilton & Preslan (1970) made an extensive survey of μ_m and ($K_s + s$) values down a line near to 105° W. from 17° N. to 15° S., taking samples at the top, middle, and bottom of the euphotic zone. The heterotrophic activity was largely a function of the general productivity of the area. About 50 % of the samples taken gave workable kinetics. Aspartic acid, glutamic acid, arginine, glycine and glucose produced useful data from eight or more stations of the 12 tested. Values for μ_m were about 2–35×10^{-3} μg. carbon/l./h. with the ten substrates tested. Values for ($K_s + s$) were from 4 to 133 μg. carbon/l., with a mean around 50.

Hobbie, Crawford & Webb (1968) studied the rates of uptake of amino acids in the York River Estuary by using labelled substrates, and compared these values with the concentrations of free amino acids obtained by direct analysis. There was no correspondence, and a sub-strate present in a given concentration could have either high or low μ value for its *in situ* value of μ. Glycine, methionine and serine had the highest uptake rates.

Very similar studies to the above have been carried out in lakes by Drs Hobbie and Wright and their collaborators (Wright & Hobbie, 1964, 1965, 1966). Quantities of glucose and acetate were generally related to the eutrophication of the lake and varied between less than 5 to over 30 μg. carbon/l.

My own attempts to measure the rates of uptake of glucose and other substrates in deep samples have convinced me that the method cannot be used below about 1,000 m. in sea water where uptake rates are 0·001 μg. carbon/l./h. or less. Only when substrates with very high specific activities can be found will we be able to extend the use of this technique to depths in the ocean below 2,000 m.

CONCLUDING REMARKS

I have tried to leave you the general impression of the oceans, or a large lake, as full of metabolizing microbial material in the first few tens of metres, where light is adequate. Production is by photosynthesis by single-cell microplants which are largely kept to a relatively low population density by herbivore predators. In most areas regeneration of nutrients due to waste processes, as matter ascends the chain from one trophic level to another, is sufficient to supply the plants with new nourishment. In some locations the upwelling of fresh cold water with high nutrient concentration, or the presence of man-made pollution, produces large blooms of plankton with the plants growing much faster than they can be grazed. The kinetics of all of these processes can be expressed remarkably well with one simple equation, which is making the prospect of computerized model-making of useful predictive ability a little nearer to reality.

Below the first 100 m. or more of water to about 500–1,000 m. we have a regime with considerable activity, largely by microbial-sized creatures and some larger animals which migrate to and from the surface region. The bulk of the oxygen which enters the sea from the atmosphere is consumed in this layer.

Below about 1,000 m., activity is very slight and in this cold water the average division time of the living material (0·05 μm. carbon/l. or less) is very long, being many tens of days. The consumption of oxygen in the deep oceans (a few microlitres a year) is trivial compared with the amount of oxygen present (several millilitres per litre) but should be detectable after many hundred years. In the Pacific where the speed and direction of deep sea currents are thought to be known, the oxygen-consumption measurement by electron-transport and actual observations of down-stream oxygen loss are in surprisingly good agreement.

Despite this impression of an immense universe of very cold water with only a miniscule concentration of substrate and microorganisms,

Professor Isaacs and his colleagues here at the Scripps Institution continue to record large sharks and hag fish at bottom depths as low as 5,000 m. The food chain leading to this apparently large population is an enigma. It is hard to believe it can originate from suspended microorganisms and must either be from the sea bed or immense vertical migrations, scavenging larger objects of decay descending relatively quickly from above. As the production of whales and large fish is only a small fraction of the production of phytoplankton a year, few of the large dead animals can get back down into deep waters.

We need many more years of good solid observations with better techniques before we will have all the answers to these interesting problems of the aquatic food chain.

REFERENCES

ALEEM, A. A. (1955). Measurement of plankton populations by triphenyltetrazolium chloride. *Kieler Meeresforsch.* **11**, 160.

ALLEN, H. L. (1968). Acetate in fresh water: natural concentrations determined by dilution bioassay. *Ecology* **49** (2), 346.

ARON, S. H. B. & STOMMEL, H. (1967). On the abyssal circulation of the world ocean. III. An advection–lateral mixing model of a tracer property in an ocean basin. *Deep-Sea Res.* **14**, 441.

BEERS, J. R., STEWART, G. L. & STRICKLAND, J. D. H. (1967). A pumping system for sampling small plankton. *J. Fish. Res. Bd Can.* **24**, 1811.

BERNARD, F. (1963). Vitesse de chute en mer désarmas palmelloides de *Cyclococcolithus*. Ses conséquences pour le cycle vital des mers chaudes. *Pelagos* **1**, 5.

BURNS, C. W. & RIGLER, F. H. (1967). Comparison of filtered rates of *Daphnia rosea* in lake water and in suspensions of yeast. *Limnol. Oceanogr.* **12**, 492.

CAPERON, J. (1967). Population growth in micro-organisms limited by food supply. *Ecology* **48**, 715.

CARLUCCI, A. F. & SILBERNAGEL, S. B. (1969). Effect of vitamin concentrations on growth and development of vitamin-requiring algae. *J. Phycol.* **5**, 64.

CONOVER, R. J. (1966). Factors affecting the feeding assimilation of organic matter by zooplankton and the question of superfluous feeding. *Limnol. Oceanogr.* **11**, 345.

CURL, H., JR. & SANDBERG, J. (1961). The measurement of dehydrogenase activity in marine organisms. *J. Mar. Res.* **19**, 123.

CUSHING, D. H. (1968). Grazing by herbivorous copepods in the sea. *J. Conseil Int. Explor. Mer.* **32**, 70.

DODSON, A. N. & THOMAS, W. H. (1964). Concentrating plankton in a gentle fashion. *Limnol. Oceanogr.* **9**, 455.

DOWD, J. E. & RIGGS, S. (1964). A comparison of estimates of Michaelis–Menten kinetic constants from various linear transformations. *J. biol. Chem.* **246**, 863.

DROOP, M. R. (1957). Vitamin B_{12} in marine ecology. *Nature, Lond.* **180**, 1041.

DROOP, M. R. (1966). Vitamin B_{12} and marine ecology. III. An experiment with a chemostat. *J. mar. biol. Ass. U.K.* **46**, 659.

DROOP, M. R. (1968). Vitamin B_{12} and marine ecology. IV. The kinetics of uptake, growth and inhibition in *Monochrysis lutheri*. *J. mar. biol. Ass. U.K.* **48**, 689.

DUGDALE, R. C. (1967). Nurtient limitation in the sea: dynamics, identification, and significance. *Limnol. Oceanogr.* **12**, 685.

EPPLEY, R. W. & COATSWORTH, J. L. (1968). Uptake of nitrate and nitrite by *Ditylum brightwellii* – kinetics and mechanisms. *J. Phycol.* **4**, 151.

EPPLEY, R. W. & SLOAN, P. R. (1966). Growth rates of marine phytoplankton: correlation with light absorption by cell chlorophyll *a*. *Physiologia Pl.* **19**, 47.

EPPLEY, R. W. & THOMAS, W. H. (1969). Comparison of half-saturation 'constants' for growth and nitrate uptake of marine phytoplankton. *J. Phycol.* **5**, 375.

EPPLEY, R. W., ROGERS, J. N. & MCCARTHY, J. J. (1969). Half-saturation constants for uptake of nitrate and ammonium by marine phytoplankton. *Limnol. Oceanogr.* **14**, 912.

FOURNIER, O. (1966). North Atlantic deep-sea fertility. *Science, N.Y.* **153**, 1250.

HAMILTON, R. D. & AUSTIN, K. E. (1967). Assay of relative heterotrophic potential in the sea: the use of specifically-labelled glucose. *Can. J. Microbiol.* **13**, 1165.

HAMILTON, R. D. & HOLM-HANSEN, O. (1967). Adenosine triphosphate content of marine bacteria. *Limnol. Oceanogr.* **12**, 319.

HAMILTON, R. D. & PRESLAN, J. E. (1970). Observations on heterotrophic activity in the Eastern Tropical Pacific. *Limnol. Oceanogr.* (In the Press.)

HAMILTON, R. D., MORGAN, K. M. & STRICKLAND, J. D. H. (1966). The glucose uptake kinetics of some marine bacteria. *Can. J. Microbiol.* **12**, 995.

HARRIS, J. G. K. (1968). A mathematical model describing the possible behaviour of a copepod feeding continuously in a relative dense randomly distributed population of algal cells. *J. Conseil. Int. Explor Mer.* **32**, 83.

HAQ, S. M. (1967). Nutritional physiology of *Metridia lucens* and *M. longa* from the Gulf of Maine. *Limnol. Oceanogr.* **12**, 40.

HENRICI, A. T. (1936). Studies on freshwater bacteria. III. Quantitative aspects of the direct microscopic method. *J. Bact.* **32**, 265.

HERBERT, D., ELSWORTH, R. & TELLING, R. C. (1956). The continuous culture of bacteria, a theoretical and experimental study. *J. gen. Microbiol.* **14**, 601.

HICKS, S. E. & CARY, F. G. (1968). Glucose determination in natural waters. *Limnol. Oceanogr.* **13**, 361.

HINSHELWOOD, C. H. (1946). *The Chemical Kinetics of the Bacterial Cell*. Oxford: Clarendon Press.

HOBBIE, J. E. & CRAWFORD, C. C. (1969). Respiration corrections for bacterial uptake of dissolved organic compounds in natural waters. *Limnol. Oceanogr.* **14**, 528.

HOBBIE, J. E., CRAWFORD, C. C. & WEBB, K. L. (1968). Amino acid flux in an estuary. *Science, N.Y.* **159**, 1463.

HOLM-HANSEN, O. (1969). Determination of microbial biomass in ocean profiles. *Limnol. Oceanogr.* **14**, 740.

HOLM-HANSEN, O. & BOOTH, C. R. (1966). The measurement of adenosine triphosphate in the ocean and its ecological significance. *Limnol. Oceanogr.* **11**, 510.

HOLM-HANSEN, O., POMEROY, L. R. & PACKARD, T. T. (1970). Recovery of ATP and ETS from membrane filters and plankton concentrates. *Limnol. Oceanogr.* (In the Press.)

HOLM-HANSEN, O., LORENZEN, C. J., HOLMES, R. W. & STRICKLAND, J. D. H. (1965). Fluorometric determination of chlorophyll. *J. Cons. perm. int. Explor. Mer* **30**, 3.

IVANOFF, M. V. (1955). Method for the determination of bacterial biomass formation in reservoirs. *Mikrobiologia* **24**, 79.

JANNASCH, H. W. (1958). Studies on planktonic bacteria by means of a direct membrane filter method. *J. gen. Microbiol.* **18**, 609.

JANNASCH, H. W. (1964). Microbial decomposition in natural waters as determined in steady state systems. *Verh. int. Ver. Limnol.* **25**, 562.

JANNASCH, H. W. (1967). Growth of marine bacteria at limiting concentrations of organic carbon in seawater. *Limnol. Oceanogr.* **12**, 264.

JANNASCH, H. W. (1969). Edgardo Baldi Memorial Lut. Current concepts in aquatic microbiology. *Verh. int. Verein. theor. angew. Limnol.* **17**, 25.

JANNASCH, H. W. & JONES, G. E. (1959). Bacterial populations in sea water as determined by different methods of enumeration. *Limnol. Oceanogr.* **4**, 128.

KIMBALL, J. F., JR., CORCORAN, E. F. & WOOD, E. J. F. (1963). Chlorophyll-containing microorganisms in the euphotic zone of the oceans. *Bull. mar. Sci. Gulf Caribb.* **13**, 574.

KRISS, A. E. (1963). *Marine Microbiology (Deep-sea)*, trans. J. M. Shewan and Z. Kabata. London: Oliver and Boyd.

KUZNEZOW, S. L. (1959). Die Rölle der Mikrooganismen in Stoffkreislauf der Seen. VEB. *Dentscher Verlag der Wissenshaften.* Berlin.

LEVIN, G. V., CLENDENNING, J. R., CHAPPELLE, E. W., HEIM, A. H. & ROCEK, E. (1964). A rapid method for the detection of micro-organisms by ATP assay; its possible application in vitrus and cancer studies. *Bioscience* **14**, 37.

MCALLISTER, C. D., SHAH, N. & STRICKLAND, J. D. H. (1964). Marine phytoplankton photosynthesis as a function of light intensity: a comparison of methods. *J. Fish. Res. Bd Can.* **21**, 159.

MACISAAC, J. J. & DUGDALE, R. C. (1969). The kinetics of nitrate and ammonium uptake by natural populations of marine phytoplankton. *Deep-Sea Res.* **16**, 415.

MCMAHON, J. W. & RIGLER, F. H. (1965). Feeding rate of *Daphnia magna* Strauss in different foods labelled with radioactive phosphorus. *Limnol. Oceanogr.* **10**, 105.

MENZEL, D. W. & SPAETH, J. P. (1962). Occurrence of vitamin B_{12} in the Sargasso Sea. *Limnol. Oceanogr.* **7**, 151.

MULLIN, M. M. (1963). Some factors affecting the feeding of marine copepods of the genus *Calanus*. *Limnol. Oceanogr.* **8**, 239.

MUNK, W. H. (1966). Abyssal recipes. *Deep-Sea Res.* **13**, 707.

PACKARD, T. T. (1969). *The estimation of oxygen utilization rate in sea water from the activity of the respiratory electron transport system in plankton.* Ph.D. thesis; University of Washington.

PACKARD, T. T. & TAYLOR, P. B. (1968). The relationship between succenic dehydrogenase activity and oxygen consumption in the brine shrimp *Artemia salina*. *Limnol. Oceanogr.* **13** (3), 552.

PARSONS, T. R., & STRICKLAND, J. D. H. (1961). On the production of particulate organic oceanic carbon by heterotrophic processes in sea water. *Deep-Sea Res.* **8**, 211.

PARSONS, T. R., LE BRASSEUR, R. J. & FULTON, J. D. (1967). Some observations on the dependence of zooplankton grazing on the cell size and concentration of phytoplankton blooms. *J. Oceanogr. Soc. Japan* **23**, 10.

POMEROY, L. R. & JOHANNES, R. E. (1968). Occurrence and respiration of ultraplankton in the upper 500 meters of the ocean. *Deep-Sea Res.*, **15**, 381.

POMEROY, L. R. & WIEBE, W. J. (1970). Respiration and biomass of micro-organisms in a transect of the Southern Ocean. *Antarctic J.* (In the Press.)

POMEROY, L. R., WIEBE, W. J., FRANKENBERG, D., HENDRICKS, C. & LAYTON W. L., JR. (1969). Metabolism of total water columns. *Antarctic J.* **4**, 188.

RIGLER, F. H. (1961). The relation between concentration of food and feeding rate of *Daphnia magna* Strauss. *Can. J. Zool.* **39**, 857.

RILEY, G. A. (1951). Oxygen, phosphate and nitrate in the Atlanic Ocean. *Bull. Bingham oceanogr. Coll.* **13**, 1.

RILEY, G. A., WANGERSKY, P. J. & VAN HERMET, D. (1964). Organic aggregates in tropical and subtropical waters of the North Atlantic Ocean. *Limnol. Oceanogr.* **9**, 546.

ROMANENKO, W. I. (1964 a). The dependence between the amounts of O_2 and CO_2 consumed by bacteria. *Dokl. Akad. Sci. USSR* **157**, 178.

ROMANENKO, W. I. (1964 b). Heterotrophic assimilation of CO_2 by the aquatic microflora. *Microbiologia* **33**, 679.

RYTHER, J. H. (1954). Inhibitory effects of phytoplankton upon the feeding of *Daphnia magna* with reference to growth, reproduction, and survival. *Ecology* **35**, 522.

SEKI, H. & ZOBELL, C. E. (1967). Microbial assimilation of carbon dioxide in the Japanese Trench. *J. Oceanogr. Soc. Japan* **23**, 182.

SOROKIN, YU. I. (1964). On the primary production and bacterial activities in the Black Sea. *J. du Conseil* **24**, 41.

STRICKLAND, J. D. H. (1965). Production of organic matter in the primary stages of the marine food chain. In *Chemical Oceanography*, vol. 1. Ed. J. P. Riley and G. Skirrow. New York: Academic Press.

TAMYIA, H., HASS, E., SHIBATA, K., MITUYA, A., NIHEI, T. & SASA, T. (1953). Kinetics of growth of *Chlorella* with special reference to its dependency on quantity of available light and on temperature. In *Algal Culture from Laboratory to Pilot Plant*, p. 202. Ed. J. S. Burlew. *Publs Carnegie Instn*, no. 600.

THOMAS, W. H. & DODSON, A. N. (1968). Effects of phosphate concentration on cell division rates and yields of a tropical diatom. *Biol. Bull. mar. biol. Lab.*, *Woods Hole* **134** (1), 199.

VACCARO, R. F. (1969). The response of natural microbial populations in sea water to organic enrichment. *Limnol. Oceanogr.* **14**, 726.

VACCARO, R. F. & JANNASCH, H. W. (1966). Studies on heterotrophic activity in sea water based on glucose assimilation. *Limnol. Oceanogr.* **11**, 596.

VACCARO, R. F., HICKS, S. E., JANNASCH, H. W. & CAREY, F. G. (1968). The occurrence and the role of glucose in seawater. *Limnol. Oceanogr.* **13**, 356.

WRIGHT, R. T. & HOBBIE, J. E. (1964). The uptake of organic solutes in lake water. *Limnol. Oceanogr.* **10**, 22.

WRIGHT, R. T. & HOBBIE, J. E. (1965). Bioassay with bacterial uptake kinetics: glucose in fresh water. *Limnol. Oceanogr.* **10**, 471.

WRIGHT, R. T. & HOBBIE, J. E. (1966). Use of glucose and acetate by bacteria and algae in aquatic ecosystems. *Ecology* **47**, 447.

YENTSCH, C. S. & MENZEL, D. W. (1963). A method for the determination of phytoplankton, plankton, chlorophyll and phaeophytin by fluorescence. *Deep-Sea Res.* **10**, 221.

ZOBELL, C. E. (1946). *Marine Microbiology*. Chronica Botanica. U.S.A.: Waltham.

ZOBELL, C. E. & ALLEN, E. C. (1933). Attachment of marine bacteria to submerged slides. *Proc. Soc. exp. Biol. N.Y.* **30**, 1409.

As this volume was going to press, we heard of the untimely death of John Strickland. We hope that this, his last publication, will be a fitting memorial to the outstanding contributions which John Strickland made in marine microbiology – The Editors.

MICROBIAL PRODUCTIVITY
IN SOIL

T. R. G. GRAY AND S. T. WILLIAMS
Hartley Botanical Laboratories, University of Liverpool

INTRODUCTION

Early studies of microorganisms in soil were concerned with collecting data on the types of organisms present and the nature of the chemical transformations they brought about. Subsequently, there was a growing realization that many of the organisms isolated from soil by conventional methods were present in a dormant condition and considerable attention was paid to developing methods for the study of the so-called active organisms. More and more attention was paid to this active component of the microflora, especially by soil mycologists, with the results that the nature and extent of the dormant organisms were often forgotten. Methods were developed which attempted to determine which organisms were present in a vegetative condition, to distinguish living from dead cells and to measure the metabolic activity of populations, usually by respirometry.

Concepts were worked out which stressed the role of competition and microbial metabolites (e.g. antibiotics) in the development of populations, while organisms were classified according to the substrates they utilized (e.g. sugar fungi, lignin fungi, autochthonous and zymogenous bacteria). At the same time, the idea that soil was a heterogeneous environment in which a diversity of microhabitats existed was advanced. It was argued that activity was concentrated in some of these microhabitats (e.g. root surfaces), and that elsewhere the activity of organisms was diminished. In this paper we shall present evidence which suggests that soil organisms grow relatively slowly and that the maintenance of rapidly growing populations for long periods probably cannot take place. We shall re-emphasize the importance of the dormant nature of much of the soil microflora and relate this to the energy supply available for growth, the occurrence of unfavourable environmental conditions and the presence of recalcitrant compounds in the soil.

EVIDENCE FOR THE INACTIVITY
OF SOIL MICROORGANISMS

Presence of resting structures in the soil

The form of many microorganisms in soil often indicates their state of activity. Many soil organisms form resting structures which much evidence suggests are common in natural soil and might often out-number active cells.

The best-known resting structures produced by bacteria are the endospores formed by *Bacillus* and *Clostridium* species and by species in a few other genera. *Bacillus* species occur in most soils and are particularly prominent in acidic peaty soils (Holding, Franklin & Watling, 1965; Goodfellow, Hill & Gray, 1968) and steppe and desert soils (Mishustin, 1956). The relatively inhospitable nature of these habitats for bacteria suggests that dormant endospores might give rise to many of the colonies formed on isolation plates. The occurrence of *Bacillus* endospores, as indicated by pasteurization of soil before plating, has been studied by Mishustin (1956). He found that endospores can account for up to 24·5 % of the total bacterial count in steppe soils. Hill & Gray (1967) demonstrated the presence of endospores in an acid forest soil by staining soil with a fluorescent antiserum specific for *Bacillus subtilis* spores. Soil is also the principal habitat of many clostridia (Skinner, 1968), and since these are obligate anaerobes some of the isolates from well-aerated soils almost certainly originate from endospores. Other bacterial resting structures are the cysts of *Azotobacter*, which can survive better in soil than young vegetative cells (Brown, Jackson & Burlingham, 1968).

Most soil bacteria do not produce resting cells obviously different in gross structure from the vegetative cells, and many probably exist in soil for much of the time as vegetative cells in a reduced state of metabolic activity. The ability of bacterial cells to survive starvation is not necessarily correlated with the ability to grow rapidly (Harrison & Lawrence, 1963), as instanced by stationary phase cells of *Aerobacter aerogenes* which remained viable for a relatively long period compared with logarithmic phase cells. Starvation resistant mutants had arisen which were smaller in size, lower in their RNA:DNA ratio, greater in light-scattering ability and more heat-resistant than the logarithmic phase cells. This has not been demonstrated in soil organisms but it could be relevant to the problem of bacterial survival in soil. Most actinomycetes produce spores, and there is evidence to show that many soil isolates arise from spores. Skinner (1951) noted the differential susceptibilities of spores and hyphae to killing when they were shaken with sand particles. By shaking soil suspensions he was able to show that most streptomycetes in soil existed as spores. This was also shown by Ruddick (1969), who applied maceration, and Watson (1970), who made use of the different susceptibilities of streptomycete spores and hyphae to heating and drying. Cross (1968) also found the endospores of the thermophilic *Thermoactinomyces vulgaris* in a wide range of soils.

Soil fungi can form a variety of resting structures which are easily distinguished from vegetative hyphae. The predominant fungal propagules in most soils are spores of such common organisms as *Aspergillus*, *Mucor*, *Penicillium* and *Trichoderma*, which are frequently found on dilution plates (Warcup, 1955, 1957). Thus, as with the actinomycetes, most viable counts of fungi are in fact counts of spores. More selective isolation techniques can demonstrate the presence of spores of other types of fungi in soil, e.g. Warcup & Baker (1963).

In some fungi, areas of cytoplasm within hyhpae become condensed and surrounded by a thick wall to form chlamydospores. These occur in such common soil forms as *Fusarium*, *Mucor* and *Trichoderma* and their presence in soil has been demonstrated for *Mucor ramannianus* (Hepple, 1958) and *Fusarium* spp. (Christou & Snyder, 1962). Detection of these spores is not always easy as they may be associated with hyphae which penetrate organic particles.

Another resting structure formed by some fungi in soil is the sclerotium. This consists of a tightly interwoven mass of live hyphae surrounded by a layer of thickened, dead cells. The best-known examples of soil fungi which form sclerotia are *Sclerotium* and *Rhizoctonia* species. Sclerotia isolated from soil by Warcup & Talbot (1962) and placed on laboratory media gave rise to a variety of fungi, including *Aspergillus*, *Cephalosporium* and *Penicillium* species, which do not often produce sclerotia in culture. This suggests that the ability to form sclerotia may be more widespread and significant in soil than is generally realized.

The various structures discussed so far are generally referred to as being inactive. However, resting spores and other bodies do have low levels of metabolic activities, which in fungal spores are less than those of hyphae (Sussman & Halvorson, 1966). Hyphae are generally considered to be active but their presence in soil cannot always be regarded as proof of activity. They may be detected in natural soil by direct observation (Jones & Mollison, 1948; Burges & Nicholas, 1961) and may be picked out of soil by micromanipulation and grown on laboratory media (Warcup, 1955, 1957). Many of these hyphae appear to be devoid of cytoplasm and have pigmented walls which may be very resistant to lysis (Bloomfield & Alexander, 1967), so one can imagine that large quantities of dead hyphae could build up in soil. Warcup (1957) found that, on average, only 23 % of hyphae picked from a wheat-field soil were viable and this figure dropped to 3–15 % during dry periods. Therefore methods devised to measure the frequency of lengths of 'active' hyphae in soil may be based partially upon the detection of dead structures.

Even when hyphae are isolated and shown to be viable, it cannot be presumed that they were fully active in the soil. Growth may originate from chlamydospores within them or the hyphae may be in a reduced state of metabolic activity (Hawker, 1957; Warcup, 1967). Little is known about the changes in fine structure of fungal hyphae in different states of metabolic activity but recently Trinci & Righelato (1970), who studied glucose-limited *Penicillium chrysogenum*, found that a proportion of cytologically normal hyphal components were present at all stages and suggested that maintenance or cryptic growth of some hyphae (or components) occurred at the expense of others.

Respiration rates of soil microorganisms

A commonly used measure of microbial activity is its respiration. Such measurements can be correlated with soil organic-matter content, extent of nitrogen and phosphorus transformations, soil pH, production of metabolic intermediates and average microbial numbers (Stotzky, 1965). However, the correlation of microbial numbers and respiration is not always high and the maximum rates of respiration often occur before the development of maximum numbers of viable microorganisms. This may be due to several factors, most important of which is that respiratory rates are related to metabolic activity and biomass rather than numbers. Numbers are not a good indication of the biomass of microbial material in the soil because the objects being counted range from lengths of fungal hyphae and large fungal spores to individual or clumped bacterial cells. Animal populations or plant roots, both of which respire actively, will further obscure relationships.

The effect of plant roots on the respiratory rates recorded for soil is probably considerable. Unfortunately, there are no methods for determining the relative amounts of respiration due to roots in undisturbed soil and the amount due to microorganisms. Wiant (1967) has shown that in an average forest soil the annual leaf fall (1,780 kg. C/ha.) is sufficient to yield 653 g. CO_2/m.^{2}/year. If respiration occurred at an even rate throughout the year, 0·075 g. CO_2/m.^{2}/h. would be evolved.

Rates considerably in excess of this are generally reported and can be explained only by assuming that the methods used are inaccurate, that much of the carbon dioxide comes from plant roots (either directly or by microbial action on root exudates) or that respiration is only maintained at high rates for a part of the year. Thus carbon dioxide evolution at the rate of 0.3 g./m.^{2}/h. could only be maintained for 91 24 h. days in the absence of root respiration. Witkamp (1966a) also found that the complete oxidation of oak litter in a forest could yield only 452 l. CO_2/m.^{2}/year, while 775 l. were actually evolved; inaccuracies in the method of measurement could account for the difference. Excessive respiration rates have been recorded by many other workers, including Rommell (1932), Feher (1933), Wallis & Wilde (1957), Witkamp & van der Drift (1961), Tamm & Krzysch (1965) and Reiners (1968).

We may conclude from such data that any microbial growth rates calculated from respiration measurements will be overestimates, but they nevertheless show that rates are probably extremely low. There have been surprisingly few studies in which both respiration and microbial numbers or biomass have been measured on the same samples. Gray & Wallace (1957) and Witkamp (1966b) found a good correlation between bacterial numbers and carbon dioxide evolution, while Latter, Cragg & Heal (1967) found that respiration levels in peat soils increased together with bacterial numbers and lengths of stained hyphal fragments. Gray & Wallace (1957) measured CO_2 by an aspiration method in cores of agricultural soils. By assuming that there was no fungal or animal respiration (probably not so), it is possible to calculate a hypothetical growth rate for bacteria in these soils from the data in Fig. 1. If a bacterial cell weighs about 1.5×10^{-9} mg., then 1 mg. bacteria would evolve 0.121 mg. CO_2/h., i.e. 0.033 mg. carbon. Assuming that this represents about 65 % of the metabolized carbon, the remaining 35 % will have contributed to an increase in biomass or excretion of metabolites. A typical bacterial cell has 50 % carbon, so the maximum increase in biomass of bacteria is 0.035 mg./h., that is, 1 mg. of bacteria will increase to 2 mg. every $28\frac{1}{2}$ h.; errors involved in this calculation suggest that the actual growth rate is slower. Dilution plate counts of cells have almost certainly underestimated bacterial numbers, while CO_2 output measures are overestimates for the reasons already outlined. The accuracy of the estimate for efficiency of utilization of substrate carbon is unknown, but Stotzky (1965) suggests it will vary from 60 % to 80 % and will be dependent upon the nature of the substrate, microbial populations and their growth rate.

In common with other experiments we have discussed, the rate of

CO_2 output in Gray and Wallace's studies was excessive, i.e. about 0.3 mg./h./100 g. soil. Assuming that most respiration takes place in the upper 10 cm. of the soil, a square metre containing 1.0×10^9 g. soil would evolve 0.3 g. CO_2/h. which would require the addition of 14,000 kg. per hectare plant material to the soil each year (see Tables 1 and 2).

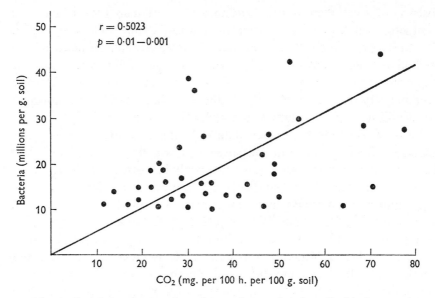

Fig. 1. Correlation between bacterial numbers and carbon dioxide for samples taken in June. (From Gray & Wallace, 1957.)

CAUSES OF MICROBIAL INACTIVITY

Two types of dormancy, constitutive and exogenous, can be recognized in microbial cells (Sussman, 1965). In constitutive dormancy the length of the resting period is controlled by innate properties of the cell such as its permeability. Exogenous dormancy is imposed on the cell by unfavourable chemical or physical factors in its external environment.

Constitutive dormancy

It has been suggested that constitutive dormancy is brought about by the presence of self-inhibitors in cells. Evidence for these substances is obtained from metabolic studies on the decreased germination of spores found in concentrated spore suspensions (Sussman, 1969). The existence of constitutive dormancy has been demonstrated in a range of microbial resting bodies, including a number of soil-inhabiting, plant-pathogenic fungi; for example, oospores of *Phytophthora cactorum* require a 3- to 4-

week period before they will germinate (Blackwell, 1943). Dormancy can be broken by a variety of triggers such as exposure to temperature extremes, alternate wetting and drying, or certain chemicals; ascospore dormancy can be broken also by heat-treatment for short periods (Sussman & Halvorson, 1966) and this may have been an important factor in the isolation of ascomycetes from soil after heating (Warcup & Baker, 1963). Similarly, bacterial endospores may require a heat shock or exposure to chemical or mechanical treatments to induce germination (Sussman, 1969). There is, however, little information on the importance of constitutive dormancy in maintaining populations of dormant propagules in the soil but it is generally accepted that exogenous dormancy is important because of the occurrence of unfavourable physical environmental factors, the presence of inhibitory substances and the low levels of available nutrients.

Exogenous dormancy

Physical environmental factors

The major factors in the soil environment which can limit microbial activity are moisture, aeration, temperature and pH, and it is important to realize these may interact. The availability of water in soil, as indicated by the suction pressures which an organism must exert to obtain it, is an important limiting factor, although the response of different species varies (Griffin, 1963). In a sandy soil containing 10 % water, a suction pressure of at least pF 1·5 has to be exerted to obtain water and a microbe would therefore have little difficulty in growing. However, in a loam the required suction pressure might be pF 4·0 and in a clay pF 5·2 at the same moisture level; consequently growth would be restricted or even prevented. In waterlogged soil growth can be restricted by poor aeration, which has a differential effect on the microbial population, which includes obligate anaerobes like clostridia and obligate aerobic fungi. The balance between anaerobic and aerobic conditions in soil is delicate and it has been estimated that a change from one metabolism to the other takes place when oxygen concentration becomes less than 3×10^{-6} M. On this basis, water-saturated crumbs of more than 3 mm. diameter would have anaerobic centres even if surrounded by air. Carbon dioxide concentrations are also important although some soil microbes are more tolerant of high concentrations than others (Stotzky & Goos, 1965). Temperatures below about 5° or above 30° will severely limit the growth of the predominantly mesophilic soil population. In many soils, temperatures seldom reach levels required by the bulk of the soil microbes for optimum growth but it is possible that local, short-term increases of temperature occur at sites of microbial activity: Clark, Jackson & Gardner (1962) have recorded rises in temperature of 5·3° in the vicinity of organic residues added to soil. If soil pH falls below about 5·5, growth of many bacteria and actinomycetes is prevented, although acidophilic strains of some organisms are known. The isolation of strains requiring higher pH values for growth from acid soils can only be explained by assuming that they grow in localized microsites of high pH in the soil or that the less-acid isolation media have only allowed the growth of atypical mutant forms. Localized sites of high or low pH may occur around soil particles, the pH differing from that of a bulk soil sample by 0·5 units or more. Negatively charged mineral particles may attract hydrogen ions, thus lowering the pH, while organic matter may accumulate ammonia released during decomposition leading to increased pH values. The situation is further complicated by the accumulation of hydrogen ions at the surfaces of the microbes themselves.

Inhibitory substances

Substances capable of inhibiting microbial growth may originate from microbes themselves or from other sources in soil. Some soils apparently contain unidentified inhibitory factors which prevent the germination of some fungal spores resulting in the phenomenon termed 'residual mycostasis' by Dobbs & Gash (1965). Organic residues in soil may contain inhibitory compounds such as phenols, aldehydes and organic acids. Germination of streptomycete spores was prevented by extracts from pine roots which contained a leucoanthocyanin (Mayfield, 1969). Breakdown products of lignin were shown to inhibit germination of fungus spores (Lingappa & Lockwood, 1962).

Metabolic products of microorganisms may inhibit growth of their producers or other organisms. Germination of conidia of *Glomerella cingulata* was inhibited by diffusible compounds produced by the conidia (Lingappa & Lingappa, 1966). An inhibitory labile staling substance was detected in cultures of *Fusarium oxysporum* by Park (1961, 1963), who suggested that such substances might be important in limiting fungal growth in soil (Park, 1960). The best-known inhibitory substances produced by microorganisms are the antibiotics, but evidence for their occurrence in natural soil is limited. Their production can be demonstrated in soil amended with nutrients and kept under laboratory conditions but not in unamended soil; some are quickly degraded when added to non-sterile soil while others (e.g. the basic and amphoteric antibiotics) become strongly adsorbed by colloidal particles (Soulides,

1965). Methods used to detect their presence in soil may not be sufficiently sensitive and the possibility remains that they may be produced
in inhibitory concentrations in highly localized sites; better techniques
such as the bioassay method of Soulides (1964) may provide more
information on this point.

Lack of energy-yielding materials for growth

It is unlikely that any of the factors that we have dealt with so far
could account for the general lack of activity apparent in most soil
populations. It has been known for some time that the addition of fresh
energy-yielding substrates to soil almost invariably leads to an increase
in one or other component of the soil microflora, although the rates of
response vary. Only in extreme conditions where some other factor is
limiting a part or whole of the population will nutrient additions fail to
induce a response. Thus Mayfield (1969) and Lowe (1969) showed that
actinomycetes and bacteria would not respond to nutrient additions to
acid forest soils but would do so if the soils were made neutral or
alkaline. Under field conditions, temperature also limits the utilization
of organic matter and it is noticeable that soil stored in a moist condition in the laboratory at higher temperatures often shows evidence of
increased hyphal growth on and around organic-matter particles.

Brock (1967 a) has suggested that environments such as soil may be
viewed as stable ecosystems in which the input of energy is exactly
balanced by the output. If we could determine the annual input of
energy, we would know the amount of energy potentially available to
the soil microflora and thus would be able to calculate the maximum
average rate of growth of the whole soil population.

Sources of energy

The most important source of energy for the growth of heterotrophic
microbes is a mixture of plant material from trees, shrubs and herbs.
Satchell (1970) has outlined the measurements which must be taken to
determine the energy input from these sources as follows.

The production of plant material can be divided into that above
ground and that below ground. Above-ground production can be
determined by calculating the mean annual increments of stems and
branches, the production of stem and branch litter, the annual increase in
butt weights, production of leaf and other litter and the amount of organic
matter washed from the soil canopy. To determine net primary production
we also need to know the amount of this material that is consumed by
herbivores and removed from the site (see Newbould, 1967; Bunce, 1968;

Milner & Hughes, 1968). Although the methods used are sometimes difficult to apply, especially in mixed communities, the estimates obtained are thought to be relatively reliable. However, estimates of below-ground production of roots are unreliable and accurate figures have never been obtained for root growth in natural communities. Nevertheless, Satchell (1970) has suggested, from a comparison of root and stem biomass data and knowledge of the rates of stem growth, that below-ground production of tree and shrub roots is about one-fifth that of the above-ground production. Such calculations ignore production of root exudates and sloughed off cells however. The rate of exudation of materials from roots growing in natural soils are unknown, but Harmsen & Jager (1963) showed that during the 6-week period following germination of vetch seeds, 1·6–2·9 mg. carbon were exuded for each 100 mg. dry root material. Assuming that exuded material is composed of 50 % carbon, then approximately 3–6 % of the root biomass was exuded within 6 weeks. These figures must be regarded as minimum excretion levels since part of the exudates were decomposed by the microorganisms present in the soil.

If the vegetation has not reached a stable climax condition, it is also necessary to determine how much of the production is being added to the standing crop. Then we may assume that the rest is added to the soil as litter (above and below ground) or as leachates.

Table 1. *Partition of primary production in different communities*

Units are equivalents of kg. dry matter/m.²/year

Community	Gross primary production	Respiration	Net primary production			Mean stock
			Total	Eaten	Decomposed	
Plankton marine temperate	0·72	0·06	0·65	0·65	0·01	0·004
Algae salt marsh	0·50	0·05	0·45	0·45	0·01	0·003
Spartina salt marsh	1·17	0·10	1·07	0·07	1·00	1·06
Grazed meadow	1·17	0·12	1·05	0·39	0·66	1·00
Beech wood	2·35	1·00	1·35	0·95	0·40	15·5
Rain forest	5·35	4·00	1·35	0·90	0·45	24·0

Other energy input sources (fertilizers, pesticides, rain run-off, etc.) are probably of relatively little significance, although there has been an inadequate amount of work done on them. Energy fixation by auto-trophic soil organisms will also take place but the rates of fixation are likely to be low except in unusual circumstances (Morrill & Dawson, 1962; Postgate, this symposium).

The net primary production of plant material does not vary much with latitude since increased rates of photosynthesis are usually balanced by increased rates of respiration (Macfadyen, 1970). An approximate value for net dry-matter production is 1 kg./m.²/year, i.e. 4,800 kcal. Some of this material is eaten by herbivores and does not reach the soil; this amount varies in different environments (Table 1). In aquatic microbial ecosystems the amount removed by herbivores is large (90 %) while in terrestrial communities it is much smaller (6–40 %). This means that varying amounts of energy-yielding material reach the soil and are available to the decomposer organisms.

Nature and amount of decomposer organism

The types and proportions of decomposer organisms are very varied and to some degree related to the soil and vegetation type. Unfortunately, there is almost complete absence of data on the biomass of all the components in any one soil, so that precise comparisons of different soils are impossible. The International Biological Programme is partly concerned with the construction of energy and nutrient budgets for different plant communities, and as the data from these studies are published some generalizations may emerge. Preliminary information (Satchell, 1970) on the biomass of organisms in a deciduous woodland soil at Meathop, Lancashire, a brown earth with a mull humus, are given in Table 2, but with more recent estimates for the microbial component included.

Table 2. *Provisional biomass estimates and annual litter production for Meathop Wood soil (after Satchell, 1970)*

Group	Dry matter biomass (kg./ha.)
Bacteria	7·3
Actinomycetes	0·2
Fungi	454·0
Protozoa	1·0
Nematodes	2·0
Earthworms	12·0
Enchytraeidae	4·0
Molluscs	5·0
Acari	1·0
Collembola	2·0
Diptera	3·0
Other arthropods	6·0
Total microflora	461·5
Total microfauna	36·0
Total biomass	497·5
Annual litter production	7640·0

It is apparent that the biomass of the microfauna is insubstantial in comparison with the microflora but this does not necessarily mean that the microflora is more important. Importance is related to the *rate* of biomass production, not to biomass itself.

Maintenance and growth of the microbial populations

Babiuk & Paul (1970) suggested that a considerable proportion of the energy input of soil was used in maintaining the microbial populations rather than permitting growth. They applied the following equation (Marr, Nilson & Clark, 1963):

$$\frac{dx}{dt} + ax = Y\frac{ds}{dt}, \tag{1}$$

where a is the specific maintenance constant (h.$^{-1}$), x the concentration of cells in grams, Y the yield coefficient (i.e. efficiency of conversion of substrate to cell material) and s the substrate available for maintenance.

Assuming that no growth is occurring, then

$$\frac{dx}{dt} = 0, \tag{2}$$

and so

$$\frac{ds}{dt} = \frac{ax}{Y}, \tag{3}$$

i.e. the rate of substrate utilization for maintenance of the population could be determined. From a consideration of the literature on the behaviour of microorganisms in continuous culture systems, Babiuk & Paul (1970) suggested that a reasonable value for a would be 0·001 h.$^{-1}$ and for Y, 0·35. Substituting these values and the available figures for bacterial biomass determined by direct counting of cells in a grassland soil they found that over half of the available energy was required for cell maintenance and that the bacteria could divide only a few times during the year. They recognized that dilution-plate-count figures might have given a more useful estimate of the metabolizing cells in the soil and suggested that the division rates obtained using these figures more closely resembled those found in laboratory conditions. However, the diversion of some of the energy to the growth and maintenance of fungi and soil animals has not been taken into account in their calculations.

Similar calculations can be made using the data obtained for microorganisms in the Meathop Wood soil already referred to in Table 2. It can be shown that to maintain a population of 7·3 kg. bacteria/ha. for a whole year requires the utilization of about 184 kg. of substrate. Annual litter production in this site is 7,640 kg./ha. so that 7,456 kg. are left for growth. If we assume that all of the substrate is used by the bacteria and that the bacterial cells formed act as secondary substrates, then the number of generations per year would be given by the equation

$$Y(S + xR) = xR, \tag{4}$$

where Y is the yield coefficient, S the substrate available for growth, x the concentration of cells in grams and R the number of divisions per year.

Substituting the value determined for Meathop Wood we find that $R = 543$, i.e. the bacteria could divide once every 16 h. However, this figure is unrealistic since it assumes that only the bacteria are utilizing the available substrates. If we make additional calculations for the fungal biomass, it can be shown that it would require 11,363 kg. of substrate per hectare per annum to maintain itself, i.e. almost 400 kg. more substrate than is actually present. In this case, no substrate would be available for fungal growth or maintenance and growth of any other soil organism. There are several possible explanations for this apparently absurd situation:

(a) The fungal biomass figures used are too high (see p. 258).

(b) The primary production figures are underestimates (see p. 264).

(c) The assumptions made concerning the maintenance constant and the yield coefficient are incorrect. If this is so then the maintenance coefficient would have to be lower or the yield coefficient higher to produce more realistic results. Some attempt to measure these parameters for soil organisms growing under low nutrient conditions at low temperatures are required.

(d) The system is not an equilibrium and more energy is being lost from the soil system than is being added to it. However, Satchell (1970) has suggested that this is not so and that the amount of organic matter in this soil is increasing very slowly each year.

Table 3 summarizes these calculations and compares them with estimates of growth rates derived in various ways for other natural habitats including fresh water, sea water and benthic sediments (modified from Hissett, 1970). In all cases long generation times have been recorded.

If growth rates of organisms in the soil are slow, then it follows that some of the chemical transformations brought about by them must also be slow. Alexander (1961) has shown such difficulties associated with rates of nitrogen fixation in soil. *Azotobacter* requires 1 g. of sugar in order to fix 5–20 mg. of nitrogen. On a field scale this means that 1122 kg. of organic matter per year would be required to fix 5–9 kg. of nitrogen per hectare, or put another way, cell turnover rates of 50,000 per day would be needed to produce an annual increase of 2.47 kg. of nitrogen per hectare. Since *Azotobacter* is rather uncommon in soil, it would be surprising if it was able to do this.

Table 3. *Growth rates of bacteria in natural habitats (modified from Hissett, 1970)*

Species	Habitat	Temperature	Basis of calculation	Generation time (h.)	Source
	Water				
Mixed	Fresh water	16°	Oxygen uptake	23·9–56·4	Belyatskaya (1958)
Mixed	Fresh water, 0·5 m.	—	Periodic counting	7–200	Drabkova (1965)
Mixed	Fresh water, 4·0 m.	—		7–200	
Mixed	Fresh water, 8–10·0 m.	—		12–200	
Spirillum	Sea water	18°	Continuous culture	75–127	Jannasch (1969)
Serratia	Sea water	18°	Continuous culture	98–130	
Leucothrix	Sea water on algae	—	Thymidine uptake	11·0–11·4	Brock (1967b)
	Benthic sediments (lakes)				
Mixed	Sand	13·5°	Growth in predator-free samples	90	Gambaryan (1965)
Mixed	Sand	4·5°	Growth in predator-free samples	18–61	
Mixed	Sand	0·5°	Growth in predator-free samples	300	
	Soil				
Mixed	Grassland	—	Energy input	c. 1,200	Babiuk & Paul (1970)
Mixed	Agricultural soil	—	CO_2 output	28·5	This paper
Mixed	Deciduous forest	—	Energy input	16	
Nitrosomonas	Agricultural soil	—	Metabolite production in perfused soil	38–100	Morrill & Dawson (1962)
Nitrobacter	Agricultural soil	—		21–58	

Effect of energy limitations on dormant propagules

Dobbs & Hinson (1953) showed that germination of the spores of many common soil fungi was prevented in soil by some unknown factor and this phenomenon was termed fungistasis or mycostasis. Similar effects have since been detected in many soils throughout the world, e.g. America (Lingappa & Lockwood, 1964), Africa (Jackson, 1958) and Malaya (Griffiths, 1966). Not all fungi are affected and the inhibitory effect, occurring in the presence of microbial activity, can be temporarily removed by addition of nutrients or sterilization of the soil. It is reintroduced in sterile soil by inoculation with soil microbes or non-sterile soil. Microbes rather than physical factors are involved (Dobbs & Hinson, 1953; Dobbs, Hinson & Bywater, 1960) and antibiotics have been implicated (Lockwood, 1959). Thus sterile agar discs are inhibitory to fungus spores if they have been in contact with non-sterile soil. However, it is also possible that essential nutrients move from the disc into the soil since soil microbes would act as a sink for the nutrients (Lockwood, 1968*b*). Charcoal is also sometimes reported as removing fungistasis but the effect has not always been found (Lingappa & Lockwood, 1961).

It is difficult to explain why the inhibitory effects are so widespread if they are due to an inhibitor and why they are removed by addition of nutrients. Alternatively it is possible that substances required for germination are not available due to the low level of energy supply to the soil: Lockwood (1964) suggested that microbial competition for limited supplies of exogenous nutrients could be an important factor in fungistasis. Since then a variety of effects of various nutrients on the germination of spores of various fungi have been found; these support this suggestion (Lingappa & Lockwood, 1964; Ko & Lockwood, 1967).

Interesting information on organic residues stimulating germination of spores subject to fungistasis has also been obtained which suggests that, apart from pieces of fresh undecomposed organic matter, the bulk of the soil is deficient in nutrients. Fungistasis may be regarded, therefore, not as a complete inhibition of growth but as an indication of the very spasmodic occurrence of active fungal growth in soil. Confirmation of this viewpoint would suggest the abandonment of the terms fungistasis or mycostasis because of their implications.

Evidence for similar effects on other soil microbes is comparatively limited. However, Lloyd (1969) found that many conidia of streptomycetes did not germinate in soil under physical conditions apparently suitable for their growth; a much higher percentage of conidia germinated in sterilized soil. Similar trends have been observed in our laboratories and the addition of glucose to non-sterile soil also increases markedly the frequency of germination. When streptomycete spores of the same species were inoculated on to pieces of organic debris and buried in natural soil, growth was detected after 2 weeks by scanning electron microscopy (Plate 1). Cysts of *Azobacter* in soil were stimulated to germinate by the presence of pea roots (Brown *et al.* 1968). Thus, the principles outlined to explain fungistasis are probably of much wider relevance to microbial populations in soil.

EFFECTS OF NUTRIENT SHORTAGE

Laboratory cultures

The nutritional situation in soil is apparently more analogous to nutrient-limited continuous culture than to batch culture where cells are growing at maximum specific growth rate with excess nutrients, which later become exhausted.

In laboratory chemostat experiments Righelato *et al.* (1968) showed that *Penicillium chrysogenum* requires 'rations' for both growth and maintenance. At maintenance levels of nutrient supply, morphological and chemical changes occur in the cells. At 1·7 times the maintenance requirement, conidia production is greatest. Induction of dormant structures by low nutrient supply has also been shown in blue-green algae (Wolk, 1965) and in bacteria (Harrison & Lawrence, 1963; Mandelstam, 1969).

A reduction in viability at low nutrient supply rates has been obtained in *Aerobacter aerogenes* by Tempest, Herbert & Phipps (1967). Under these conditions the nutrient situation becomes complex as dead auto-lysing cells contribute to the food supply of the viable organisms and estimates of overall growth rates cease to be meaningful (Postgate, 1965). The discrepancies between viable and total counts of bacteria in soil is probably due partly to the presence of many dead cells (Skinner, Jones & Mollison, 1952). Therefore bacteria in soil may be growing at minimum rates and replacing both viable and non-viable fractions of the population. Tempest *et al.* (1967) also noted that the minimum growth-rate value was temperature-dependent and postulated that in natural environments with sub-optimal temperatures this might allow greater maintenance of viability.

If microbes grow only slowly in soil, it is necessary to ask how they obtain energy to synthesize new proteins and enzymes which are required for adaptation to new conditions. It has been shown that in culture, synthesis of new proteins may be possible under starvation conditions

(Pardee, 1961; Mandelstam & Halvorson, 1960). Energy-rich storage products produced during active growth (e.g. poly-β-hydroxybutyric acid, glycogen) may also provide energy and Tempest & Hunter (1965) showed that bacteria may produce more carbohydrate at sub-optimal temperatures.

However, soil is a heterogenous structured system in which localized changes can occur which are unrelated to the soil as a whole, and continuous culture represents a gross over-simplification. Thus organisms may be growing rapidly on a root for short periods of time while starving and dying elsewhere. Environmental differences may occur which profoundly affect the growth of organisms around soil particles but not in the ambient soil solution. Thus what may be applicable at the soil–plant ecosystem level becomes far more complicated at the micro-environmental level where conditions vary spatially and temporally.

Effects on the sites and patterns of growth in soil

The results from studies of fungistasis and related phenomena suggest that zones of microbial activity are often associated with particulate organic substrates in soil. The existence of such microsites of growth, although generally accepted, has not been demonstrated often in natural soil. This probably reflects to some extent the infrequency of their occurrence, as will be discussed later.

The most studied particulate organic substrate in soil is undoubtedly the plant root, although even here there is a limited amount of work on the direct observation of roots grown in natural soil. The root provides a welcome, relatively long-term supply of nutrients in the nutrient-poor soil system. The stimulation of microbial activity in the rhizosphere soil around roots is probably analogous to the stimulation of spores in the vicinity of fresh organic residues added to soil.

Direct observation of root surfaces usually reveals the presence of fungal hyphae. These are most pronounced in older parts of the root where external moribund tissues occur (Parkinson, 1967). It has been suggested that colonization of roots by fungi occurs by successive lateral ingrowth from adjacent soil (Taylor & Parkinson, 1961). Actinomycetes and bacteria are less easily detected by direct observation but their presence on roots has been reported (Starkey, 1938). Rovira (1956) examined stained roots and noted bacteria growing both in colonies (ranging from a few to hundreds of cells) and in more diffuse patterns; Jenny & Grossenbacher (1963) detected bacterial colonies embedded in a mucigel layer surrounding the roots by electron microscopy.

Organic substrates in litter layers above the soil have been studied and the growth of fungi both on and inside pine needles observed by Kendrick & Burges (1962) and Hayes (1965). Smaller organic particles within the soil have received less attention, but there is evidence that in sandy soils occurrence of bacteria is greater on organic fragments than on mineral grains (Gray *et al.* 1968); similar evidence has been obtained for fungi and actinomycetes (Williams, 1962; Mayfield, 1969). Microbial cells themselves provide particulate substrates for other microbes. There is much indirect evidence for the colonization of fungal hyphae by bacteria and actinomycetes (Lockwood, 1968 *a*). Bodies of soil animals also provide substrates for soil microbes and growth of bacteria, actinomycetes and fungi on wings of arthropods buried in soil was observed by Okafor (1966).

In addition to such particulate organic substrates, growth on mineral grains may also occur. Although this is relatively infrequent on larger sand grains, soil crumbs can be heavily colonized. The surfaces of soil crumbs from a forest soil stored in the laboratory were found to be quite heavily colonized by microorganisms when examined by scanning electron microscopy. It is likely that nutrients are adsorbed on to the surfaces of clay crumbs and Quastel (1955) found nitrifying bacteria in perfused systems to be most active at the surface of crumbs where ammonium ions were held.

Occurrence of microbes on any surface in soil does not in itself prove that nutrients are being obtained from the substrate. Utilization of nutrients more generally distributed in the soil solution is always possible. The dissolved inorganic and organic components of the soil solution are in a constantly shifting equilibrium with the solid, liquid and gas phases of soil, as well as with soil organisms and plant roots (Chapman, 1965).

It is possible, to some extent, to discern different growth patterns of microorganisms which reflect their sites of growth in the soil. One of the earliest attempts to relate growth pattern to substrate utilization was made by Winogradsky (1924). He distinguished the zymogenous flora, mainly in a resting phase with short periods of activity in the presence of suitable substrates, from the autochthonous flora which was more or less continually active. Garrett (1951) made a similar distinction between the 'sugar fungi', which were essentially zymogenous, and the 'lignin fungi', which were autochthonous. These two groups of fungi could be equated on the basis of their growth pattern with the '*Penicillium* pattern' and 'Basidiomycete pattern' respectively (Burges, 1960). By modification, of the Burges scheme, the following growth patterns may be recognized (Gray & Williams, 1971).

(*a*) *Non-migratory, unicellular pattern.* Unicellular microbes, such as yeasts and non-motile bacteria, exist in soil as single cells or colonies, the number of cells in which varies with the substrate and the organism. Such colonies can be detected on surfaces in soil but it is not possible to decide to what extent the cells move passively in the soil water (Plate 1, fig. 2).

(*b*) *Migratory, unicellular pattern.* Some bacteria, fungi and algae possess limited powers of movement when grown in culture but conclusive proof of active movement in soil is lacking. Brown *et al.* (1968) showed that *Azotobacter* could travel along root surfaces but hardly at all through soil. Starkey (1968) found that plants inoculated with *Rhizobium* did not transmit these bacteria to adjacent, uninoculated roots. Thus if movement does occur in soil, it is probably mostly along continuous surfaces such as roots.

(*c*) *Plasmodial pattern.* Myxomycetes consist of naked masses of protoplasm which move in an amoeboid fashion until transformed into a mass of walled spores which form a fruit body. These organisms can migrate over the surface of soil but their capacity to do so over surfaces within the soil is undetermined. The myxobacteria may have a similar but more restricted growth pattern in soil.

(*d*) *Restricted hyphal pattern.* A reasonable amount of direct evidence for the occurrence of this pattern in soil is available. Fungi, such as *Penicillium* and other common soil forms, together with streptomycetes, form restricted hyphal growth on and within substrates. They rapidly produce spores above the substrate in large quantities which are then dispersed passively in the soil. The 'sugar fungi' (Garrett, 1951) have a pattern of this type, spending long periods as resting spores. Observations of natural soil by scanning electron microscopy indicated that streptomycetes grew on particulate substrates in restricted areas ranging from about 100 μm. across to a few isolated spore chains covering an area of no more than 20 μm. (Plate 1, fig. 3). Burges (1960) suggested that this type of growth pattern might be associated with the ability to produce antibiotics and the inclusion of streptomycetes emphasizes this possibility. The demonstration of antibiotic production by *Trichoderma viride* inoculated on to organic material buried in soil also supports this suggestion (Wright, 1956*a*, *b*). Failure to detect antibiotics in natural soil is probably partly due to their localized occurrence in both space and time in such microsites of growth.

(*e*) *Locally spreading hyphal growth.* The fungus *Mucor ramannianus* can colonize a particulate substrate and form aplanospores (as in the above pattern) or spread out into the surrounding soil where it forms chlamydospores within its hyphae (Burges, 1960). The hyphae can then disintegrate, leaving chlamydospores unassociated with the original substrate. The same may be true of other fungi, such as *Cylindrocarpon* and *Fusarium* which form chlamydospores when introduced into soil (Matturi & Stenton, 1964; Sequeira, 1962).

(*f*) *Mycelial strand or rhizomorph pattern* (Plate 1, fig. 4). The hyphae of some fungi can form strands which grow through soil from one substrate to another. More organized aggregate hyphal structures, rhizomorphs, are also formed by some fungi (e.g. *Armillaria mellea*) and it is clear that these grow faster than unorganized hyphae (Garrett, 1956) and spread through soil (Marsh, 1952). Many fungi in this category are pathogens of plant roots but some also have a reasonable capacity for saprophytic growth in soil.

(*g*) *Diffuse spreading hyphal pattern.* Examination of soil often reveals isolated fungal hyphae which are apparently growing unassociated with a particulate substrate. Burges (1960) suggested that *Zygorhynchus* typified this pattern and that it used nutrients in the soil solution. However, as stated previously, many hyphae seen in soil are dead but persistent, and so occurrence of lengths of hyphae unattached to a substrate need not necessarily indicate their growth pattern.

It is next appropriate to consider how frequently periods of active growth occur in the various sites within soil. Most of the evidence presented so far has indicated that it is difficult to detect active microbial cells in soil. Only on substrates such as plant roots, fresh leaf-litter or other recently added organic substrates is there much likelihood of detecting sites of growth, as opposed to indications of past growth, such as spores or dead hyphae. At the present time it is impossible to make an accurate assessment of how often cells in soil grow and reproduce (see calculations, based on energy input, p. 266). A further indication can be obtained from considering the length of survival of dormant propagules in soil if it is presumed that the total population level remains reasonably constant. When streptomycete conidia were added to an acidic soil which was unsuitable for their growth, their number was reduced by 50 % in a month (Mayfield, 1969). Lloyd (1969) found that about 33 % of streptomycete conidia added to soil remained viable after 16 days. Even restricted growth of a streptomycete in a microsite in natural soil can lead to the production of about 80 new spores (Plate 1, fig. 3). Therefore with a decrease in numbers of 50 % in a month and the occurrence of only restricted growth, it would be necessary for only about 1 % of the conidia in soil to germinate within 1 month to maintain the spore population at a constant level. Unfortunately there is little information on the survival of microbial cells in natural soil. In sterile soil, propagules of bacteria, actinomycetes, fungi and protozoa can survive for periods of 5 years or more (Sussman & Halvorson, 1966). In natural soil, lysis of cells induced by the presence of other microbes will reduce chances of survival. Thus Sequeira (1962) found that conidia of *Fusarium oxysporum* added to soil were lysed within 12–15 days; few conidia of *Trichoderma viride* survived for a year in a garden soil (Caldwell, 1958). Nevertheless, as already pointed out, even loss of a high percentage of viable propagules does not necessarily require frequent periods of activity to replenish numbers. There is also evidence that some propagules are more persistent in natural soil and the above workers found that chlamydospores of *Fusarium* and *Trichoderma* were not lysed. It is also clear that propagules of some plant pathogens survive for several years in the absence of the host, although with these organisms too it is difficult to assess survival rates accurately (Menzies, 1963). It seems likely, therefore, that many saprophytic soil microbes encounter the problems faced by obligate plant pathogens in the absence of their host, but on a time scale of days and weeks rather than months or years.

Effects on microorganisms added to soil

Unsuccessful attempts have been made to alter the fertility of soil by adding microbes such as *Azotobacter*, rhizobia, phosphate-solubilizing organisms and predacious fungi. These failures may be due to the unsuitability of the intrinsic physical and chemical conditions or to changes induced by the additives. Thus autolysis of added fungal hyphae is triggered by antibiotics from indigenous bacteria and actinomycetes stimulated by their presence (Lloyd & Lockwood, 1966). Added organisms, less well adapted to the environment, are also at a disadvantage in competition for limited supplies of nutrients. Because of these problems, heavy inocula must be used to increase their chances of survival (Johnson, Means & Weber, 1965; Brown *et al.* 1968; Postgate, Harley, this symposium).

AVAILABILITY OF ORGANIC MATTER

Evidence has been presented to show that growth of heterotrophic microbes in soil is severely limited by shortage of energy-yielding substrates, and it would therefore be reasonable to expect organic material added to soil, naturally or by man, to be rapidly decomposed. While this is true of some substances, it is clear that others can persist in soil for long periods and the reasons for this are now considered.

The nature and decomposition of persistent organic matter

Some components of naturally occurring organic residues added to soil are assimilated rapidly, while others persist and contribute directly or indirectly to the formation of humus. Thus when ^{14}C-labelled rye-grass was added to soil, 60 % of the carbon had left the soil after 6 months but 20 % still remained after 4 years (Jenkinson, 1965). When relatively pure compounds, such as glucose or cellulose, are added to soil, some of their carbon also enters the humic fraction (Mayaudon & Simonart, 1958; Sørenson, 1963). The persistence of some humic components in soil is given by radiocarbon dating techniques. By this method the age of humus in podzols was between 1,580 and 2,860 years (Perrin, Willis & Hodge, 1964); humus in a chernozemic soil was 1,000 years old (Paul *et al.* 1964), the humic acid fraction being oldest.

Humus is a complex mixture of substances, containing a small fraction of water-soluble substances such as amino acids and sugars, and a bulk of insoluble dark-coloured material consisting of humic acid, fulvic acid and humin. It seems likely that humic acid is formed by the polymerization of compounds derived from lignins, proteins, etc., of plants, animals and microbes. Its structure has not been exactly defined but it is generally thought to have a nitrogen-poor aromatic core with a nitrogen-rich aliphatic periphery; its exact composition probably varies in different soils.

Among the most persistent components of natural organic residues are phenols, lignins and waxes. Phenolic compounds are thought to contribute to the formation of humic acid by polymerization (Flaig, 1960). Lignins consist of an aromatic polymer of phenylpropane, but aliphatic hydroxyl and carbonyl groups are also present (Brauns & Brauns, 1960) and are extremely resistant to decomposition (Sørenson, 1963). So also are the surface components of plants, such as cutin, which is thought to consist of 16–18 fatty acids joined by ester linkages and occasional oxygen bridges (De Vries, Bredemeijer & Heinen, 1967).

In addition to these natural substances, a variety of man-made chemicals, such as fungicides, herbicides and pesticides enter the soil and persist for periods comparable with those of naturally occurring recalcitrant substances. Among the most persistent are the chlorinated hydrocarbon pesticides, the herbicides simazine (2-chloro-4,6-bis (ethylamino)-*s*-triazine) and monuron (3-(p-chlorophenyl)-1,1-dimethylurea), and fungicides such as zineb (zinc ethylene bisdithiocarbamate). It is not easy to decide why a particular substance is recalcitrant in soil. Many such substances have complex molecules, often aromatic; introduction of nitro, methyl or halogen groups can increase recalcitrance (Alexander, 1965*a*, *b*).

Simpler substances, readily available to microbes in the laboratory, may also be recalcitrant in the soil. This may be due to their adsorption on clay particles (Pinck, Dyal & Allison, 1954), their forming metal–organo complexes (Alexander, 1965*a*) or their protection by recalcitrant substances such as tannins (Basaraba & Starkey, 1966).

However, most recalcitrant substances are eventually decomposed and their decomposition rates largely determine rates of disappearance and accumulation of organic matter in soil (Minderman, 1968). Attempts to detect microbes responsible for their degradation are often unsuccessful; a summary of evidence obtained, mainly from laboratory cultures, is given in Table 4.

Table 4. *Examples of soil microorganisms involved in the decomposition of recalcitrant compounds*

Compounds	Microorganisms	Reference
Cutin	*Penicillium spinulosum*	Heinen & De Vries (1966)
	Rhodotorula spp.	
	Bacillus subtilis	
	Streptomyces spp.	Gray & Lowe (1967)
	Mortierella marburgensis	
Lignins	*Arthrobotrys* sp.	
	Cephalosporium sp.	Jones & Farmer (1967)
	Humicola sp.	
	Pseudomonas spp.	
	Flavobacterium spp.	Sørensen (1962)
	Basidiomycetes (white rots)	Generally known
Tannins	*Aspergillus* spp.	
	Penicillium spp.	Lewis & Starkey (1969)
	Other Fungi Imperfecti	
Humic acid	*Penicillium frequentans*	Mathur & Paul (1967*a*)
	Penicillium luteum	
	Polystictus sp.	Latter & Burges (1960)
Fulvic acid	*Poria subacida*	Mathur (1969)
Simazine	Bacteria	
	Streptomyces spp.	
	Aspergillus fumigatus	Kaufman, Kearney & Sheets (1963)
	Fusarium spp.	
Phenoxy herbicides	*Flavobacterium* sp.	Macrae & Alexander (1963)

The concept of zymogenous and autochthonous components of soil bacterial populations has been applied to other groups of the soil microflora. Thus Garrett (1951) distinguished between the 'sugar' and 'lignin fungi' and Burges (1960) defined the penicillia and basidiomycete growth patterns according to their zymogenous and autochthonous modes of existence. In such groupings, distinction is made between forms using readily available but ephemeral substrates and those using more persistent forms of organic matter. It has been relatively easy to demonstrate the presence of forms which are inactive in soil for long periods, having short intense periods of activity in the presence of fresh substrates. However, the presence of more or less continually active forms using residual components of the soil organic matter has proved more difficult to demonstrate. It is difficult to observe active microbes in soil, except in sites where fresh supplies of organic matter are available. Most of the microbes shown to be able to use recalcitrant forms of organic matter (Table 4) would normally be regarded as having the attributes of zymogenous rather than autochthonous species. In this context, Sundman *et al.* (1964) found that a wide range of soil microbes, including yeasts and bacteria, were involved in lignin breakdown. The wood-rotting basidiomycete *Polystictus versicolor* could not compete with the soil microflora and contributed little to lignin decomposition.

It seems possible that microbes with long inactive periods may contribute to the breakdown of persistent organic matter during their brief periods of activity. After being stimulated into activity by the arrival of readily available nutrients, more persistent materials in the microsite may also be attacked and evidence from a number of sources gives support to this suggestion.

It is clear from many laboratory experiments that complete or partial decomposition of recalcitrant compounds is more likely to occur when other more available nutrients are also present. Thus a number of streptomycetes and fungi attacked side chains of humates if an available carbon source was present (Szegi & Gulyas, 1968) and many Fungi Imperfecti tested by Lewis & Starkey (1969) could attack tannins if the medium was supplemented with glucose. A wide range of fungi can decolorize humic acid, but its decomposition is dependent on an accessory energy source (Burges, 1965). Hurst, Burges & Latter (1962) found that one of the enzymes involved in humic acid decomposition required NADPH and hence an accessory carbon source would be needed before decomposition in soil occurred.

There is also evidence that soil microorganisms are sometimes able to attack resistant substances only after adaptation to form new enzymes during a period of growth in their presence. When studying the metabolism of aromatic compounds related to lignin, Henderson (1961) noted a lag period of 2 days before utilization of these substances by fungi reached a maximum. This lag period could be eliminated if fungi were first incubated in the presence of various intermediate breakdown products of lignin. Mathur & Paul (1967*b*) found that enzyme systems involved in the degradation of humic acid by *Penicillium frequentans* were formed in contact with humic acid for 18 h.; cf. Loos, Roberts & Alexander (1967) and the oxidation of the herbicide 2,4-dichlorophenoxyacetate.

Finally, it is well known that addition of fresh organic residues or simple compounds such as glucose to soil results in the decomposition of some of the native soil organic matter and produces the so-called 'priming effect' (Jansson, 1960; Macura, Szolnoki & Vancura, 1963). Small, frequent additions of carbonaceous materials results in more pronounced decomposition of soil organic matter than do large, less frequent additions (Hallam & Bartholomew, 1953; Macura *et al.* 1963).

Therefore it is possible to view the bulk of the soil population as being relatively inactive for a large proportion of the time due to lack of available nutrients. While this is supported by a wide variety of experimental evidence, the existence of more or less continually active forms remains to be conclusively demonstrated.

We wish to thank the Natural Environmental Research Council and the North Atlantic Treaty Organisation for grants enabling us to participate in some of the work described in this paper. We also thank Dr R. Hissett for permission to quote some unpublished data.

REFERENCES

ALEXANDER, M. (1961). *Introduction to Soil microbiology*. New York: Wiley.

ALEXANDER, M. (1965*a*). Biodegradation: problems of molecular recalcitrance and microbial fallibility. *Adv. appl. Microbiol.* **7**, 35.

ALEXANDER, M. (1965*b*). Persistence and biological reactions of pesticides in soils. *Proc. Soil Soc. Am.* **29**, 1.

BABIUK, L. A. & PAUL, E. A. (1970). The use of fluorescein isothiocyanate in the determination of the bacterial biomass of grassland soil. *Can. J. Microbiol.* **16**, 57.

BASARBA, J. & STARKEY, R. L. (1966). Effect of plant tannins on decomposition of organic substances. *Soil Sci.* **101**, 17.

BELYATSKAYA, Y. S. (1958). Generation time and energy utilization of aquatic bacteria. *Nauch. Dokl. vȳssh Shk.* (Biology) **2**, 147.

BLACKWELL, E. (1943). The life history of *Phytophthora cactorum* (Lob. & Cohn) Schroet. *Trans. Br. Mycol. Soc.* **26**, 71.

BLOOMFIELD, B. J. & ALEXANDER, M. (1967). Melanins and resistance of fungi to lysis. *J. Bact.* **93**, 1276.

BRAUNS, F. E. & BRAUNS, D. A. (1960). *The Chemistry of Lignin* (supplementary volume). New York: Academic Press.

BROCK, T. D. (1967*a*). The ecosystem and the steady state. *Bioscience* **17**, 166.

BROCK, T. D. (1967*b*). Bacterial growth rate in the sea: direct analysis by thymidine autoradiography. *Science, N.Y.* **155**, 81.

BROWN, M. E., JACKSON, R. M. & BURLINGHAM, S. K. (1968). Growth and effects of bacteria introduced into soil. In *The Ecology of Soil Bacteria*, p. 531. Ed. T. R. G. Gray and D. Parkinson. Liverpool University Press.

BUNCE, R. G. H. (1968). Biomass and production of trees in a mixed deciduous woodland. 1. Girth and height as parameters for the estimation of tree dry weight. *J. Ecol.* **56**, 759.

BURGES, A. (1960). Dynamic equilibria in soil. In *The Ecology of Soil Fungi*, p. 185. Ed. D. Parkinson and J. S. Waid. Liverpool University Press.

BURGES, A. (1965). Biological processes in the decomposition of organic matter. In *Experimental Pedology*, p. 189. Ed. E. G. Hallsworth and D. V. Crawford. London: Butterworth.

BURGES, A. & NICHOLAS, D. P. (1961). Use of soil sections in studying amounts of fungal hyphae in soil. *Soil Sci.* **92**, 25.

CALDWELL, R. (1958). Fate of spores of *Trichoderma viride* introduced into soil. *Nature, Lond.* **181**, 1144.

CHAPMAN, H. D. (1965). Chemical factors of the soil as they affect microorganisms. In *Ecology of Soil-borne Plant Pathogens*, p. 120. Ed. K. F. Baker and W. C. Snyder. Los Angeles: University of California Press.

CHRISTOU, T. & SNYDER, W. C. (1962). Penetration and host–parasite relationships of *Fusarium solani* f. *phaseoli* in the bean plant. *Phytopathology* **52**, 219.

CLARK, F. E., JACKSON, R. D. & GARDNER, H. R. (1962). Measurement of microbial thermogenesis in soil. *Proc. Soil Sci. Am.* **26**, 155.

CROSS, T. (1968). Thermophilic actinomycetes. *J. Appl. Bact.* **31**, 36.

DE VRIES, H., BREDEMEIJER, G. & HEINEN, W. (1967). The decay of cutin and cuticular components by soil micro-organisms in their natural environment. *Acta bot. neerl.* **16**, 102.

DOBBS, C. G. & GASH, M. J. (1965). Microbial and residual mycostasis in soils. *Nature, Lond.* **207**, 1354.

DOBBS, C. G. & HINSON, W. H. (1953). A widespread fungistasis in soils. *Nature, Lond.* **175**, 500.

DOBBS, C. G., HINSON, W. H. & BYWATER, J. (1960). Inhibition of fungal growth in soils. In *The Ecology of Soil Fungi*, p. 130. Ed. D. Parkinson and J. S. Waid. Liverpool University Press.

DRABKOVA, V. G. (1965). Dynamics of the bacterial number generation time and production of bacteria in the matter of Red Lake (Punnus-jarvi). *Microbiology* **34**, 933.

FEHER, D. (1933). *Mikrobiologie des Waldbodens*. Berlin.

FLAIG, W. (1960). Comparative chemical investigations on natural humic compounds and their model substances. *Proc. Roy. Soc., Dublin* A **1**, 149.

GAMBARYAN, M. E. (1965). Method of determining the generation time of microorganisms in benthic sediments. *Microbiology* **34**, 939.

GARRETT, S. D. (1951). Ecological groups of soil fungi: a survey of substrate relationships. *New Phytol.* **50**, 149.

GARRETT, S. D. (1956). Rhizomorph behaviour in *Armillaria mellea* (Vahl.) Quel. II. Logistics of infection. *Ann. Bot.* N.S. **20**, 193.

GOODFELLOW, M., HILL, I. R. & GRAY, T. R. G. (1968). Bacteria in a pine forest soil. In *The Ecology of Soil Bacteria*, p. 500. Ed. T. R. G. Gray and D. Parkinson. Liverpool University Press.

GRAY, P. H. H. & WALLACE, R. H. (1957). Correlation between bacterial numbers and carbon dioxide in a field soil. *Can. J. Microbiol.* **3**, 191.

GRAY, T. R. G., BAXBY, P., HILL, I. R. & GOODFELLOW, M. (1968). Direct observation of bacteria in soil. In *Ecology of Soil Bacteria*, p. 171. Ed. T. R. G. Gray and D. Parkinson. Liverpool University Press.

GRAY, T. R. G. & LOWE, W. E. (1967). Techniques for studying cutin decomposition in soil. *Bact. Proc.*, 3.

GRAY, T. R. G. & WILLIAMS, S. T. (1971). *Soil Micro-organisms*. Edinburgh: Oliver and Boyd.

GRIFFIN, D. M. (1963). Soil moisture and the ecology of soil fungi. *Biol. Rev.* **38**, 141.

GRIFFITHS, D. A. (1966). Vertical distribution of mycostasis in Malayan soils. *Can. J. Microbiol.* **12**, 149.

HALLAM, M. J. & BARTHOLOMEW, W. V. (1953). Influence of rate of plant residue addition in accelerating the decomposition of soil organic matter. *Proc. Soil Sci. Soc. Am.* **17**, 365.

HARMSEN, G. W. & JAGER, G. (1963). Determination of the quantity of carbon and nitrogen in the rhizosphere of young plants. In *Soil Organisms*, p. 245. Ed. J. Doeksen and J. van der Drift. Amsterdam: North Holland Publishing Co.

HARRISON, A. P. & LAWRENCE, F. R. (1963). Phenotypic, genotypic and chemical changes in starving populations of *Aerobacter aerogenes*. *J. Bact.* **85**, 742.

HAWKER, L. E. (1957). Ecological factors and survival of fungi. In *Microbial Ecology*, p. 238. Ed. R. E. O. Williams & C. C. Spicer. Cambridge University Press.

HAYES, A. J. (1965). Studies on the decomposition of coniferous leaf litter. II. Changes in external features and succession of microfungi. *J. Soil Sci.* **16**, 242.

HENDERSON, M. E. K. (1961). The metabolism of aromatic compounds related to lignin by some Hyphomycetes and yeast-like fungi of soil. *J. gen. Microbiol.* **26**, 155.

HEINEN, W. & DE VRIES, H. (1966). Stages during the breakdown of plant cutin by soil micro-organisms. *Arch. Mikrobiol.* **54**, 331.

HEPPLE, S. (1958). *Mucor rammannianus in a podsolised soil*. Ph.D. thesis, University of Liverpool.

HILL, I. R. & GRAY, T. R. G. (1967). Application of the fluorescent antibody technique to an ecological study of bacteria in soil. *J. Bact.* **93**, 1888.

HISSETT, R. (1970). *The bacterial population of leaf litter and soil in a deciduous woodland*. Ph.D. thesis, University of Liverpool.

HOLDING, A. J., FRANKLIN, D. A. & WATLING, R. (1965). The microflora of peat–podzol transitions. *J. Soil Sci.* **16**, 44.

HURST, H. M., BURGES, A. & LATTER, P. (1962). Some aspects of the biochemistry of humic acid decomposition by fungi. *Phytochemistry* **1**, 227.

JACKSON, R. M. (1958). An investigation of fungistasis in Nigerian soils. *J. gen. Microbiol.* **18**, 248.

JANNASCH, H. W. (1969). Estimations of bacterial growth rates in natural waters. *J. Bact.* **99**, 156.

JANSSON, S. K. (1960). On the establishment and use of tagged microbial tissue in soil organic matter research. *Trans. 7th Int. Congr. soil Sci.* **2**, 635.

JENKINSON, D. S. (1965). Studies on the decomposition of plant material in soil. I. Losses of carbon from ^{14}carbon labelled ryegrass inoculated with soil in the field. *J. Soil Sci.* **16**, 104.

JENNY, H. & GROSSENBACHER, K. (1963). Root–soil boundary zones as seen by the electron microscope. *Proc. Soil Sci. Am.* **27**, 273.

JOHNSON, H. W., MEANS, U. M. & WEBER, C. R. (1965). Competition for nodule sites between strains of *Rhizobium japonicum* applied as inoculum and strains in the soil. *Agron. J.* **57**, 179.

JONES, D. & FARMER, V. C. (1967). The ecology and physiology of soil fungi involved in the degradation of lignin and related aromatic compounds. *J. Soil Sci.* **18**, 74.

JONES, P. C. T. & MOLLISON, J. E. (1948). A technique for the quantitative estimation of soil micro-organisms. *J. gen. Microbiol.* **2,** 54.

KAUFMAN, D. D., KEARNEY, P. C. & SHEETS, T. J. (1963). Simazine: degradation by soil micro-organisms. *Science, N.Y.* **142,** 405.

KENDRICK, W. B. & BURGES, A. (1962). Biological aspects of decay of *Pinus sylvestris* leaf litter. *Nova Hedwigia* **4,** 313.

KO, W. H. & LOCKWOOD, J. L. (1967). Soil fungistasis: relation to fungal spore nutrition. *Phytopathology* **57,** 894.

LATTER, P. M. & BURGES, A. (1960). Experimental decomposition of humic acid by fungi. *Trans. 7th Int. Congr. Soil Sci.* **2,** 643.

LATTER, P. M., CRAGG, J. B. & HEAL, O. W. (1967). Comparative studies on the microbiology of four moorland soils in the Northern Pennines. *J. Ecol.* **45,** 445.

LEWIS, J. A. & STARKEY, R. L. (1969). Decomposition of plant tannins by some soil micro-organisms. *Soil Sci.* **107,** 235.

LINGAPPA, B. T. & LINGAPPA, Y. (1966). The nature of self-inhibition of germination of conidia of *Glomerella cingulata. J. gen. Microbiol.* **43,** 91.

LINGAPPA, B. T. & LOCKWOOD, J. L. (1961). The nature of the widespread soil fungistasis. *J. gen. Microbiol.* **26,** 473.

LINGAPPA, B. T. & LOCKWOOD, J. L. (1962). Fungitoxicity of lignin monomers, model substances and decomposition products. *Phytopathology* **52,** 295.

LINGAPPA, B. T. & LOCKWOOD, J. L. (1964). Activation of soil microflora by fungus spores in relation to soil fungistasis. *J. gen. Microbiol.* **35,** 215.

LLOYD, A. B. & LOCKWOOD, J. L. (1966). Lysis of fungal hyphae in soil and its possible relation to autolysis. *Phytopathology* **56,** 595.

LLOYD, A. B. (1969). Behaviour of streptomycetes in soil. *J. gen. Microbiol.* **56,** 165.

LOCKWOOD, J. L. (1959). *Streptomyces* spp. as a cause of natural fungitoxicity in soils. *Phytopathology* **49,** 327.

LOCKWOOD, J. L. (1964). Soil fungistasis. *Ann. Rev. Phytopath.* **2,** 341.

LOCKWOOD, J. L. (1968*a*). The fungal environment of soil bacteria. In *The Ecology of Soil Bacteria,* p. 44. Ed. T. R. G. Gray and D. Parkinson. Liverpool University Press.

LOCKWOOD, J. L. (1968*b*). Discussion on the environment of soil bacteria. In *The Ecology of Soil Bacteria,* p. 94. Ed. T. R. G. Gray and D. Parkinson. Liverpool University Press.

LOOS, M. A., ROBERTS, R. N. & ALEXANDER, M. (1967). Phenols as intermediates in the decomposition of phenoxyacetates by an arthrobacter species. *Can. J. Microbiol.* **13,** 679.

LOWE, W. E. (1969). *An ecological study of coccoid bacteria in soil.* Ph.D. thesis, University of Liverpool.

MACFADYEN, A. (1970). Soil metabolism in relation to ecosystem energy flow and to primary and secondary production. In *Methods for the Study of Production and Energy Flow in Soil Communities,* ed. J. Phillipson. Paris: UNESCO.

MACRAE, I. C. & ALEXANDER, M. (1963). Metabolism of phenoxyalkyl carboxylic acids by a *Flavobacterium* species. *J. Bact.* **86,** 1231.

MACURA, J., SZOLNOKI, J. & VANCURA, V. (1963). Decomposition of glucose in soil. In *Soil Organisms,* p. 231. Ed. T. Doeksen and J. van der Drift. Amsterdam: North Holland Publishing Co.

MANDELSTAM, J. (1969). Regulation of bacterial spore formation. In *Microbial Growth,* p. 377. Ed. P. Meadows and S. J. Pirt. Cambridge University Press.

MANDELSTAM, J. & HALVORSON, H. (1960). Turnover of protein and nucleic acid in soluble and ribosome fractions of non-growing *Escherichia coli. Biochim. biophys. Acta* **40,** 43.

MARR, A. G., NILSON, E. H. & CLARK, D. J. (1963). Maintenance requirements of *Escherichia coli. Ann. N.Y. Acad. Sci.* **102,** 536.

MARSH, R. W. (1952). Field observations on the spread of *Armillaria mellea* in apple orchards and in blackcurrant plantations. *Trans. Br. Mycol. Soc.* **35,** 201.

MATHUR, S. P. (1969). Microbial use of podzol Bh fulvic acids. *Can. J. Microbiol.* **15,** 677.

MATHUR, S. P. & PAUL, E. A. (1967*a*). Microbial utilization of soil humic acids. *Can. J. Microbiol.* **13,** 573.

MATHUR, S. P. and PAUL, E. A. (1967*b*). Partial characterization of soil humic acids through biodegradation. *Can. J. Microbiol.* **13,** 581.

MATTURI, S. T. & STENTON, H. (1964). The behaviour in soil of spores of four species of *Cylindrocarpon. Trans. Br. Mycol. Soc.* **47,** 589.

MAYAUDON, J. & SIMONART, P. (1958). Study of the decomposition of organic matter in soil by means of radioactive carbon. II. The decomposition of radioactive glucose in soil and distribution of radioactivity in the humus fractions of soil. *Pl. Soil* **9,** 376.

MAYFIELD, C. I. (1969). *A study of the behaviour of a successful soil streptomycete.* Ph.D. thesis, University of Liverpool.

MENZIES, J. D. (1963). Survival of microbial plant pathogens in soil. *Bot. Rev.* **29,** 79.

MILNER, C. & HUGHES, R. E. (1968). *Methods for the Measurement of the Primary Production of Grassland. IBP handbook*, no. 6. Oxford: Blackwell.

MINDERMAN, G. (1968). Addition, decomposition and accumulation of organic matter in forests. *J. Ecol.* **56,** 355.

MISHUSTIN, E. N. (1956). The law of zonality and the study of the microbial associations of the soil. *Soils Fertil., Harpenden* **19,** 385.

MORRILL, L. G. & DAWSON, J. E. (1962). Growth rates of nitrifying chemoautotrophs in soil. *J. Bact.* **83,** 205.

NEWBOULD, P. J. (1967). *Methods for Estimating the Primary Production of Forests, IBP handbook*, no. 2. Oxford: Blackwell.

OKAFOR, N. (1966). The ecology of micro-organisms on, and the decomposition of, insect wings in soil. *Pl. Soil* **25,** 211.

PARDEE, A. B. (1961). Response of enzyme synthesis and activity to environment. In *Microbial Reaction to Environment*, p. 19. Ed. C. G. Meynell and H. Gooder. Cambridge University Press.

PARK, D. (1960). Antagonism – the background to soil fungi. In *The ecology of soil fungi*, p. 148. Ed. D. Parkinson and J. S. Waid. Liverpool University Press.

PARK, D. (1961). Morphogenesis, fungistasis and cultural staling in *Fusarium oxysporum* Snyder and Hansen. *Trans. Br. Mycol. Soc.* **44,** 377.

PARK, D. (1963). Evidence for a common fungal growth factor. *Trans. Br. Mycol. Soc.* **46,** 541.

PARKINSON, D. (1967). Soil micro-organisms and plant roots. In *Soil Biol.* p. 449. Ed. A. Burges and F. Raw. London and New York: Academic Press.

PAUL, E. A., CAMPBELL, C. A., RENNIE, D. A. & McCALLUM, K. J. (1964). Investigations of the dynamics of soil humus utilizing carbon dating techniques. *Trans. 8th Int. Congr. Soil Sci.* **3,** 201.

PERRIN, R. M. S., WILLIS, E. H. & HODGE, C. A. H. (1964). Dating of humus podzols by residual radiocarbon activity. *Nature, Lond.* **202,** 165.

PINCK, L. A., DYAL, R. S. & ALLISON, F. E. (1954). Protein–montmorillonite complexes, their preparation and the effects of soil micro-organisms on their decomposition. *Soil Sci.* **78,** 109.

POSTGATE, J. (1965). Continuous culture: attitudes and myths. *Lab. Prac.* **14**, 1140.

QUASTEL, J. H. (1955). Soil metabolism. *Proc. Roy. Soc. Lond.* B **143**, 159.

REINERS, M. (1968). Carbon dioxide evolution from the floor of three Minnesota forests. *Ecology* **49**, 471.

RIGHELATO, R. C., TRINCI, A. P. J., PIRT, S. J. & PEAT, A. (1968). The influence of maintenance energy and growth rate on the metabolic activities; morphology and conidiation of *Penicillium chrysogenum. J. gen. Microbiol.* **50**, 399.

ROMMELL, L. G. (1932). Mull and duff as biotic equilibria. *Soil Sci.* **34**, 161.

ROVIRA, A. D. (1956). A study of the development of the root surface microflora during the initial stages of plant growth. *J. appl. Bact.* **19**, 72.

RUDDICK, S. M. (1969). *Some aspects of the structure and behaviour of actinomycete spores in soil.* Ph.D. thesis, University of Liverpool.

SATCHELL, J. (1970). Feasibility study of an energy budget for Meathop Wood. In *Production of the World's Forests*, ed. J. Duvingneaud. Brussels.

SEQUEIRA, L. (1962). Influence of organic amendments on survival of *Fusarium oxysporum* f. *cubense* in the soil. *Phytopathology* **52**, 976.

SKINNER, F. A. (1951). A method for distinguishing between viable spores and mycelial fragments of actinomycetes in soil. *J. gen. Microbiol.* **5**, 159.

SKINNER, F. A. (1968). The anaerobic bacteria in soil. In *The Ecology of Soil Bacteria*, p. 573. Ed. T. R. G. Gray and D. Parkinson. Liverpool University Press.

SKINNER, F. A., JONES, P. C. T. & MOLLISON, J. E. (1952). A comparison of a direct- and a plate-counting technique for quantitative estimation of soil micro-organisms. *J. gen. Microbiol.* **6**, 261.

SØRENSON, H. (1962). Decomposition of lignin by soil bacteria and complex formation between autoxidised lignin and organic nitrogen compounds. *J. gen. Microbiol.* **27**, 21.

SØRENSON, H. (1963). Studies on the decomposition of C^{14} labelled barley straw in soil. *Soil Sci.* **95**, 45.

SOULIDES, D. A. (1964). Antibiotics in soils. VI. Determination of micro-quantities of antibiotics in soil. *Soil Sci.* **97**, 286.

SOULIDES, D. A. (1965). Antibiotics in soil. VIII. Production of streptomycin and tetracyclines in soil. *Soil Sci.* **100**, 200.

STARKEY, R. L. (1938). Some influence of the development of higher plants upon the micro-organisms in soil. VI. Microscopic examination of the rhizosphere. *Soil Sci.* **45**, 207.

STARKEY, R. L. (1968). The ecology of soil bacteria: discussion and concluding remarks. In *The Ecology of Soil Bacteria*, p. 635. Ed. T. R. G. Gray and D. Parkinson. Liverpool University Press.

STOTZKY, G. (1965). Microbial respiration. In *Methods of Soil Analysis*, Vol. II. *Chemical and Microbiological Properties*, p. 1150. Ed. C. A. Black. Madison: American Society of Agronomy.

STOTZKY, G. & GOOS, R. D. (1965). Effect of high CO_2 and low O_2 tensions on the soil microbiota. *Can. J. Microbiol.* **11**, 853.

SUNDMAN, V., KUUSI, T., KUHANEN, S. & CARLBERG, G. (1964). Microbial decomposition of lignin. IV. Decomposition of lignin by various micro-populations. *Acta Agr. scand.* **24**, 229.

SUSSMAN, A. S. (1965). Dormancy of soil micro-organisms in relation to survival. In *Ecology of Soil-borne Plant Pathogens*, p. 99. Ed. K. F. Baker and W. C. Snyder. Los Angeles: University of California Press.

SUSSMAN, A. S. (1969). The prevalence and role of dormancy. In *The Bacterial Spore*, p. 1. Ed. G. W. Gould & A. Hurst. London: Academic Press.

SUSSMAN, A. S. & HALVORSON, H. O. (1966). *Spores, their Dormancy and Germination*. New York and London: Harper Row.

SZEGI, J. & GULYAS, F. (1968). Data on the humus-decomposing activity of some streptomycetes and microscopic fungi. *Agrochemica Talajtan* **17**, 109.

TAMM, E. & KRZYSCH, G. (1965). Zur dynamik der Bodenatmung und des CO_2-Gehaltes der bodennahen Luftschnicht während der Vegetationsruhe. *Z. Acker- Pflanzenbau* **122**, 209.

TAYLOR, G. S. & PARKINSON, D. (1961). Studies on fungi in the root region. IV. Fungi associated with the roots of *Phaseolus vulgaris*. L. *Pl. Soil* **15**, 261.

TEMPEST, D. W., HERBERT, D. & PHIPPS, P. J. (1967). Studies on the growth of *Aerobacter aerogenes* at low dilution rates in a chemostat. In *Microbial Physiology and Continuous Culture*, p. 240. Ed. E. O. Powell *et al.* H.M.S.O.

TEMPEST, D. W. & HUNTER, J. R. (1965). The influence of temperature and pH value on the macromolecular composition of magnesium-limited and glycerol-limited *Aerobacter aerogenes* growing in a chemostat. *J. gen. Microbiol.* **41**, 267.

TRINCI, A. P. J. & RIGHELATO, R. C. (1970). Changes in constituents and ultra-structure of hyphal compartments during autolysis of glucose-starved *Peniclilum chrysogenum*. *J. gen. Microbiol.* **60**, 239.

WALLIS, G. W. & WILDE, S. A. (1957). Rapid method for the determination of carbon dioxide evolved from forest soils. *Ecology* **38**, 359.

WARCUP, J. H. (1955). Isolation of fungi from hyphae present in soil. *Nature, Lond.* **175**, 953.

WARCUP, J. H. (1957). Studies on the occurrence and activity of fungi in a wheat field soil. *Trans. Br. Mycol. Soc.* **40**, 237.

WARCUP, J. H. (1967). Fungi in soil. In *Soil Biology*, p. 51. Ed. A. Burges and F. Raw. London: Academic Press.

WARCUP, J. H. & BAKER, K. F. (1963). Occurrence of dormant ascospores in soil. *Nature, Lond.* **197**, 1317.

WARCUP, J. H. & TALBOT, P. H. B. (1962). Ecology and identity of mycelia isolated from soil. *Trans. Br. Mycol. Soc.* **45**, 495.

WATSON, E. T. (1970). *Synecological studies of actinomycetes in a coastal sand belt*. Ph.D. thesis, University of Liverpool.

WIANT, H. V., JR. (1967). Has the contribution of litter decay to forest 'soil respiration' been overestimated? *J. For.* **65**, 408.

WILLIAMS, S. T. (1962). *The soil washing technique and its application to a study of fungi in a podsolised soil*. Ph.D. thesis, University of Liverpool.

WINOGRADSKY, S. (1924). Sur la microflore autochthone de la terre arable. *C. r. hebd. Séanc. Acad. Sci., Paris* **178**, 1236.

WITKAMP, M. (1966a). Decomposition of leaf litter in relation to environment, microflora and microbial respiration. *Ecology* **47**, 194.

WITKAMP, M. (1966b). Rates of carbon dioxide evolution from the forest floor. *Ecology* **47**, 492.

WITKAMP, M. & VAN DER DRIFT, J. (1961). Breakdown of forest litter in relation to environmental factors. *Pl. Soil* **15**, 295.

WOLK, P. (1965). Heterocyst formation under defined conditions. *Nature, Lond.* **205**, 201.

WRIGHT, J. M. (1956a). The production of antibiotics in soil. IV. Production of antibiotics in coats of seeds sown in soil. *Ann. appl. Biol.* **44**, 561.

WRIGHT, J. M. (1956b). The production of antibiotics in soil. III. Production of gliotoxin in wheat straw buried in soil. *Ann. appl. Biol.* **44**, 461.

EXPLANATION OF PLATE

Scanning electron micrographs of microorganisms in soil

Fig. 1. Growth of streptomycete spores inoculated on to organic matter buried in natural soil. × 2,000.

Fig. 2. Colony of bacteria on the surface of a mineral grain. × 6,000.

Fig. 3. Restricted growth of a streptomycete producing approximately 80 spores. × 6,000.

Fig. 4. Rhizomorph of a fungus. × 1,000.

PLATE 1

RELEVANT ASPECTS OF THE PHYSIOLOGICAL CHEMISTRY OF NITROGEN FIXATION

JOHN POSTGATE

A.R.C. Unit of Nitrogen Fixation,
University of Sussex, Brighton, BN1 9QJ, Sussex, England

INTRODUCTION

Biological nitrogen fixation has been the subject of a Discussion meeting (Chatt & Fogg, 1969) and an extensive review (Hardy & Burns, 1968) as well as a condensed review (Postgate, 1970*a*), all published within a few months of preparing this contribution. Research in this area is 'moving' rapidly, so I shall not review the subject yet again, nor provide an exhaustive bibliography. I shall summarize the position at the end of May 1970 before discussing practical implications. The summary will necessarily be biased in the direction of matters which seem of interest in the context of the present symposium.

I shall use the term 'dinitrogen' to refer to the N_2 molecule, in accordance with present-day chemical convention.

BACKGROUND

The enzyme

Nitrogenase, the enzyme complex which reduces dinitrogen to ammonia, has now been extracted from numerous microorganisms. These include aerobic, anaerobic, photosynthetic bacteria as well as Cyanophyceae and the leguminous symbiosis (Table 1). The crude preparations of nitrogenase vary considerably in their sensitivity to oxygen: extracts of *Clostridium pasteurianum*, though the first to be obtained (Carnahan *et al.* 1960), are very sensitive and must be handled anaerobically, yet extracts from *Azotobacter vinelandii* or *A. chroococcum* obtained by sonication or decompression are stable in air. Such *Azotobacter* extracts are particulate (they sediment at 150,000 *g* over 2–4 h.) and they can be resolved by anaerobic ion-exchange chromatography into two protein components, one of which is very readily damaged by oxygen and the other of which is not. The two protein components of nitrogenase will be referred to as proteins '1' and '2' respectively; the parent organisms will be referred to by the appropriate

Table 1. *Organisms which have yielded preparation or extracts containing active nitrogenase*

The 'code' abbreviation after the organisms is used in the text.

Clostridium pasteurianum	*Cp*	Carnahan *et al.* (1960)
Klebsiella pneumoniae	*Kp*	Hamilton, Burris & Wilson (1964)†
Bacillus polymyxa	*Bp*	Grau & Wilson (1962)
Rhodospirillum rubrum	*Rr*	Bulen, Burns & LeComte (1965)
Mycobacterium flavum 301	*Mf*	Biggins & Postgate (1969)
Anabena cylindrica	*Anc*	Stewart, Haystead & Pearson (1969)
Azotobacter vinelandii	*Av*	Bulen, Burns & LeComte (1965)
Azotobacter chroococcum	*Ac*	Kelly (1968)
Desulfovibrio desulfuricans	*Dd*	M. G. Yates, J. R. Postgate unpublished
Chromatium‡	*Cv*	Arnon *et al.* (1960)
Soya beans + *Rhizobium japonicum*		Bergersen (1966); Koch, Evans & Russell (1967)

† *Kp* was thought to be an achromobacter at this time.
‡ Probably 'Chromatium D', a strain of *C. vinosum*.

italic initials as in Table 1, so that Ac_1 refers to protein 1 of *Azotobacter chroococcum*, Kp_2 to protein 2 of *Klebsiella pneumoniae*, and so on.

An oxygen-sensitive protein appears to be a universal component of nitrogenase and 'crude' preparations do not necessarily show a correlation between sedimentability and oxygen sensitivity. The fact that crude microbial extracts are often less sensitive to oxygen than the purified sensitive protein implies that this protein is protected by other components. This is particularly true of extracts of *Ac* and *Av*. The situation regarding oxygen sensitivity of these proteins may well be complicated by protective effects of a non-specific kind: Kelly, Klucas & Burris (1967) reported that cytochromes protected stored *Av* nitrogenase proteins and Yates (1970b) showed non-specific protection of $Ac_1 + Ac_2$ against oxygen damage by crude NADH dehydrogenase. Other proteins could simulate involvement in the electron transport part of the N_2 fixation process; this matter is discussed again later.

Proteins 1

Proteins 1 contain Mo as well as non-haem Fe and 'labile' sulphur; Cp_1 has been called 'molybdoferredoxin'. Proteins 1 are not specially sensitive to oxygen and aeration has been used to free Ac_1 from Ac_2 (Kelly, 1969a). Systematic data on the oxygen sensitivities of proteins 1 from various organisms are needed; Biggins & Postgate (1969) had evidence that *Mf* nitrogenase was relatively oxygen-sensitive and D. R. Biggins (to be published) showed that Mf_1 was less stable than Kp_1, Ac_1, or Bp_1, all of which were undamaged after 30 min. stirring in

Table 2. *Selected data on nitrogenase proteins*

Code	Mol.wt ($\times 10^{-5}$)	Atomic ratios of constituents			Remarks	Reference
		Fe	Mo	Labile S		
Cp_1	2	7·5	1	7·5	2 Mo/mole	(1)
Av_1	—	12	1	—	—	(2)
Av_1	2·7–3	20	1	15	Crystalline, 2 Mo/mole	(3)
Ac_1	c. 2·5	11	1	—	Mol.wt by gel exclusion	(4)
Kp_1	1·8	7·4	1	7·5	One electrophoretic component may be a 2 Mo dimer	(5)
Cp_2	0·4	2	—	2	2 Fe/mole	(6)

References: (1) Dalton & Mortenson (1970); (2) Bulen & LeComte (1966); (3) Burns, Holsten & Hardy (1970); (4) Mr K. A. Cook and Dr. M. Kelly (unpublished); (5) Kelly (1969*b*) and Mr K. A. Cook (unpublished); (6) Moustafa & Mortenson (1969).

air according to Kelly (1969*b*). Data on the molecular weight and other properties of proteins 1 differ widely according to laboratory and organisms; electrophoretic inhomogeneity caused by aggregation and disaggregation may account for some differences (cf. Table 2). Burns, Holsten & Hardy (1970) obtained crystals of Av_1 which came out of solution of low ionic strength; it was not homogeneous to electrophoresis. A report of 6 sub-units, but only two Mo atoms (Dalton & Mortenson, 1970) in Cp_1, if authenticated, indicates that the protein may be an aggregate of dissimilar molecular species. In these circumstances the widespread belief that nitrogenase is a two-component system (Jeng, Devanathan & Mortenson, 1969) will need reconsideration, and reports of three components (Taylor, 1969; Kajiyama, Maksuki & Nosoh, 1969) which have not been substantiated elsewhere may require taking more seriously.

Proteins 2

Proteins 2 are the oxygen-sensitive components of nitrogenase and are non-haem iron proteins, without molybdenum, of mol.wt in the region of 50,000. Data for Cp_2, which has been called 'azoferredoxin', are included in Table 2; data for proteins 2 purified from other microorganisms have not been published, though reports of cruder preparations of Av_2 (Bulen & LeComte, 1966) and Ac_2 (Kelly, 1969*a*) suggest that their properties are similar. Oxygen damage to proteins 2 from all microorganisms is rapid and irreversible; in Cp_2 it leads to bleaching and denaturation (Moustafa & Mortenson, 1969).

Need for ATP

Nitrogenase from all sources studied so far require ATP and Mg^{2+} for activity, as well as a reductant. Thus 'reductant-activated ATP-ase' activity accompanies nitrogenase function. In theory the formation of ammonia from dinitrogen would be exergonic, yet the amounts of ATP required are stoichiometric rather than catalytic. Values for the (ATP consumed)/(electron transferred) (ATP/e) range from 2·5 to less than 1 in the literature; with undissociated nitrogenase from Av or Cp a value approximating to 2·5 is consistently obtained (see Hadfield & Bulen, 1969), but with purified Ac_1 plus Ac_2 Kelly (1969b) found that the value varied considerably, with substrate being reduced (below) and with ratio of protein 1 to 2. The reasons for these discrepancies are obscure, but Jeng et $al.$ (1969) and Bui & Mortenson (1969) have evidence that even highly purified $Cp_1 + Cp_2$ show non-reductant-activated ATP-ase activity at about pH 5·2, though the enzyme complex has no nitrogenase activity at this pH value. Such side effects could lead to high values for the ATP/e ratio if they persist at physiological pH values, particularly if, as Kelly (1969a) proposed, substrates may have different effects in diverting ATP into such 'useless' hydrolysis (i.e. hydrolysis not contributing to nitrogenase function).

The amount of ATP needed and its role in nitrogenase function is thus still obscure. The empirical necessity for ATP is, however, unquestioned. This need apparently persists in living organisms and is reflected, particularly with anaerobic bacteria, in the low yields of bacteria obtained per mole of substrate when the organisms are fixing nitrogen compared with when they are assimilating ammonium.

Substrate specificity

Nitrogenase from all sources reduces substrates other than dinitrogen and, if no substrate is available, discharges the H^+ ion of the aqueous environment as H_2. Thus nitrogenase shows a 'reductant-activated hydrogenase' function, but this property does not explain the enigmatic association of nitrogenase with conventional hydrogenase. Reducible analogues of dinitrogen include acetylene, the cyanide ion, the azide ion, nitrous oxide, methyl isocyanide and various homologues of these compounds. Reduction of any of these substrates is accompanied by conversion of ATP to ADP and by evolution of hydrogen; it is specifically associated with nitrogenase and is absent from organisms in which nitrogenase synthesis has been repressed by growth with ammonia. Hardy & Burns (1968) have listed the classes of reducible substrates and their homologues and discussed the significance of the reactions; a late addition to the list is cyanogen (Dr M. Kelly, personal communication). The reduction of acetylene is of great practical value because the product, ethylene, may be detected with great sensitivity using vapour-phase chromatography; acetylene reduction is now widely used as a test for nitrogenase, a matter discussed further below. The reduction of methyl isocyanide is of special theoretical interest because the products, mainly methane plus methylamine,

$$CH_3NC \rightarrow CH_3NH_2 + CH_4,$$

are those to be expected if the substrate is bound to a transition metal (Kelly, Postgate & Richards, 1967). Some C_2 and even C_3 by-products are formed both by the enzyme and by inorganic models; formation of C_1 radicals or insertion reactions with 'alkylnitrogenase' intermediates have been proposed to account for these products (Hardy & Burns, 1968). Metal binding is also suggested from the chemistry of the reduction of allene and of acrylonitrile (Fuchsman & Hardy, 1970). Reducible substrates usually interact competitively with respect to each other and to the hydrogen evolution reaction, but further study of the kinetics of mixed substrate reduction should be rewarding in view of a curious discovery in the A.R.C. Unit that acetylene may stimulate cyanide reduction at low concentrations (Biggins & Kelly, 1970). Carbon monoxide, a structural analogue of dinitrogen, inhibits the reduction of N_2 and of all other substrates, but not the hydrogen evolution reaction or the accompanying ATP hydrolysis. Oxygen is theoretically an analogue of dinitrogen, but experiments to study its possible reduction are not easily designed because of the oxygen sensitivity of protein 2 and also the need to use autoxidizable electron donors such as ferredoxin or sodium dithionite in the assay systems.

Separate functions

The need for a reductive step and an ATP-utilization process in nitrogenase function has tempted scientists to seek separate functions for proteins 1 and 2. From exclusion chromatography studies with gels equilibrated with radioactive substrates, Bui & Mortenson (1968) obtained evidence for specific interaction of Cp_1 with radioactive ATP and Cp_2 with radioactive cyanide. Biggins & Kelly (1970) performed comparable experiments with Kp_1 and Kp_2 and cast doubt on the significance of such tests: proteins such as ferredoxin, serum albumin and haemoglobin showed 'preference' for ATP or cyanide. Kelly & Lang (1970) used equilibrium dialysis at low temperatures to show that

Kp_1 and Kp_2 together were necessary for binding of labelled cyanide and that CO then antagonized the process.

Cross reactions

Proteins 1 and 2 show cross-reactivity (Detroy et al. 1968; Kelly, 1969b). For example, $Ac_1 + Kp_2$ are as active as $Ac_1 + Ac_2$, and the reverse combination, $Kp_1 + Ac_2$, is equally active. Partial cross-activities exist: $Bp_1 + Ac_2$ is only 55 % as active as $Bp_1 + Bp_2$, and crosses such as $Cp_{1,\,2} + Ac_{2,\,1}$ are inactive. Yet crosses of the type $Bp_{1,\,2} + Ac_{2,\,1}$ and $Bp_1 + Cp_2$ are active, so a matrix of cross-activities may be constructed which may reflect evolutionary relationships among the appropriate microbes. The pattern of cross-reactions is, to a first approximation, independent of substrate being reduced, but the extent of ATP hydrolysis does not always conform (Kelly, 1969b), perhaps for reasons similar to those (above) which affect precise estimation of the ATP/e ratio. Differences in the ratios of secondary products from methyl isocyanide, depending on the type of cross tested, suggest that both proteins are involved in the stereochemistry of the functional site of nitrogenase.

Reductant and electron transport

Sodium dithionite was introduced by Bulen and his colleagues (Bulen, Burns & LeComte, 1965) as reductant for Av nitrogenase preparations. Excess dithionite can be harmful: preparations from the blue-green alga Anc appear relatively sensitive to dithionite (Smith & Evans, 1970). Nevertheless, dithionite is much used in enzymological studies but is hardly a 'physiological' reductant. The natural substrate for Cp nitrogenase is pyruvate, which acts as both reductant and ATP-generator, and the principal electron transport factor is ferredoxin. Crude preparations from Kp and Bp function with pyruvate; pyruvate also acts with Anc extracts, which suggests that transfer of electrons generated by the photosynthetic system to nitrogenase is not essential in this phototroph. Inhibition experiments indicate that light-generated ATP is used however (Cox & Fay, 1969). In Av, Ac and Mf, pyruvate does not link with nitrogenase and the nature of the natural electron donor, as well as the electron transport system, is problematical. Klucas & Evans (1968) presented evidence that an NADH-generating system was involved, which could be linked to Av or rhizobial N_2-ase with a viologen dye; in nodule preparations a flavin and an acetone powder factor would replace the viologen dye (Evans, 1970). Yates & Daniel (1970) studied preparations of Ac of different degrees of disruption and observed particulate preparations which coupled

N_2-ase to NADH dehydrogenase via a viologen dye. They also obtained membrane preparations ('large particles') which reduced acetylene with NADH or NADH-generating substrates (glucose, sucrose, etc.) provided endogenous substrates were dialysed away. These preparations and that described by Evans (1970) from nodules are the first from aerobic systems which make use of wholly 'natural' substrates and components from the organisms themselves. The ATP relationships of Yates & Daniel's 'large particles' from *Azotobacter* were peculiar: no added ATP was needed and it was, in fact, inhibitory. Anaerobic conditions were required for maximum activity in all their preparations.

Bulen, Burns & LeComte's (1964) earliest consistently active preparation of *Av* used a hydrogen-donating system from *Cp* (hydrogenase + ferredoxin) as electron donor for dinitrogen reduction. Adopting the same principle, Arnon and his colleagues (Benemann *et al.* 1969; Yoch *et al.* 1970) used chloroplast preparations from spinach (heated to suppress O_2 evolution) to donate electrons to *Av* preparations. They report that proteins of low molecular weight isolated from *Av* (azotoflavin and 'azotobacter ferredoxin') also couple to *Av* nitrogenase. Koch *et al.* (1970) obtained a non-haem iron protein from soybean nodules which coupled such chloroplasts to the N_2-ase extracted from soybean nodules. Yates (1970*b*) showed that *Ac* cytochrome-c_4, 'ferredin' and an unidentified dialysable low-molecular-weight nonprotein factor all augmented acetylene reduction by a dilute $Ac_1 + Ac_2$ but had no effect with strong preparations. They also augmented oxygen damage to the nitrogenase preparation slightly and Yates proposed that they influenced the proportion or the conformation of an Ac_{1+2} complex in such a manner as to favour formation of the prosthetic site. Thus they stimulated acetylene reduction but did not directly influence electron transport at all. This view represents a valid criticism of the present evidence for electron transport factors in this system: factors which promote association of the components of nitrogenase will simulate electron transport function in tests of this kind. But it would be unjustified to dismiss Arnon's electron transport factors as operational artifacts of this kind; the detailed nature of their activity needs studying with more purified systems.

Reports that the normal cytochrome pathway of *Av* and *Ac* are involved in nitrogen fixation are the subject of a dispute which need not be rehearsed here (Yates, 1971); at the time of writing I remain unconvinced that the normal respiratory chain of these organisms plays any part at all.

CONSEQUENCES

Distribution and character of nitrogen fixation

The ability of N_2-ase to reduce acetylene has provided workers with a rapid and extremely sensitive test for the enzyme in living organisms as well as in extracts, because the product, ethylene, may be detected and estimated simply, in nanomolar concentrations, by gas–liquid chromatography. Though ethylene does appear as a mould or plant metabolite in nature, progressive acetylene reduction, suppressed by growth with NH_4, is presumptive evidence for nitrogenase in an organism. A consequence of the availability of this test has been a drastic reassessment of the nature and distribution of the agents of the nitrogen fixation step of the biological nitrogen cycle. Parejko & Wilson (1968) and Hill & Postgate (1969) cast doubt on the status of *Azotomonas insolita* and *A. fluorescens* as N_2-fixers; the latter authors could also obtain no evidence for N_2 fixation by strains of *Nocardia, Pseudomonas azotocolligans* and *P. azotogensis*, which had been deposited with the NCIB as able to fix dinitrogen. A strain of *P. azotogensis* did fix N_2, but only when growing anaerobically; it differed in several respects from its discoverers' description. Millbank (1969) has reassessed the ability of yeast and *Pullularia* to fix N_2 and concluded that they do not do so. His published tests did not exclude the possibility of anaerobic micro-aerophilic fixation, but more recent experiments have taken care of this (Dr J. Millbank, personal communication). Millbank's experiments seem to exclude the only representatives of eukaryotes from the list of authentic nitrogen-fixing microbes.

If these findings are accepted as generally valid, the ability to fix dinitrogen *aerobically* seems to be restricted to the Cyanophyceae, the family Azotobacteriaceae (genera *Azotobacter, Azotococcus, Azomonas, Beijerinckia* and *Derxia*) and two other obligate aerobic bacteria: *Mycobacterium flavum* 301 (an organism which does not fit into non-Russian criteria for *Mycobacterium*: D. R. Biggins, unpublished) and '*Pseudomonas methanitrificans*' (Coty, 1967), a methane-oxidizing organism which is not a pseudomonas since it multiplies by budding (Whittenbury *et al.* 1970). Though the acetylene test has not been the only evidence for the exclusion of the aerobes listed, it has played an important part in such studies by virtue of its simplicity. At the time of writing there is a clear need for critical re-examination of reports of fixation by arthrobacters, spirilla, pseudomonas and other aerobes. One reason for previous errors regarding these bacteria is their ability to scavenge fixed N from the atmosphere or from impurities in medium constituents; some of them also have a remarkable ability to 'make do' with extremely low N-contents and this may make growth tests misleading: a yeast which simulated N_2-fixation in my laboratory contained just over 1 % N compared with an average content for most microorganisms of between 10 % and 15 % N. Incorporation of isotopic dinitrogen in earlier experiments is less easily dismissed, though handling errors (residual $^{15}NO_2$ or $^{15}NH_3$ in the gas) and cross-contamination from glassware (Newman, 1966), tap grease and so on could conceivably lead to error.

In contrast to the situation with aerobes, the variety and distribution of recognized anaerobic nitrogen-fixing bacteria has increased. Nitrogen fixation among the non-sporulating sulphate-reducing bacteria has been a matter of dispute since Sisler & ZoBell's (1951) report. Their finding was not confirmed in other laboratories and Le Gall, Senez & Pichinoty (1959) felt it necessary to report anew two authentic strains, the 'Berre' strains of *Desulfovibrio desulfuricans*, which fixed nitrogen both in their laboratories and in other people's. Reiderer-Henderson & Wilson (1970) reported fixation by strains of *Desulfovibrio vulgaris*, *D. gigas* and *D. desulfuricans* which had, in unpublished experiments in other laboratories, proved negative; their claim was substantiated by positive acetylene tests. Presumptive nitrogen fixation by other strains of *D. desulfuricans* based on the acetylene test was reported by Postgate (1970b), though type strains of *D. salexigens* and *D. africanus* did not fix. Mesophilic members of the genus *Desulfotomaculum*, the sporulating sulphate-reducing bacteria, all fixed dinitrogen (Postgate, 1970b); following an apparently universal rule, a thermophilic strain of this genus did not. *Klebsiella*, once regarded as a 'doubtful' fixer, is now recognized to be an active and widespread organism which, like *Bacillus polymyxa*, only fixes in anaerobic conditions. The so-called '*Pseudomonas azotogensis*' of Voets & Debacher (1956) belongs to the class of facultative fixers, though it is apparently unrelated to either *Klebsiella* or *Bacillus* (Miss S. Hill, unpublished).

A virtue of the acetylene test is that it permits rapid surveys of strains or samples. The present state of evidence with the genera *Desulfovibrio*, *Bacillus* and *Klebsiella* suggests that not all species within a genus and, indeed, not all strains within a species, can fix N_2. The photosynthetic bacteria in particular require more exhaustive study because the original N_2-fixing strain of *Chlorobium thiosulfatophilum* (Lindstrom, Burris & Wilson, 1949; the strain was mis-named '*Chlorobacterium*': Larsen, 1953) was one strain ('G 2a') out of several non-fixing *Chlorobium* strains and has been lost (Professor

Table 3. *Annotated table of putative and authenticated free-living nitrogen-fixing microbes*

Numbers in parenthesis refer to references listed at foot of table.

	Comments
Obligate aerobic bacteria	
The family Azotobacteriaceae (genera: *Azotobacter, Azotococcus, Azomonas, Beijerinckia, Derxia*) (1)	All strains fix
Mycobacterium flavum 301, *Mycobacterium* sp. 571, *M. rose oalbum* 368, *M. azotabsorptum* (2)	Probably not mycobacteria according to 'western' taxonomy (3)
'Methylosinus trichosporium' (4)	Earlier reported as *Pseudomonas methanitrificans* (5)
Azotomonas sp., *Pseudomonas, Azotocolligans, P. azotogensis* 9277, *Nocardia cellulans, N. calcarea*	Type strains do not fix, see (6); one strain did, was not pseudomonas
Achromobacter, Spirillum, etc.	Need rechecking
Aerobic phototrophs (Cyanophyceae)	
Order Nostocales (*Anabena, Anabenopsis, Aulosira, Calothrix, Chlorogloea, Cylindrospermum, Nostoc, Scytonema Tolypothrix*)	Form heterocysts
Order Stigonemataoes (*Fischerella, Hapalosiphon, Mostigocladus, Stigonema, Westiellopsis*)	Form heterocysts
Order Chroococcales (*Gleocapsa*)	Does not form heterocysts (7)
Facultative anaerobic bacteria	
Klebsiella pneumoniae, K. rubiacearum	*K. rubiacearum* is probably involved in leaf nodule symbiosis (8). Many strains and species of *Klebsiella* do not fix
Bacillus polymyxa, B. macerans	
Unidentified rod	'*Pseudomonas azotogensis* Strain V' (see 6)
Fungi	
Saccharomyces, Rhodotorula, Pullularia	Probably do not fix (9)
Phototrophic bacteria	
The family Athiorhodaceae: *Rhodospirillum rubrum, Rhodopseudomonas spheroides, Rhodomicrobium* (10)	Fixation shown with single strains; is light-dependent and anaerobic
The family Thiorhodaceae: *Chlorobium thiosulfatophilum, Chromatium vinosum, C. minutissimum, Chloropseudomonas ethylicum*	Fixation seems rare in this group (see text). Two strains of *C. ethylicum* fix (Dr M. C. W. Evans, personal communication)
Obligate anaerobes	
Clostridium pasteurianum, C. butyricum	Not all strains of *C. pasteurianum* fix readily
Methanobacillus (*Methanobacterium*), *Omelianskii* (11)	Now known to be a stable association of two bacteria (14). Which fixes?
Desulfovibrio desulfuricans, D. vulgaris, D. gigas	(12, 13) Not all strains within species fix; *D. africanus* and *D. salexigens* do not fix
Desulfotomaculum ruminis, D. orientis	(12, 13) A thermophilic species (*D. nigrificans*) did not fix.

References: (1) De Ley & Park (1966); (2) Federov & Kalininskaya (1961); (3) D. R. Biggins, (unpublished); (4) Whittenbury *et al.* (1970); (5) Coty (1967); (6) Hill & Postgate (1969); (7) Wyatt & Silvey (1969); (8) Centifanto & Silver (1964); (9) Millbank (1969); (10) Lindstrom, Tove & Wilson (1950); (11) Pine & Barker (1954); (12) Reiderer-Henderson & Wilson (1970); (13) Postgate (1970*b*); (14) Bryant *et al.* (1967).

H. Larsen, personal communication); fixation by red sulphur bacteria appears only to be established for single strains of *Chromatium vinosum* and *Chromatium minutissimum*. A wider examination of the Athiothodaceae than that carried out by Kamen & Gest (1949) and Lindstrom, Tove & Wilson (1950) is now feasible and desirable. A list of free-living, nitrogen-fixing bacteria is given in Table 3; the Cyanophyceae are taken, with one addition, from Stewart's (1969) list, a list which was exemplary in giving the numbers of strains examined.

Use of the tubes devised by Pankhurst (1966) for anaerobes has facilitated the detection and enumeration of anaerobic and facultatively anaerobic nitrogen-fixing bacteria in natural samples, making use of the acetylene test (Campbell & Evans, 1969), and Evans, Campbell & Hill (1970) showed that organisms of this kind are widespread in the rhizosphere of leguminous plants. A role for *Klebsiella* in the leaf-nodule symbiosis is now well known, as is the part played by Azotobacteriaceae in the phyllosphere (Ruinen, 1965). An ecological role for free-living anaerobes in the rhizosphere is becoming increasingly probable. Perhaps the most remarkable association suggested by the acetylene test is Bergersen & Hipsley's (1970) demonstration of active, N_2-fixing *Klebsiella*-like organisms in the intestine of a guinea pig on a low nitrogen diet and in the faeces of men. Early reports of nitrogen-fixing bacteria in the blood of whales (Laurie, 1933) or the tissues of insects (Toth, 1952) may need re-evaluation. An almost facetious (and probably unoriginal) proposal that fixation might occur in the cow rumen (Postgate, 1968) seems not to be substantiated by a test with $^{15}N_2$ though appropriate bacteria are present (Moisio, Kreula & Virtanen, 1969). Dr P. Hobson, Mr Ware and I have also demonstrated appropriate bacteria in the rumina of sheep.

Physiology of nitrogen fixation

The need for ATP in biological N_2 fixation cannot be explained on thermodynamic grounds (Bayliss, 1956) but is nevertheless an empirical fact. In consequence, nitrogen-fixing populations are obliged to divert a proportion of their carbon and energy resources to nitrogen fixation with consequent diminution of molar growth yields (Senez, 1962; Daesch & Mortenson, 1968; Dalton & Postgate, 1969 b). The physiological inefficiency of N_2 fixation probably accounts for the rarity of constitutive nitrogenase-postive strains: there must be considerable selection pressure against their persistance. Reports of such mutants of *Av* (Sorger, 1968) need checking with the acetylene test; nitrogenase-defective mutants of *Av* are well established (Fisher & Brill, 1969;

Sorger & Trefimenkoff, 1970). NH_4 represses nitrogenase synthesis but NH_4-limited populations form nitrogenase under argon or helium without the need of gross amounts of N_2 in the atmosphere. The question whether nitrogenase synthesis is a derepression or an induction process has not been advanced since the publication of reviews cited at the beginning of this article.

The oxygen sensitivity of nitrogenase presents aerobic organisms with a difficult physiological problem. Though Azotobacteriaceae are obligate aerobes, Dalton & Postgate (1969a) showed that they were readily inhibited by excessive aeration if they were fixing N_2 but not if there were utilizing NH_4. Biggins & Postgate (1969) and Kalininskaya (1967) both presented evidence that Mf grows best, when fixing nitrogen, at low P_{O_2} values despite being an obligate aerobe; Hill & Postgate (1969) showed that Derxia gummosa preferred a low P_{O_2}. Dalton & Postgate (1969a) showed that oxygen-sensitivity was greatest among carbon- or phosphate-limited populations. It seems reasonable to regard oxygen sensitivity during nitrogen fixation as a universal property of aerobic N_2-fixing bacteria; one which is influenced by the nutritional status of the population. (This property may also influence dilution counts on Azotobacteriaceae and be an important source of error in ecological studies: Billson, Williams & Postgate, 1970). Oxygenation may lead to reversible 'switch off' of nitrogenase activity (Yates, 1970a; Drozd & Postgate, 1970a; this process could interfere with field assays using the acetylene test) and populations 'adapted' to growth at high P_{O_2} values are less prone to 'switch off', have enhanced respiration but have normal contents of nitrogenase (Drozd & Postgate, 1970b). A rationalization of this kind of study is emerging in terms of two regulatory processes which are believed to occur in Azotobacter.

(a) Respiratory protection: actively N_2-fixing populations are regarded as adjusting their respiration to keep up with the external availability of oxygen and thus to prevent access of O_2 to the functioning nitrogenase. N_2 fixation therefore becomes very inefficient under high aeration and correspondingly more efficient at low dissolved oxygen concentrations.

(b) Conformational protection: chemostat cultures, unless grown in special conditions, show all the properties of populations which are limited by their ability to fix N_2, even at growth rates well below μ_{max}. This finding, together with the reversible 'switch off' of nitrogenase mentioned above and the fact that one can extract oxygen-insensitive forms of nitrogenase from azotobacters, have led to the view that simple competition for electrons between nitrogenase and respiratory oxidases does not adequately interpret the behaviour of these bacteria. The proposal has been made that the activity of nitrogenase is regulated and, in circumstances in which respiratory protection fails, a conformational change in the enzyme occurs which protects its sensitive sites from damage by oxygen but which renders it unable to fix N_2 or to reduce acetylene.

Arguments for this view, which has the status of a working hypothesis, have been presented elsewhere (Postgate, 1971) and will not be repeated here; Oppenheim & Marcus (1970a) and Oppenheim et al. (1970) have shown that N_2-grown Av possess an extensive internal membrane network which is not present in NH_4-grown populations and the formation of which is repressed by NH_4. Oppenheim & Marcus (1970b) assert that the membrane protects an insoluble nitrogenase from damage by oxygen. This membrane system would obviously provide a morphological locus for the conformational protection proposed earlier. Conformational protection may be regarded as a special case of a more general compartmentation required by aerobic, N_2-fixing organisms. Such compartmentation is physiologically necessary to separate oxygen metabolism from nitrogen fixation and, among the blue-green algae, presents a special problem because their photosynthesis yields oxygen. There is considerable evidence that nitrogenase is restricted in the majority of these organisms to specialized cells called heterocysts (Stewart, Haystead & Pearson, 1969). (This view has been questioned by Smith & Evans (1970), who claim to have obtained nitrogenase from vegetative Anc tissue, but the extent of leakage from heterocysts during their sonication procedure is not clear from their report.) An exception to the heterocyst rule is provided by the coccoidal blue-green alga Gleocapsa (Wyatt & Silvey, 1969), which fixes N_2 but does not form heterocysts; it is a remarkably photo-sensitive alga (Professor R. Y. Stanier, personal communication), which would be consistent with the low oxygen tolerance that one would predict for an organism possessing limited possibilities for physiological compartmentation.

On a grosser scale the confinement of symbiotic nitrogen-fixing microbes in root nodules or in nodules in the leaves of plants performs a similar biological function to the intracellular compartmentation just discussed, because both the leaf nodule and the root nodule systems are known to involve enzymes which require strictly anaerobic conditions. Earlier evidence that leghaemoglobin was directly involved in binding of dinitrogen now seems highly doubtful (Appleby, 1969) and the question of its function arises again. Its affinity for oxygen is very high and a plausible function might be to act as a redox buffer, protecting the bacteroids in their nodules from oxygen; a preliminary report that fixation could be detected in a tissue culture of soyabean roots infected with *Rhizobium* (Holsten *et al.* 1970) opens interesting possibilities for research on this question as on other aspects of the legume symbiosis.

As far as microbial productivity is concerned, recent studies on the physiology of nitrogen fixation have three obvious consequences.

(*a*) Most laboratory cultures of aerobic nitrogen-fixing bacteria behave as if they were limited by their ability to fix N_2. This situation arises because the functioning of nitrogenase is regulated and it follows that, in many circumstances, the organisms do not fix nitrogen at the maximum possible rate.

(*b*) Even in anaerobic bacteria, considerable diversion of the substrate from biosynthesis appears to take place during nitrogen fixation. Hence these organisms are likely to be far more productive when they are able to grow in conditions in which the availability of carbon and energy source determines their growth rate.

(*c*) Aerobic bacteria are obliged to develop complex physiological systems – either morphological compartmentation or the 'conformationally protected' state of the enzyme – to guard against damage caused by oxgyen and, in addition, must 'burn up' considerable amounts of their carbon substrates for respiratory protection. It follows that, in well-aerated laboratory cultures, rich in carbon-energy source, they will give a false impression of their efficiency as nitrogen fixers. In the natural environment, where the P_{O_2} may be low and the supplies of carbon-energy source may be slight, they probably contribute much more seriously to the productivity of that environment.

Role of transition metals

The discovery of the dinitrogen complexes of transition metals in the last four years has provided the first sound chemical basis for proposing a direct prosthetic role for iron or molybdenum in the functioning of nitrogenase. Though accident more than design entered into the discovery of the earliest examples, systematic procedures for preparing these compounds have now been developed and homologous series are available for purely chemical study. Table 4 gives examples of compounds of this class. Reasons for believing that they may be regarded as models, for certain aspects at least, of biological nitrogen fixation have, in the main, been circumstantial. For example, the products of isocyanide reduction by nitrogenase (methane and C_2 products) are

Table 4. *Examples of dinitrogen complexes*

	$\nu(N\equiv N)$	Reference
$[Ru(NH_3)_5N_2]^{+2}$	2,105 cm^{-1}	Allen & Senoff (1965)
$[IrCl(P\varnothing_3)N_2]$	2,095 cm^{-1}	Collman & Kang (1966)
$[CoH(P\varnothing_5)N_2]$	2,082 cm^{-1}	Yamamoto *et al.* (1967)
$[(\varnothing Me_2P)_4ClReN_2MoCl_4(PEt\varnothing_2)]$	1,810 cm^{-1}	Chatt *et al.* (1969)
N_2	2,331 cm^{-1}	—

those which would reasonably be expected if the substrate was initially bound to a transition metal; the bond migrations which occur in the reduction of acrylonitrile or allene most probably involve associations with a metal; so does the activation of acetylene. Dinuclear dinitrogen complexes can be obtained with decreased triple bond character in the N—N link, which would be more susceptible to reduction than $N\equiv N$ itself. But chemical reductants either displace nitrogen from transition metal complexes or leave it undisturbed, and a dinitrogen complex in which the N_2 group can be reduced chemically has yet to be reported. Direct evidence for involvement of a metal in the enzyme function has been slow in coming, but Kelly & Lang (1970) obtained Mössbauer spectra indicating that iron is involved. They purified Kp_1 and Kp_2 labelled with ^{57}Fe and showed that their spectra were similar but distinguishable. They were additive when Kp_1 and Kp_2 were mixed together, unaffected by addition of ATP or of dithionite separately, but underwent a significant change when both ATP and dithionite were added. By permuting labelled and unlabelled Kp proteins they showed that the major change was in the direction of reduction of the iron in Kp_1 by Kp_2 + the other components. Acetylene, CO and, to some extent, N_2 influenced the intensity of the combined resonances but not their positions. The detailed interpretation of the spectra is beyond the scope of the present contribution and is to some extent speculative, but two broad conclusions may be drawn.

(*a*) A change in the ligand environment of the iron in Kp_1 occurs when all the components necessary for nitrogenase activity are present.

(*b*) Substrates and the specific inhibitor influence the intensity of absorption in a manner which would, at first sight, suggest that an iron–dinitrogen complex is not formed, but that the ligand environment of the iron changes in response to substrate or inhibitor binding elsewhere. However, since there is a great deal of intramolecular iron in the enzyme complex (some twelve Fe atoms/active unit of $Kp_1 + 2Kp_2$), and only one is likely to combine with N_2, the appropriate resonances may have been masked.

The second conclusion is therefore extremely tentative, but these experiments provide direct evidence for the involvement of iron in a fairly intimate way in the biological N_2-fixation process.

CONCLUSIONS

The biological productivity of an eco-system is measured by the rate at which the biological elements change from biological to purely chemical combination and back again: the rate at which the biological cycles of the elements 'turn'. In many parts of this planet the primary productivity of the earth's surface is limited by the nitrogen cycle, and the rate-determining step within this cycle is biological nitrogen fixation. In the context of the limited protein resources of this planet, and considering its present and predicted population, any steps which may be taken to increase the rate of nitrogen fixation may be considered desirable, at least in the first instance. In the long term the demand must obviously be diminished rather than the productivity increased, but this requires a degree of enlightenment which is not practicable when vast masses of people live on the borders of protein starvation. It can be argued that advances in knowledge of biological nitrogen fixation may make little if any contribution to the present protein crisis, because established processes are available, and have been available for many years, able to produce ammonium fertilizers extremely cheaply by purely chemical means using effectively amortized capital investment. Moreover, when nitrogen fertilizers are freely available, other deficiencies become manifest; notably the widespread appearance of sulphur-deficient soils in many parts of this planet in the last decade. Use of chemically fixed nitrogen has, in some countries, given rise to areas in which the sulphur cycle, not the nitrogen cycle, is the limiting primary production process. These considerations are true of relatively highly industrialized countries, but the countries where the need for protein is greatest are frequently the least industrialized and also those with the least ability to pay for transport of chemical fertilizer, however cheaply produced, from countries able to make it. Hence, on a global scale, understanding of this process from both biological and chemical points of view is extremely important. In particular, knowledge of the sort of organisms that contribute to productive ecosystems, knowledge of the most productive kinds of associations, can be expected to be extremely valuable. We have learnt in recent years why the area around blowholes of natural gas proves to be extra fertile; why leaf fall increases the fertility of soils over and above the nitrogen content of leaves;

we now have suggestions of associations with animals, insects and roots which were not recognized earlier; we now realize that the sulphur cycle can be self-sufficient as regards nitrogen; we are learning the distinctive features of arctic, tropical and even marine ecosystems. We have also learnt techniques for detecting and studying nitrogenase function, which have opened possibilities for mutant studies and genetic manipulation; we have discovered new properties of the enzyme, and of 'model' chemicals, which will have repercussions in the areas far removed from agriculture and even from any reasonable extension of the nitrogen cycle. Does this symposium regard contributions to general technical and scientific knowledge derived from the study of nitrogen-fixing microbes as forms of biological productivity? I hope so.

The moral, as always, is that the major rewards of basic research are often the most unexpected ones. In the present context, it is clear that more exhaustive study and reassessment of the contributors to the N_2 fixation step of the nitrogen cycle in a variety of ecosystems is still badly needed. This thought leads us to what may seem a most mundane conclusion. Ecological studies require the enumeration of the organisms contributing to the character of the ecosystem. We now have trustworthy ways, making use of the acetylene test, of assessing numbers and activities of anaerobic microorganisms capable of fixing nitrogen. What we need badly, still, is a trustworthy method of enumerating aerobic nitrogen fixers.

REFERENCES

ALLEN, A. D. & SENOFF, C. V. (1965). Nitrogenopentammineruthenium (II) complexes. *Chemical Communications*, p. 621.

APPLEBY, C. A. (1969). The nature of the supposed N_2 complex of ferroleghaemoglobin. *Biochim. biophys. Acta* **180**, 202.

ARNON, D. I., LOSADA, M., NOZAKI, M. & TAGAWA, K. (1960). Photoproduction of hydrogen, photofixation of nitrogen and a unified concept of photosynthesis. *Nature, Lond.* **190**, 601.

BAYLISS, N. S. (1956). The thermochemistry of biological nitrogen fixation. *Aust. J. Biol. Sci.* **9**, 364.

BENEMANN, J. R., YOCH, D. C., VALENTINE, R. C. & ARNON, D. I. (1969). The electron transport system in nitrogen fixation by *Azotobacter*. I. Azotoflavin as an electron carrier. *Proc. natn. Acad. Sci. U.S.A.* **64**, 1079.

BERGERSEN, F. J. (1966). Some properties of nitrogen-fixing breis prepared from soybean root nodules. *Biochem. biophys. Acta* **130**, 304.

BERGERSEN, F. J. & HIPSLEY, E. H. (1970). The presence of N_2-fixing bacteria in the intestines of man and animals. *J. gen. Microbiol.* **60**, 61.

BIGGINS, D. R. & KELLY, M. (1970). Reaction of nitrogenase from *Klebsiella pneumoniae* with ATP or cyanide. *Biochim. biophys. Acta* **205**, 288.

BIGGINS, D. R. & POSTGATE, J. R. (1969). Nitrogen fixation by cultures and cell-free extracts of *Mycobacterium flavum* 301. *J. gen. Microbiol.* **56**, 181.

BILLSON, S., WILLIAMS, K. & POSTGATE, J. R. (1970). A note on the effect of diluents on the determination of viable numbers of Azotobacteriaceae. *J. appl. Bact.* **33**, 270.

BRYANT, M. P., WOLIN, E. A., WOLIN, M. J. & WOLFE, R. S. (1967). *Methanebacillus omelianskii*, a symbiotic association of two species of bacteria. *Arch. Mikrobiol.* **59**, 20.

BUI, P. T. & MORTENSON, L. E. (1968). Mechanism of the enzymic reduction of N_2: the binding of adenosine S-triphosphine and cyanide to the N_2-reducing system. *Proc. natn. Acad. Sci. U.S.A.* **61**, 1021.

BUI, P. T. & MORTENSON, L. E. (1969). The hydrolysis of adenosine triphosphate by purified components of nitrogenase. *Biochemistry* **8**, 2462.

BULEN, W. A. & LeCOMTE, J. R. (1966). The nitrogenase system from Azotobacter: two-enzyme requirement for N_2-reduction, ATP-dependent H_2 evolution and ATP hydrolysis. *Proc. natn. Acad. Sci. U.S.A.* **56**, 979.

BULEN, W. A., BURNS, R. C. & LeCOMTE, J. R. (1964). Nitrogen fixation: cell-free system with extracts from azotobacter. *Biochem. biophys. Res. Commun.* **17**, 3.

BULEN, W. A., BURNS, R. C. & LeCOMTE, J. R. (1965). Nitrogen fixation: hydrosulfite as electron donor with cell-free preparations of *Azotobacter vinelandii* and *Rhodospirillum rubrum*. *Proc. natn. Acad. Sci. U.S.A.* **53**, 532.

BURNS, R. C., HOLSTEIN, R. D. & HARDY, R. W. F. (1970). Isolation by crystallization of the Mo-Fe protein of *Azotobacter* nitrogenase. *Biochem. biophys. Res. Commun.* **39**, 90.

CAMPBELL, N. E. R. & EVANS, H. J. (1969). Use of Pankhurst tubes to assay acetylene reduction by facultative and anaerobic nitrogen-fixing bacteria. *Can. J. Microbiol.* **15**, 1342.

CARNAHAN, J. E., MORTENSON, L. E., MOWER, H. F. & CASTLE, J. E. (1960). Nitrogen fixation in cell-free extracts of *Clostridium pasteurianum*, *Biochim. biophys. Acta* **44**, 520.

GENTIFANTO, Y. M. & SILVER, W. S. (1964). Leaf-nodule symbiosis. I. Endophyte of *Psychotria bacteriophila*. *J. Bact.* **88**, 776.

CHATT, J. & FOGG, G. E. (1969). A discussion on nitrogen fixation. *Proc. Roy. Soc. Lond.* B **172**, 317.

CHATT, J., DILWORTH, J. R., RICHARDS, R. L. & SANDERS, J. R. (1969). Chemical evidence concerning the function of molybdenum in nitrogenase. *Nature, Lond.* **224**, 1201.

COLLMAN, J. P. & KANG, J. W. (1966). Iridium complexes of molecular nitrogen. *J. Am. Chem. Soc.* **88**, 3459.

COTY, V. F. (1967). Atmospheric nitrogen fixation by hydrocarbon-oxidizing bacteria. *Biotech. Bioeng.* **9**, 25.

COX, R. M. & FAY, P. (1969). Special aspects of nitrogen fixation by blue-green algae. *Proc. Roy. Soc. Lond.* B **172**, 357.

DAESCH, G. & MORTENSON, L. E. (1968). Sucrose catabolism in *Clostridium pasteurianum* in relation to N_2 fixation. *J. Bact.* **96**, 346.

DALTON, H. & POSTGATE, J. R. (1969*a*). Effect of oxygen on growth of *Azotobacter chroococcum* in batch and continuous culture. *J. gen. Microbiol.* **54**, 463.

DALTON, H. & POSTGATE, J. R. (1969*b*). Growth and physiology of *Azotobacter chroococcum* in continuous culture. *J. gen. Microbiol.* **56**, 307.

DALTON, H. & MORTENSON, L. E. (1970). Studies on the iron-molybdenum complex of molybdoferredoxin. *Bact. Proc.* p. 148.

DE LEY, J. & PARK, I. W. (1966). Molecular biological taxonomy of some free-living nitrogen-fixing bacteria. *Antonie van Leeuwenhoek* **32**, 6.

DETROY, R. W., WITZ, D. F., PAREJKO, R. A. & WILSON, P. W. (1968). Reduction of N_2 by complementary functioning of the components from nitrogen fixing bacteria. *Proc. natn. Acad. Sci. U.S.A.* **51**, 537.

DROZD, J. & POSTGATE, J. R. (1970a). Interferency by oxygen in the acetylene reduction test for aerobic nitrogen-fixing bacteria. *J. gen. Microbiol.* **60**, 427.

DROZD, J. & POSTGATE, J. R. (1970b). Effects of oxygen on acetylene reduction, cytochrome content and respiratory activity of *Azotobacter chroococcum.* *J. gen. Microbiol.* **63**, 63.

EVANS, H. J. (1970). How legumes fix nitrogen. In *How Crops Grow.* Ed. J. G. Horsefall. *Bull. Conn. Agric. Exp. Stn* no. 708, p. 110.

EVANS, H. J., CAMPBELL, N. E. R. & HILL, S. (1970). Asymbiotic nitrogen-fixing bacteria from the surfaces of nodules and roots of legumes. *J. Bact.* (In the Press).

FEDEROV, M. V. & KALININSKAYA, T. (1961). A new species of nitrogen-fixing mycobacterium and its physiological properties. *Mikrobiologiya* **30**, 9.

FISHER, R. J. & BRILL, W. J. (1969). Mutants of *Azotobacter vinelandii* unable to fix nitrogen. *Biochim biophys. Acta* **184**, 99.

FUCHSMAN, W. H. & HARDY, R. W. F. (1970). Model studies on nitrogenase (N_2-ase): acrylonitrile reduction. *Bact. Proc.* p. 148.

GRAU, F. H. & WILSON, P. W. (1962). Hydrogenase and nitrogenase in cell-free extracts of *Bacillus polymyxa. J. Bact.* **85**, 446.

HADFIELD, K. L. & BULEN, W. A. (1969). Adenosine triphosphate requirement of nitrogenase from *Azotobacter vinelandii. Biochemistry* **8**, 5103.

HAMILTON, I. R., BURRIS, R. H. & WILSON, P. W. (1964). Hydrogenase and nitrogenase in a nitrogen-fixing bacterium. *Proc. natn. Acad. Sci. U.S.A.* **52**, 637.

HARDY, R. W. F. & BURNS, R. C. (1968). Biological nitrogen fixation. *A. Rev. Biochem.* **37**, 331.

HILL, S. & POSTGATE, J. R. (1969). Failure of putative nitrogen-fixing bacteria to fix nitrogen. *J. gen. Microbiol.* **58**, 277.

HOLSTEN, R. D., HEBERT, R. R., BURNS, R. C. & HARDY, R. W. F. (1970). Symbiotic N_2-fixation in plant cell cultures. *Bact. Proc.* p. 149.

JENG, P. Y., DEVANATHAN, T. & MORTENSON, L. E. (1969). Components of cell-gree extracts of *Clostridium pasteurianum* W 5 required for acetylene reduction and N_2 fixation. *Biochem. biophys. Res. Commun.* **35**, 625.

JENG, D. Y., MORRIS, J. A., BUI, P. T. & MORTENSON, L. E. (1969). Influence of hydronium ions on ATP utilization by purified nitrogenase of *Clostridium pasteurianum. Fedn Proc.* **28**, 667.

KAJIYAMA, S., MAKSUKI, T. & NOSOH, Y. (1969). Separation of the nitrogenase system of azotobacter into three components and purification of one of the components. *Biochem. biophys. Res. Commun.* **37**, 711.

KALININSKAYA, T. (1967). The role of symbiotic microbes in the fixation of nitrogen by free-living micro-organisms. In *Biological Nitrogen Fixation and Its Role in Agriculture.* Moscow: U.S.S.R. Academy of Sciences.

KAMEN, M. D. & GEST, H. (1949). Evidence for a nitrogenase system in the photosynthetic bacterium *Rhodospirillum rubrum. Science, N.Y.* **109**, 560.

KELLY, M. (1968). The kinetics of the reduction of isocyanides, acetylene and the cyanide ion by nitrogenase preparations from *Azotobacter chroococcum* and the effect of inhibitors. *Biochem. J.* **107**, 1.

KELLY, M. (1969a). Some properties of purified nitrogenase of *Azotobacter chroococcum. Biochim. biophys. Acta* **171**, 9.

KELLY, M. (1969b). Comparisons and cross-reactions of nitrogenase from *Klebsiella pneumoniae, Azotobacter chroococcum* and *Bacillus polymyxa. Biochim. biophys. Acta* **191**, 527.

KELLY, M. & LANG, G. (1970). Evidence from Mössbauer spectroscopy for the role of iron in nitrogen fixation. *Biochim. biophys. Acta.* (In the Press.)

KELLY, M., KLUCAS, R. V. & BURRIS, R. H. (1967). Fractionation and storage of nitrogenase from *Azotobacter vinelandii*. *Biochem. J.* **105**, 3c.

KELLY, M., POSTGATE, J. R. & RICHARDS, R. L. (1967). Reduction of cyanide and isocyanide by nitrogenase of *Azotobacter chroococcum*. *Biochem. J.* **102**, 1c.

KLUCAS, R. V. & EVANS, H. J. (1968). An electron donor system for nitrogenase-dependent acetylene reduction by extracts of soybean nodules. *Pl. Physiol.* **43**, 1458.

KOCH, B., EVANS, H. J. & RUSSELL, S. (1967). Reduction of acetylene and nitrogen gas by breis and cell-free extracts of soybean root nodules. *Pl. Physiol.* **42**, 466.

KOCH, B., WONG, P., RUSSELL, S., HOWARD, R. & EVANS, H. J. (1970). Purification and some properties of a non-haem iron protein from the bacteroids of soybean (Glycine max. merr.) nodules. *Biochem J.* (In the Press.)

LARSEN, H. (1953). On the microbiology and biochemistry of the photosynthetic green sulfur bacteria. *Det. Kgl. Norska Videnshaben Selskaln Skrifter* 1953 *NRI.* Trondheim: F. Bruns.

LAURIE, A. H. (1933). Some aspects of respiration in blue and fin whales. '*Discovery*' *Rep.* **7**, 365.

LE GALL, J., SENEZ, J. C. & PICHINOTY, F. (1959). Fixation de l'azote par les bactéries sulfato-réductrices. *Annls Inst. Pasteur, Paris* **96**, 223.

LINDSTROM, E. S., BURRIS, R. H. & WILSON, P. W. (1949). Nitrogen fixation by photosynthetic bacteria. *J. Bact.* **58**, 313.

LINDSTROM, E. S., TOVE, S. R. & WILSON, P. W. (1950). Nitrogen fixation by the green and purple sulfur bacteria. *Science, N.Y.* **112**, 197.

MILLBANK, J. W. (1969). Nitrogen fixation in moulds and yeasts – a reappraisal. *Arch. Mikrobiol.* **68**, 32.

MOISIO, T., KREULA, M. & VIRTANEN, A. I. (1969). Experiments on nitrogen fixation in cow's rumen. *Acta chem. fenn.* B **42**, 432.

MOUSTAFA, E. & MORTENSON, L. E. (1969). Properties of azoferredoxin purified from nitrogen-fixing extracts of *Clostridium pasteurianum*. *Biochim. biophys. Acta.* **172**, 106.

NEWMAN, A. C. D. (1966). A distillation of ammonia for isotopic analysis. *Chemy Ind.* (3), 115.

OPPENHEIM, J. & MARCUS, L. (1970a). Correlation of ultra-structure in *Azotobacter vinelandii* with nitrogen source for growth. *J. Bact.* **101**, 286.

OPPENHEIM, J. & MARCUS, L. (1970b). Induction and repression of nitrogenase and internal membranes in *Azotobacter vinelandii*. *Bact. Proc.* p. 148.

OPPENHEIM, J., FISHER, R. J., WILSON, P. W. & MARCUS, L. (1970). Properties of a soluble nitrogenase in *Azotobacter*. *J. Bact.* **101**, 292.

PANKHURST, E. S. (1966). A simple culture tube for anaerobic bacteria. *Lab. Pract.* **16**, 58.

PAREJKO, R. A. & WILSON, P. W. (1968). Taxonomy of Azotomonas species. *J. Bact.* **95**, 143.

PINE, M. J. & BARKER, H. A. (1954). Studies in the methane bacteria. XI. Fixation of atmospheric nitrogen by *Methanobacterium omelianskii*. *J. Bact.* **68**, 589.

POSTGATE, J. R. (1968). How microbes fix nitrogen. *Sci. J.* **4**, 69.

POSTGATE, J. R. (1970a). Biological nitrogen fixation. *Nature, Lond.* **226**, 25.

POSTGATE, J. R. (1970b). Nitrogen fixation by sporulating sulphate-reducing bacteria including rumen strains. *J. gen. Microbiol.* **63**, 137.

POSTGATE, J. R. (1971). Fixation by free-living microbes: physiology. In *The Chemistry and Biochemistry of Nitrogen Fixation*. Ed. J. R. Postgate. London: Plenum Press.

REIDERER-HENDERSON, M. A. & WILSON, P. W. (1970). Nitrogen fixation by sulphate-reducing bacteria. *J. gen. Microbiol.* **61**, 27.

RUINEN, J. (1956). The phyllosphere. III. Nitrogen fixation in the phyllosphere. *Pl. Soil* **22**, 375.

SENEZ, J. C. (1962). Some considerations on the energetics of bacterial growth. *Bact. Rev.* **26**, 14.

SISLER, F. D. & ZOBELL, C. E. (1951). Nitrogen fixation by sulfate-reducing bacteria indicated by nitrogen/argon ratios. *Science, N.Y.* **113**, 511.

SMITH, R. V. & EVANS, M. C. W. (1970). Soluble nitrogenase from vegetative cells of the blue green algae *Anabena cylindrica*. *Nature, Lond.* **225**, 1253.

SORGER, G. J. (1968). Regulation of nitrogen fixation in *Azotobacter vinelandii* OP and in an apparently partially constitutive mutant. *J. Bact.* **95**, 1721.

SORGER, G. J. & TREFIMENKOFF, D. (1970). Nitrogenaseless mutants of *Azotobacter vinelandii*. *Proc. natn. Acad. Sci. U.S.A.* **65**, 74.

STEWART, W. D. P. (1969). Biological and ecological aspects of nitrogen fixation by free-living micro-organisms. *Proc. Roy. Soc. Lond.* B **172**, 367.

STEWART, W. D. P., HAYSTEAD, A. & PEARSON, H. W. (1969). Nitrogenase activity in heterocysts of blue-green algae. *Nature, Lond.* **224**, 226.

TAYLOR, K. B. (1969). The enzymology of nitrogen fixation in cell-free extracts of *Clostridium pasteurianum*. *J. Biol. Chem.* **244**, 171.

TOTH, L. (1952). The role of nitrogen-active micro-organisms in the nitrogen metabolism of insects. *Tijdschr. Ent.* **95**, 43.

VOETS, J. P. & DEBACHER, J. (1956). *P. azotogensis* nov.spp. A new free-living nitrogen-fixing bacterium. *Naturwissenschaften* **43**, 40.

WHITTENBURY, R., PHILLIPS, K. C. & WILKINSON, J. F. (1970). *J. gen. Microbiol.* **61**, 205.

WYATT, J. T. & SILVEY, J. K. G. (1969). Nitrogen fixation by Gleocapsa. *Science, N.Y.* **165**, 908.

YAMAMOTO, A., KITAZUME, S., PU, L. S. & IKEDA, S. (1967). Study of the fixation of nitrogen. Isolation of tris (triphenylphosphase) cobalt complex co-ordinated with molecular nitrogen. *Chem. Commun.* p. 79.

YATES, M. G. (1970a). Control of respiration and nitrogen fixation by oxgyen and adenosine nucleotides in N_2-grown *Azotobacter chroococcum*. *J. gen. Microbiol.* **60**, 393.

YATES, M. G. (1970b). Effect of non-haem iron proteins and cytochromes *c* from *Azotobacter* upon the activity and oxygen sensitivity of *Azotobacter* nitrogenase. *FEBS Letters* **8**, 281.

YATES, M. G. (1971). A review of research by Soviet scientists in the biochemistry of nitrogen fixation. In *The Chemistry and Biochemistry of Nitrogen Fixation*. Ed. J. R. Postgate. London: Plenum Press.

YATES, M. G. & DANIEL, R. M. (1970). Acetylene reduction with physiological electron donors by extracts and particulate fractions from nitrogen-fixing *Azotobacter chroococcum*. *Biochim. biophys. Acta* **197**, 161.

YOCH, D. C., BENEMANN, J. R., VALENTINE, R. C. & ARNON, D. I. (1970). Azotobacter ferredoxin as an electron carrier in N_2 fixation by extracts of *Azotobacter vinelandii*. *Bact. Proc.* p. 148.

ASSOCIATIONS OF
MICROBES AND ROOTS

J. L. HARLEY

Department of Forestry, University of Oxford

INTRODUCTION

The surfaces and environs of the roots of higher plants are specialized microhabitats for soil organisms. Perhaps the earliest indication of this is the work of Reissek (1847), who observed the colonization of the roots of angiosperms by fungal hyphae. His observations were soon followed and overshadowed by the discovery and study of the more conspicuous microbe and root associations such as the nodules of legumes and the mycorrhizas of forest trees and other plants, in which some degree of permanent morphological or histological change results from the cooperative growth of the two dissimilar organisms. It was not therefore till later (Hiltner, 1904) that it was realized that all root systems possess an entourage of microorganisms around and upon their surfaces. The rhizosphere, as it has come to be called, has been shown to be colonized in all plants by populations of bacteria, actinomycetes and fungi more numerous and active per unit weight or volume of soil than the soil at a distance from the root.

Questions have therefore arisen as to why these populations exist in the root region and whether and in what ways they affect the growth and development of the host plant. These questions have been considered separately from those concerning the properties of microbial nodules and mycorrhizas although the two subjects have much in common.

Rhizosphere populations

The root regions of plants, besides possessing extremely numerous and active populations of microorganisms, are selective. Their populations differ in the relative numbers of the different kinds of organisms from the general soil population and contain larger numbers and proportions which require special substances for optimum growth. The study both of the special nutrient requirements of their organisms and also of the materials released from roots as exudates or by the sloughing of senescent cells has afforded some explanation of the 'rhizosphere effect'. (See review by Katznelson, 1965; Rovira, 1965*a*, *b*.)

A most impressive list of substances released from growing roots has now been compiled. It includes diverse carbohydrates, amino acids, organic acids, vitamins, enzymes and other organic substances, the presence of which in the root region could well afford explanation for the presence of nutritively selective microorganisms. In addition the quality and quantity of the substances released by the roots have been shown to be affected by the age and condition of the host and to vary with the position within the root systems. Factors such as moisture stress, light intensity, nutrient supply, as well as the activities of the microorganisms themselves, all exert effects on the quantity and quality of substances released by the root and present in the root region. In addition to the known compounds a variety of unidentified substances have been shown to exist in the root region of certain species of plant which attract eelworms, stimulate hatching of eelworm cysts, encourage spore formation or spore germination of fungi, attract zoospores, inhibit mycelial or bacterial growth, etc. Hence the chemically and physically complex habitat of the root region, although affording a general explanation of the observations, as yet proved too complicated for detailed analyses.

THE EFFECT OF RHIZOSPHERE POPULATIONS ON THE HOST

In the face of these complexities the apparently simpler questions of whether and in what ways the presence of the rhizosphere populations affect the growth of the host might seem more attractive. Two aspects have been specially investigated, namely their effect upon disease incidence and their effect on nutrient absorption into the host plant.

Control of disease

Work such as that of Henry (1931) (Table 1) indicated that the infection of plants by soil-borne pathogens occurred with much greater facility in sterilized than in unsterilized soil. A trace of unsterilized soil added to sterilized soil in which wheat was germinated and grown was sufficient to afford almost complete protection to the wheat in the face of an inoculum sufficient to cause almost total disease incidence in sterilized soil. Such results, together with an increasing knowledge of biological antagonism and later of antibiotic substances, emphasized the fact that pathogens have limited specific niches in natural ecosystems and suggested that the rhizosphere and root-surface populations could especially constitute a barrier to infection by pathogens by competing for materials and possibly also by producing antagonistic substances.

Table 1. *The effect of sterility of the soil upon the pathogenicity of Helminthosporium sativum (data of Henry 1931)*

Amount of unsterile soil added per pot	Recovery of pathogen from soil per 100 attempts	Infection (%)
0	100	47·6
Trace	30	7·8
1 g.	0	7·0
5 g.	0	3·1
50 g.	0	3·5

Many experimental results have upheld these suggestions in a general way (Garrett, 1956). For example, the difficulty of inoculation of wheat, even with heavy doses of spores of *Ophiobolus graminis*, unless the plants were in sterilized soil was first described by Garrett (1939). Brooks (1965) further showed that wheat seedlings germinated on the surface of damp soil were easily infected. Indeed, infection by *Ophiobolus* ascospores was not difficult provided that young roots not yet fully equipped with a population of microorganisms in their root regions were inoculated. However, the full explanation of how the rhizosphere populations afford a protection to their hosts is not easy to obtain. Considerable efforts have been made to describe the differences between the populations of disease-resistant and susceptible varieties of crop plants but they do not afford a fully satisfying explanation. Often, however, the presence of species known to produce antibiotic substances have been invoked. The more rapid growth of *Fomes annosus* over pine roots in alkaline rather than acid soils was believed by Rishbeth (1950) to be due to the antagonistic effects of the root-surface fungi of acid soils especially of *Trichoderma viride*. This explanation has been recently queried (see Garrett, 1970) on the grounds that Rifai's (1969) revision of *Trichoderma viride* Pers. ex Fr. (*sensu* Bisby), together with the studies of the variation of Rifai's nine species in antagonistic properties by Gibbs (1967) and Mughogho (1968), throw doubt whether strains antagonistic to *Fomes* are really more prevalent on acid soils. *Trichoderma viride* was observed to be more abundant in the rhizosphere of flax resistant to *Fusarium* and in the artificial rhizospheres surrounding porous tubes of root exudate of resistant flax than in those of susceptible varieties by Timonin (1941). Resistant 'Bison' flax was recorded both by Timonin and by Reynolds (1931) to release cyanide into its rhizosphere, which reduced the growth of pathogens such as *Fusarium* and *Helminthosporium* but stimulated *Trichoderma*. Wright (1956) was able to obtain a degree of control of *Pythium* upon seedlings of mustard by dusting the seed with species of common saprophytes such as *Penicillium* species and *Trichoderma viride* (Table 2). Of these the gliotoxin-producing strains of *T. viride* were the most effective. Here the establishment of saprophytes early in the development of the young root brought about a significant control of disease and once again *Trichoderma* was specially effective. As will be mentioned later, mycorrhizal fungi of trees may protect their host from attack by disease organisms. Marx (1966) has shown a control of *Phytophthora* this way.

Table 2. *Effect of dusting seeds with fungal spores on the production disease symptons by Pythium species* (*data of Wright 1956*)

Organism used	Healthy plants after 21 days (%)
Control	12
Trichoderma viride	36
Penicillium nigricans	10
P. frequentans	29
P. godlewskii	55

Such specific examples are not very numerous and perhaps many need further examination and verification. In any event, except in a very limited way it has not yet proved possible by manipulation of rhizosphere populations by artificial inoculation or other means to bring about any certain control of disease.

Amongst the organisms of the rhizosphere are found potential pathogens which in normal conditions do not cause disease, but disturbance of the host or its environment may allow them to do so. Simmonds & Ledingham (1937), for instance, recorded that up to 50 % of the fungi of wheat roots might be potentially parasitic. These organisms clearly find an ecological niche in the root region which can therefore constitute a reserve of infectivity.

Nutrient absorption

The effects of the rhizosphere populations on nutrient absorption by their host roots are again not simple. There are two constellations of opposite effects. On the one hand, the microorganisms have essentially similar demands for nutrient elements as their host and therefore they may compete with it. On the other hand, many important reactions which result in the greater availability of nutrients take place more rapidly in the root region than in the remainder of the soil. Katznelson (1965) pointed out that in the rhizosphere of some species of plant large numbers of organisms capable of bringing phosphate into solution from insoluble organic and inorganic phosphate compounds, of trans-

Table 3. *Growth and phosphate content of plants grown in sterile and infected quartz sand with $Ca_3(PO_4)_2$ as phosphate source*

	Length of shoot (cm.)	Dry wt of plant (g.)	P_2O_5 absorbed (mg.)
	Sinapis alba		
Sterile	98	12·5	97·5
	72	9·0	70·3
Infected	111	22·0	154·0
	134	23·0	171·0
	Helianthus annuus		
Sterile	61	12·0	50·4
	65	22·5	83·2
Infected	97	35·0	119·0
	105	45·0	124·0

forming nitrogen compounds, of chelating nutrient ions, and of altering the available concentrations of many plant nutrients, were frequently present. Their effects might be either stimulatory or inhibitory to the growth of their host but few completely satisfactory experimental demonstrations have been made that their effects are significant. The work of Gerretsen (1948) is often quoted for he showed that the activities of the rhizosphere microorganisms of a number of crop plants increased the uptake of phosphate and the growth of the host when phosphate was presented in insoluble form (Table 3). Katznelson (1965) reported that similar experiments performed in his laboratory yielded negative results, although others have been more successful. It should be noted also that Gerretsen himself describes some experimental results where the rise in soluble phosphate in inoculated cultures resulted in iron deficiency and poor growth owing to the formation of insoluble iron phosphates. Microbial complexities in deficiency diseases have also been reported. 'Grey speck', manganese deficiency disease of oats, has been shown to be aggravated by the rhizosphere population (Gerretsen, 1937; Timonin, 1947; Leach, Balman & Krocker, 1954). Similarly soil microorganisms have been shown to compete for zinc with *Citrus*, so encouraging the development of the Little Leaf disease (Ark, 1937).

Experimental investigation of the effects of microorganisms on the uptake of inorganic nutrients by plants growing in solution culture has recently been made by several groups of workers (see Barber, 1968). Barber (1966, 1969) and Barber & Loughman (1967) using labelled phosphate compared uptake, esterification and translocation of soluble phosphates by barley plants growing in sterile solutions with others

Table 4. *Effect of microorganisms on the uptake, translocation to shoot, and esterification of phosphate by young barley plants at pH 4·0 during 3 h. (data of Barber, 1969)*

	Phosphate applied (μg. P/l.)			
	0·001	0·01	0·1	1·0
P absorbed (μg.)				
Sterile	0·40	3·12	24·68	95·7
Non-Sterile	0·156	2·81	28·06	100·6
P translocated (%)				
Sterile	7·5	8·7	50·2	48·6
Non-Sterile	3·8	6·4	34·3	49·6
P as soluble esters and inorganic P (%)				
Sterile	81·1	82·5	86·3	87·7
Non-Sterile	43·5	65·3	62·8	75·2

grown with casual air-borne contaminants on their root systems (Table 4). They showed that, except in the lowest concentrations, the uptake by non-sterile plants exceeded that of sterile plants; but in every case the quantity of phosphate translocated from the roots to the shoots was diminished by the presence of microorganisms. A greater proportion of the absorbed phosphate of the root system was converted into nucleic acids, phospholipids and phosphoproteins in the non-sterile roots than in the sterile ones. Hence they concluded (*a*) that the increased uptake observed in the non-sterile plants could be ascribed to accumulation and rapid esterification in the bodies of the micro-organisms and (*b*) that true uptake into the host and translocation to its shoot were diminished in unsterile conditions, especially in low external phosphate concentrations. Bowen & Rovira (1966) and Rovira & Bowen (1966) inoculated plants with a mixed population of micro-organisms isolated from the soil, for a comparison of the uptake and translocation of phosphate by sterile and non-sterile clover and tomato (Tables 5, 6). They observed both increased uptake and increased total translocation to the shoot in the non-sterile plants although the proportions translocated might be diminished. The contrast between the results of the two groups of investigators was ascribed by Barber (1969) to the differences in nature of the populations used in inocula-tion; 'when a wide spectrum of organisms is present the stimulatory effects of microbial secretions may predominate over the competitive effects'. Whatever the explanation of the discrepancy it is clear that further researches of this kind, but using actual rhizosphere organisms, are necessary both to examine the effects of true rhizosphere populations on their hosts and also to determine to what extent lack of sterility

Table 5. *Effect of microorganisms on the uptake of* ^{32}P *by wheat roots* (*data of Rovira & Bowen, 1966*)

	Sterile plants	Non-sterile plants
Fresh weight of roots (mg.)	251	201**
P uptake (μmoles/mg.)	17·6	23·8*
Soluble P (0·1 HClO$_4$)	93·1	89·0*
Insoluble P	6·9	11·0

$* P > 0.01.$ $** P > 0.001.$

Table 6. *Uptake and translocation of phosphate by sterile and non-sterile plants* (*data of Rovira & Bowen, 1966*)

	Uptake ($\times 10^{-13}$ moles/plant)	Translocation ($\times 10^{-13}$ moles/plant)
Tomato		
Sterile	81·9	4·4
Non-sterile	14·53	10·91
Subterranean clover		
Sterile	12·25	9·7
Non-sterile	19·06	6·5

may affect the results and interpretation of physiological experiments on nutrient absorption.

It should be noted, however, that experiments using soluble sources of phosphate tend to overemphasize in particular the competitive effects of microorganisms for available nutrients in solution (see also Gray, this Symposium). In natural soil situations much of the phosphate is in insoluble organic or inorganic form and the net effect of the microbial balance between dissolution absorption might be quite different. Not only may insoluble phosphates be brought into solution by microbial activity but also it is possible that absorption and translocation of phosphate by the fungal hyphae might increase the total soil volume over which insoluble soil phosphate might be exploited by the system.

MYCORRHIZA AND ROOT NODULES

The functioning of microbial nodules in fixation of atmospheric nitrogen requires no detailed comment here beyond pointing out the considerable list of plants involved and stressing the fact that it is a property of the composite organ of host and microbe, for neither fixes nitrogen in the free state. Besides the leguminosae, species of *Alnus* (Betulaceae), *Casuarina* (Casuarinaceae; only genus), *Ceanothus*, *Discaria* (Rhamnaceae), *Eleagnus*, *Hippophae*, *Shepherdia* (Eleagnaceae), *Coriaria* (Coriariaceae; only genus), *Myrica* including *Gale* and *Comptonia*

(Myricaceae), and *Dryas* (Rosaceae), all produce microbial nodules and in one or more species of all of them nitrogen-fixing ability has been confirmed and their ecological distribution, which will receive comment later, is in line with this property (see Postgate, this symposium).

The functioning of mycorrhizal organs has in the past been a subject of much discussion, but in recent years knowledge of that of the main kinds has so far improved that there is no reason to doubt the general conclusions. In contrast with microbial nodules it is not possible here to give a list of mycorrhizal host-plants because a very large number of the angiosperms, gymnosperms, pteridophytes and hepatics so far investigated have mycorrhizal infection sometime in their life-histories or in some environmental conditions.

For present purposes we may consider three kinds of mycorrhizal infection. On the one hand there are two kinds, the ectotrophic mycorrhizas of forest trees and the endotrophic mycorrhiza called vesicular–arbuscular or phycomycetous, in both of which the fungi are specialized in their nutrition and typically seem to have no other ecological niche except the root region of their hosts. On the other hand the endotrophic mycorrhiza of orchids and other partial or holosaprophytes, from which the fungi are more easily grown in culture, may often and perhaps usually have other ecological niches and may always be self-sufficient for carbon absorption.

The first two kinds are very distinct from one another histologically and morphologically. In the ectotrophic mycorrhizas the fungi, usually basidiomycetes belonging to Agaricaeceae or Boletaceae, form a tissue or mantle enclosing the host rootlets completely and from it there is restricted penetration between (and very little into) the cells of the outer cortex. The mantle also has hyphal connexions with the soil. Associated with infection there are considerable changes in the relative dimensions of the cells and their differentiation so that a definite composite organ of host and fungus, a mycorrhiza, is recognizable. By contrast, in the vesicular–arbuscular mycorrhizas the hyphae of the fungi belonging usually to the genus *Endogone*, penetrate the cortex intercellularly and intracellularly. In the cells branched haustoria called arbuscules are formed, and between or within the cells spherical or oval vesicles are formed at some stage in infection. Outside the root there is no mantle or sheath but a considerable development of a loose weft of hyphae in the root region on which may be borne vesicles or spores and sporocarps typical of *Endogone*. A further important difference is that digestion of the hyphae takes place within the invaded cells so that exploitation is controlled and the contents of the hyphae are released into the cells.

In the Orchidaceous type of mycorrhiza the fungi, usually basidiomycetes capable of cellulose and lignin utilization, penetrate into the cells of the host in which they at first are vegetatively active and later become digested and disintegrated by the host activity. The fungi are also attached to a living or dead food base external to the host (see Burgeff, 1932, 1936; Harley, 1969). A variety of digestion patterns in orchids result in the limitation of exploitation of the host tissue by the fungi and the release of hyphal contents into the host cells. This kind of mycorrhiza will not be discussed further in detail in this paper. The work upon it has stressed the part played by the fungal symbionts in supplying carbohydrates to the host especially in the early stages of seedling growth. The fungi are very active in breakdown of insoluble carbon compounds in wood, litter and humus, and the products absorbed have been shown to pass into the host. S. E. Smith (1966, 1967), for instance, showed that strains of *Rhizoctonia solani* could absorb carbohydrate from cellulose, converting it into fungal carbohydrates. In other experiments, ^{14}C-labelled carbohydrates absorbed as glucose was shown to be translocated as trehalose (sometimes as mannitol), released into growing saprophytic seedlings of *Orchis purpurella* and converted there to sucrose. Very little is known of the influence of the fungi in the inorganic nutrition of orchids.

It should be noted that similar problems arise with mycorrhizal fungi and nodule-forming microbes, as with other organisms, concerning development in the root region. The same constellation of possible factors – secretion of nutrients, accessory substances and inhibitors – have been invoked to explain their presence and activity. In addition the production of morphogenic factors by the fungi or microbes which affect the growth of the host must be involved and has been studied with some success. These cannot be further discussed in detail here but are reviewed for instance by Nutman (1963, 1965), Melin (1963), Harley (1969) and Harley & Lewis (1969).

The functioning of ectotrophic and vesicular–arbuscular mycorrhizas

In spite of their striking differences these two kinds of mycorrhiza have both been shown to be important in mineral nutrition (Mosse, 1963; Nicolson, 1967; Baylis, 1967; Harley & Lewis, 1969). Tables 7 and 8 illustrate experiments in which the growth of infected and uninfected host plants were compared. It has been found that in nutrient-deficient sites growth is increased very significantly by mycorrhizal

Table 7. *Some recent examples of the effect of inoculation with Endogone on the dry weight (g.) of host plants (compiled from various sources)*

Plant	Uninfected	Mycorrhizal
Apple	2·9	3·8
Maize	2·4	3·0
	3·37	13·30
	6·1	14·4
Tobacco	1·65	2·72
Griselinia	5·14	12·48
	5·87	7·85
	1·66	6·10
Podocarpus totara	0·38	2·63
P. dacrydiodes	0·12	0·48
Strawberry	6·69	16·04
	18·83	17·04†
Coprosma robusta	0·08	1·66
Myrsine australis	0·45	2·09
Pittosporum eugenioides	0·03	0·08
	0·07	0·45
Liriodendron tulipipera	4·2	9·7
	1·6	7·7
Tomato	0·09	0·39
	0·70	0·74†

† Soil with very high phosphate level.

Table 8. *Examples showing effects of ectotrophic mycorrhizal infection on the growth of the host plant (compiled from various sources)*

	Dry weight (g.)	
Plant	Uninfected	Mycorrhizal
Pinus strobus	0·303	0·405
	0·093	0·223
P. virginiana	0·152	0·323
P. radiata	1·99	3·71
	1·27	1·64
Oak	2·88	3·41
	3·15	3·34
	2·92	3·45
Eucalyptus dives	3·2	5·3
E. pauciflorus	3·3	6·2
E. macrorhiza	7·8	11·3
Pinus radiata		1·58
		1·45
	1·56	2·65
		2·15
		3·70
		3·60
	3·65	4·42
		4·55

infection. Analyses of the plants in comparative experiments have further shown that there is often a disproportional increase of absorption of some minerals as compared with dry weight, and although several nutrients may be involved in certain cases, the most consistent effects are observed in phosphate absorption with both kinds of mycorrhizas (Table 9).

Table 9. *Examples of the relative rate of uptake of phosphate by comparable samples of excised mycorrhizal (M) and uninfected roots (U)* (*compiled from various sources*)

Type of mycorrhiza	Plant	Relative rate (M/U)
Ectotrophic mycorrhiza	*Fagus sylvatica*	5·40
	Quercus sp.	2·12
	Pinus radiata	2·14
		4·43
		2·73
		1·00
	Alnus viridis	4·8 (10°)
		5·7 (20°)
Endotrophic mycorrhiza	*Agathis australis*	1·70
	Liriodendron tulipipera	1·90
	Subterranean clover	2·46

Explanations of the apparently greater efficiency in uptake by mycorrhizal roots as compared with uninfected have involved the possibilities that infection leads to changes in the absorbing area of the root system or of individual roots, that the hyphae emanating into the soil provide an increased absorbing area and that the hyphae (as in the rhizosphere work of Barber mentioned earlier) have a great affinity for the available ions. In any event in both kinds of mycorrhiza (see Harley, 1969) the initial site of accumulation, verified with ^{32}P, is in the fungal hyphae. In the case of ectotrophic mycorrhiza good evidence is available that the hyphal mantle acts as the primary organ of accumulation and that onward translocation of phosphate accumulated in the fungal sheath occurs, by a mechanism which is oxygen and temperature-sensitive, in conditions where external supplies of phosphate are low (see Harley, 1969; Harley & Lewis, 1969; see Figs. 1, 2, 3). It is therefore fitted to soil circumstances where there is a seasonal or periodic release of available nutrients which may be rapidly accumulated in the tissues in the face of competition, and mobilized later.

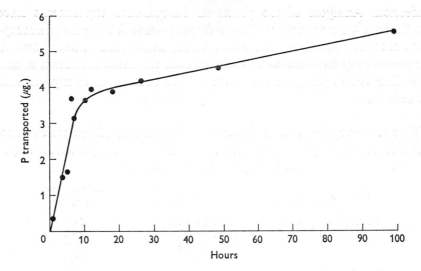

Fig. 1. The progress of movement of phosphate from the fungal sheath into the core of *Fagus* mycorrhizas at room temperature in aerated phosphate-free buffer.

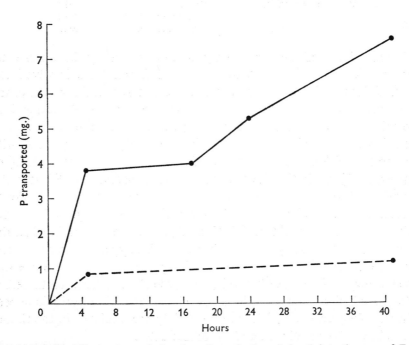

Fig. 2. Progress of movement of phosphate from the fungal sheath into the core of *Fagus* mycorrhiza at 19° (full line) and 1° (dotted line) in phosphate-free buffer.

Mycorrhiza and disease

Just as the populations of the root region of plants have been held to be barriers against infection by pathogens so have also mycorrhizal infections (Zak, 1964). The mantle of ectotrophic mycorrhiza may be thought of as a protective layer preventing the attack by pathogens, and certainly it has been stressed (Mikola, 1965) that in the absence of mycorrhizal fungi weak pathogens may occupy the root surface. Marx

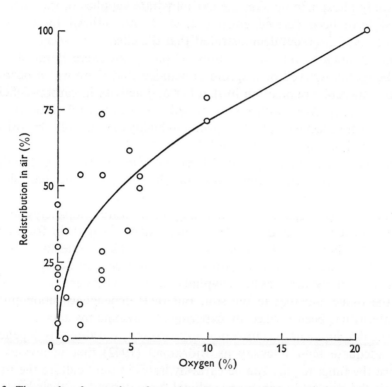

Fig. 3. The results of a number of experiments each of which compares the amount of phosphate moved from the fungal sheath to the core of *Fagus* mycorrhizas during 24 h. in phosphate-free buffer at 18° at different oxygen tensions. The results are expressed as percentages of the quantity moved in air so that the results of all the experiments may be compared.

(1966) isolated a diatretyne antibiotic from the mycorrhiza–fungus *Leucopaxillus cerealis* which appeared to confer resistance upon mycorrhizal roots to zoospores of *Phytophthora cinnamonii* and to be translocated to neighbouring uninfected roots. It is clear that properties of this sort are certainly likely to occur in symbiotic systems and are not incompatible with the known nutritive function of mycorrhizas.

Carbon nutrition of microbial associations with roots

Even in the relatively unorganized associations of microbes in the rhizosphere of roots, carbon compounds derived from the host are important. The explanation of the high activity and specificity of the rhizosphere populations all involve the release of organic nutrients and accessory growth factors which control or affect the microorganisms. Undoubtedly in the experiments performed in inorganic solution culture to study phosphate uptake, the carbohydrate supplies of the microbes must have been derived entirely from photosynthesis by the host. Indeed Barber (1969) demonstrated that the effects of the microorganisms on uptake and esterification of phosphate were diminished in carbohydrate-starved plants, and concluded that there was a decrease in the microbial population in number and activity in carbon-deficient surroundings. Work with plants in soil has shown a relatively smaller rhizosphere effect with plants growing in highly organic soils or horizons where carbon supplies are plentiful.

In many of the more specialized symbiotic associations, root nodules or mycorrhizas, the dependence of the heterotrophic partner on photosynthetic products of its host is more clear-cut. The microbes of the nodules of legumes and other angiosperms are completely enclosed in the host tissues and are clearly dependent upon them for carbon substrates both for the formation and for the maintenance of the growth and activity of the nodular structure (see Van Scheven, 1958). The fungi of mycorrhizas have hyphal connexions between the mycelium in the tissue and that in the soil, but their dependence upon photosynthesis has been verified by experiments in some instances.

The development of ectotrophic mycorrhiza in seedlings is dependent on adequate light intensity as Björkman (1942) first demonstrated. Since the fungi require simple carbohydrates in pure culture the hypothesis that the ectotrophic mycorrhizal fungi depend on their hosts for carbon substrates has been generally accepted and indeed is consistent with most observations. Melin & Nilsson (1957) showed that *Pinus sylvestris* seedlings supplied with $^{14}CO_2$ in two-membered cultures with *Boletus variegatus* or *Rhizopogon roseolus* translocated the products of photosynthesis to their mycorrhizal fungi. Lewis & Harley (1965c), using excised mycorrhizas of *Fagus sylvatica*, following this up by showing that if ^{14}C sucrose (the carbohydrate translocated in *Fagus* phloem) was applied to the cut stump of a mycorrhiza, carbohydrate was translocated into the fungal sheath. Analysis of the mycorrhizal tip showed not only that about two-thirds of the total ^{14}C passing into the tip had moved into the sheath after 24 h. but that it was mainly in the form of mannitol, trehalose and glycogen, typically fungal carbohydrates.

Further, although both the fungal sheath and the host tissue use exogenase glucose, fructose and sucrose, the fungal layer uses mannitol and trehalose (Lewis & Harley, 1965*a, b*). The fungus could therefore be viewed as a metabolic sink which could absorb the soluble carbohydrates of the host (glucose, fructose and sucrose) but which converted them to trehalose and mannitol unusable by the host. There was no evidence of considerable reciprocal movement of carbohydrate. This does not, of course, imply that there can be no movement of any carbon compound from fungus to host. Indeed there is evidence to the contrary (see for instance Reid & Woods, 1969), but it suggests that the fungal partner possesses an effective mechanism of carbon intake at the expense of its host. This is confirmed in the study of translocation in mycorrhizal seedlings. Shiroya *et al.* (1962), Nelson (1964), Lister *et al.* (1968) have studied the patterns of translocation of mycorrhizal and non-mycorrhizal *Pinus resinosa* and *P. strobus*. In these experiments, as would be expected from the foregoing, translocation of ^{14}C-labelled photosynthate to mycorrhizal root systems was more rapid than to non-mycorrhizal roots.

There is therefore a good case for supposing that a considerable part, perhaps all, of the carbon required by the fungi of ectotrophic mycorrhiza is derived from the host and it is worth while to consider its possible magnitude. Dissections of mycorrhizal apices of *Fagus sylvatica* (Harley & McCready, 1952) shows that a little less than 40 % of their dry weight consists of fungal sheath. Hence, including the hyphae within the cortical tissue, at least 40 % of the total dry weight of the mycorrhizal roots which constitute the major part (about 90 %) of the absorbing roots in woodland soil is composed of fungus. Respiratory measurements also indicate that more than half of the respiration (CO_2 emission and O_2 uptake) of each mycorrhizal rootlet is due to the fungal sheath. We must conclude therefore that mycorrhizal infection in this case must impose a very large and significant drain on carbohydrate supplies from photosynthesis.

The results of this work on carbohydrate physiology of ectotrophic mycorrhizas and on the movement of carbohydrate in mycorrhizal plants show important similarities to those obtained with other kinds of associated growth of heterotroph and autotroph. In obligate parasitic fungi such as Erisiphales and Uredinales there is a movement of carbohydrates from the photosynthesizing tissue into the fungal pustule where the typically fungal carbohydrates, trehalose, mannitol and arabitol are formed. Indeed, Smith, Muscatine & Lewis (1969) have stressed the similarity of the behaviour of these with lichens and parasitic angiosperms and have pointed out the importance of sugar alcohols in many of them. In addition in his recent work with D. H. Lewis and other colleagues Smith has further shown that the carbo-hydrate movement between algal cells and their animal host in corals and other coelenterates, is essentially similar.

Of course the processes as described here for mycorrhiza are only partly explained by the source–sink mechanism. Questions whether there is a hormonal factor associated with the change of direction of the photosynthetic products towards the fungal layer, whether the fungus interferes with cellulose or wall formation in the cortex of the host and so is able to accumulate the simpler precursors, whether the fungus affects the permeability of the cell membranes – all these have not yet been investigated with mycorrhizas, although work is in pro-gress with lichens, coelenterates and other symbiotic systems by D. C. Smith, D. H. Lewis and their colleagues.

The broad similarity of carbohydrate metabolism of all these kinds of symbiosis makes it credible that vesicular–arbuscular mycorrhiza, the formation of which depends on adequate light intensity and photo-synthesis (Schrader, 1958; Peuss, 1958; Boullard, 1960), might operate in an essentially similar fashion.

It must again be emphasized that the magnitude of the carbohydrate drain by the fungus in ectotrophic mycorrhiza is very great; a drain of similar dimensions of the photosynthetic partner undoubtedly occurs in lichens and other symbiotic systems. It is inescapable that there must be an equally large positive influence on the photosynthetic partner if the symbiosis is to be stable and of selective advantage.

Specificity of symbiotic microorganisms

The host plants of both nodular and mycorrhizal symbioses cannot be viewed as obligately symbiotic except in an ecological sense. In sub-strates rich in available nitrogen nodule formation is reduced and nitrogen-fixing activity diminished although the host plant may flourish (Hallsworth, 1958). Similarly, mycorrhiza formation, both ectotrophic and vesicular–arbuscular, is greatly diminished or abolished in nutrient-rich conditions. The host plants can be grown quite successfully uninfected (Harley, 1969), given an adequate supply of nutrients.

The symbiotic microorganisms are much more dependent on their hosts. The species of *Endogone* of vesicular–arbuscular mycorrhiza have indeed never been grown in pure culture, and although there are reports of the hyphae of some species permeating decaying vegetation in bogs (Dowding, 1959) they have no other known permanent habitat but the host plant. The basidiomycetous fungi of ectotrophic mycorrhiza can be grown in culture, but again there is some good evidence that most of them are unlikely to be able to exist apart from their hosts in natural environments except as spores, although their presence in the rhizosphere or the root-surface of plants other than host plants is a possibility worth investigation. The same is broadly true of *Rhizobia*, which may exist in addition to their symbiotic life, in soil or in the rhizosphere to a limited extent. Amongst each kind of symbiotic organism there is a range of species and of strains. These differ from one another both physiologically and in their specificity to a range of hosts.

About 100 species of basidiomycetes have been proved to form ectotrophic mycorrhiza. These range in specificity between *Boletus elegans*, which appears to be restricted to species of *Larix*, and others of very wide tolerance such as the imperfect fungus *Conococcum graniforme*, which is known from mycorrhizas with hosts belonging to at least 20 genera. Within each species there are strains differing in detail of nutritional requirements, external enzymic activity, growth in culture, and mycorrhizal potential. It is indeed one of the important present problems of ectotrophic mycorrhiza to determine something of the relative efficiency of species and strains on a particular host and in particular conditions. It is known that at the one extreme some fungi are virtually parasitic and detrimental to growth of young seedlings and in contrast at the other efficient symbionts materially improving growth.

In recent years much progress has been made in the study of variation in mycorrhizal *Endogone* (Mosse & Bowen, 1968). As far as is known, the species are not closely specific, as cross-inoculation from one host to another has demonstrated. For instance, Magrou (1936) showed that the endophyte of *Arum italicum* could infect *Arum maculatum*. Koch (1961) showed that the endophyte of *Atropa belladonna* might form mycorrhiza with about 40 other species of angiosperm, and Stahl (1949) that the endophytes of liverworts were not closely specific. Gerdemann showed that the *Endogone* endophyte of red clover also infected maize, strawberry and several clovers, but not oats. From these and other researches it may be concluded that there is a range of species of *Endogone* of which none are specific to a single host and some are relatively unspecific. Moreover, Daft & Nicolson (1966) have shown that morphologically different strains or species vary in intensity of infection and have different effects on the growth of tomato in two-membered cultures.

Strain specificity in the *Rhizobia* of legumes has been much studied because of its great importance in artificial inoculation of agriculturally important crops. Once again, there are strains and perhaps species differing in many characteristics, including ability to infect, ability to form nodules and effectiveness in nitrogen fixation. In the nodulate non-leguminous plants also cross-inoculation has been shown to be possible within *Alnus*, *Myrica*, *Eleagnus* and *Hippophaë* (see Bond, 1963). Moreover cross-inoculation between *Eleagnus* and *Hippophaë* and *Hippophaë* and *Shepherdia* has been demonstrated, so that the same kind of specificity is likely.

Mutualistic symbiosis in natural conditions

The adaptations of both constituent organisms in the permanent symbiotic systems mycorrhizal and nodular are complex. The composite structures produced by the joint growth differ from anything produced by the separate organisms when these are grown alone. Some indication of their evolutionary origin can however, be perceived from the fact that some of the physiological processes, such as the carbohydrate metabolism of ectotrophic mycorrhizas discussed above, are held in common with obligate parasites. Moreover the permanent drain on the photosynthetic processes of the host indicate that a reciprocal benefit to the host must exist if the structures are of selective advantage. It must be accepted that this can only be true where growth of nodulate systems is limited by nitrogen supply or of mycorrhizas by that of soil-derived nutrients especially of phosphate.

A striking feature of the ecological distribution of nodulate, nitrogen-fixing plants in the part they play, and have played in the past, in the colonization of situations where the soil is both deficient of nitrogen and carbon compounds. For instance, *Dryas octopetala* and *Hippophaë rhamnoides* have been recorded (see Godwin, 1956) as widespread in Northern Europe in the transition from late glacial to post-glacial times before the main spread of the forests. Similarly, at the present day *Dryas* and *Shepherdia* are pioneers in glacial outwashes in North America, together with species of legume genera such as *Oxytropis* and *Hedysarum*, and are followed during early forest colonization by *Alnus*. In such sites nitrogen-fixation by photosynthetic symbiotic systems or by photosynthesizing autotrophs such as the blue-green algae is essential to the exploitation of the habitat. In similar situations in New Zealand where the flora is largely endemic and none of the genera mentioned is present, species of *Coriaria, Discaria tommatou*, species of *Gunnera* (symbiotic with blue-green algae) and a few legume shrubs fulfil the same ecological roles. Dune sands of the sea coasts constitute another site where symbiotic nitrogen-fixation is essential. The legumes, *Ulex, Sarothamnus* also *Hippophaë* are familiar there; indeed those legumes, sometimes lupins also, are used in the various methods of stabilization of sand and the establishment of forest upon dune soils in many temperate parts of the world. In Polynesia and Australia species of *Casuarina* are found naturally on coastal sands and they are widely planted there. *Eleagnus* includes coastal species also. Other nitrogen-fixing plants are essential components of semi-desert and steppe vegetation; *Casuarina, Acacia, Prosopis, Cassia, Caesalpinia, Lupinus, Astragalus, Oxytropis, Ceanothus* include well-known examples. Other species, including in addition those of *Alnus, Ceanothus* and *Myrica*, are scree, mountain, river gravel and bog plants.

Mutualistic symbiosis and productivity

The family Leguminosae includes within its three subfamilies about 700 genera and some 14,000 species, of which about 90 % of those tested are nodulate. The family is world-wide except that it is not well represented in New Zealand. The herbaceous temperature members of the Papilionaceae have been used for thousands of years to increase the productivity of agricultural land. It is therefore indeed interesting that some 51 species have been imported into New Zealand usually from Britain (Cockayne, 1967). It is indeed impossible to think of agricultural or forest productivity without legumes. They are used in the maintenance of nitrogen content of soils, in improvement of grassland, grown as nitrogenous food crops for man and fodder crops for animals. None of the symbiotic nitrogen fixers of other families, most of which are woody, replace them in any of these important functions, and some indeed are poisonous to mammals.

Mycorrhizal infections are adaptations to nutrient, especially phosphate, deficiency. Baylis (1967) stated 'the phycomycetous endophytes assist uptake of phosphate from soils far below the minimum agricultural standard of fertility', and again (1968): 'no seedling grows in unfertilized forest soil [in New Zealand] until an endophyte is established on its roots'. Although such conclusions agree with the results of experiments on the effect of vesicular–arbuscular mycorrhiza on the host in natural ecological conditions they have never been considered in agricultural crop production. A large number of important crop plants – grasses and cereals, tomato, tobacco, rubber, coffee, vines, the legumes and many others – develop vesicular–arbuscular mycorrhiza in nutrient-deficient soils and have, as shown in Tables 7 and 8 improved efficiency in nutrient uptake in the mycorrhizal state. They are customarily grown in heavily or adequately fertilized soils at the sort of nutrient level where mycorrhizal infection has been shown experimentally to have little effect. This is not true of forest trees, where very many of the important species are either ectotrophically or endotrophically mycorrhizal. For the establishment of forests and plantations only marginal soils, unsuitable for agricultural exploitation, are usually available and it is for this reason that foresters, but not agriculturists, have interested themselves in mycorrhiza and much has been learned of ectotrophic mycorrhiza. However, because of the difficulties of isolation of the fungi of endotrophic mycorrhiza little progress has been made till recently in the investigation of vesicular–arbuscular mycorrhiza of trees, so that very little is yet known of the physiology of this kind of association in the valuable angiosperm and coniferous trees which possess it.

Extensive researches on ectotrophic mycorrhizas, as well as experience in the problems of afforestation with Pinaceae and Amentaceae, have shown that on nutrient-deficient sites the establishment of exotics is hazardous unless steps are taken to ensure inoculation with mycorrhizal fungi. At present much effort is being given to the different effects of different strains or species with a single host on different sites. In the laboratory too the physiological properties of the different types of ectotrophic mycorrhiza found on a single host have shown that great variation, worthy of extended study, in nutrient absorption occurs.

It is of interest that many nodulate species are also mycorrhizal. *Alnus*, for instance, has ectotrophic mycorrhizas which have been shown by Mejstrik & Benecke (1969) to absorb phosphate at rates considerably greater than uninfected roots. In the Leguminosae endotrophic vesicular–arbuscular mycorrhiza is very common along with nodulations. Asai (1944) concluded from his experiments that mycorrhizal infection was important for both the growth of the host and the formation of nodules. Such cases in which both nodules and mycorrhizal infection occur clearly merit further interest.

CONCLUSIONS

The root system of plants, form with soil microorganisms, ecosystems of interaction and competition. The local release of organic compounds in the root region of photosynthetic green plants provides substrates and accessory nutrients for large populations containing high numbers of nutritively specialized microorganisms. It does not seem that as a general rule the populations of the root region are especially beneficial to their hosts. The nutrient situation can be worsened or improved according to the soil condition. However, since the root microbial ecosystem is in many respects a closed community, adventive pathogens are less likely to infect the host than in sterile conditions.

Specialized associates of microorganisms and roots which result in the formation of composite organs having an integrated physiology are very common. They are symbioses between a carbon autotrophic and a carbon heterotrophic partner which have special properties of inorganic nutrition. Two kinds are especially important in natural and artificial vegetation: nitrogen-fixing nodules and nutrient-absorbing mycorrhizas.

REFERENCES

ARK, P. A. (1937). Little Leaf or rosette of fruit trees: VII. *Proc. Am. Soc. Hort. Sci.* 1936, p. 216.

ASAI, T. (1944). Über die mykorrhizabildung der Leguminosen-pflanzen. *Jap. J. Bot.* **13**, 463.

BARBER, D. A. (1966). Effect of microorganisms on nutrient absorption by plants. *Nature, Lond.* **212**, 638.

BARBER, D. A. (1968). Microorganisms and the inorganic nutrition of higher plants. *A. Rev. pl. Physiol.* **19**, 71.

BARBER, D. A. (1969). The influence of the microflora on the accumulation of ions by plants. In *Ecological Aspects of the Mineral Nutrition of Plants*, p. 191. Ed. I. Rorison. Oxford: Blackwell.

BARBER, D. A. & LOUGHMAN, B. C. (1967). The effect of microorganisms on the absorption of inorganic ions by plants. II. Uptake and utilization of phosphate by barley plants grown under sterile and non-sterile conditions. *J. exp. Bot.* **18**, 170.

BAYLIS, G. T. S. (1967). Experiments on the ecological significance of phycomycetous mycorrhizas. *New Phytol.* **66**, 231.

BAYLIS, G. T. S. (1968). Paper read to A.N.Z.A.S., Christchurch, 1968.

BJÖRKMAN, E. (1942). Über die Bedingungen der Mykorrhizabildung bei Kiefer und Fichte. *Symb. bot. upsal.* **6**, 1.

BOND, G. (1963). The root nodules of non-leguminous angiosperms. In *Symbiotic Associations*. Ed. B. Morse and P. S. Nutman. Cambridge University Press.

BOULLARD, B. (1960). La lumière et les mycorhizes. *Ann. Biol.* **36**, 231.

BOWEN, G. D. & ROVIRA, A. D. (1966). Microbial factor in short-term phosphate uptake studies with plant roots. *Nature, Lond.* **211**, 665.

BROOKS, D. H. (1965). Root infection by ascospores of *Ophiobolus graminis* as a factor in the epidemiology of the take-all disease. *Trans. Br. Mycol. Soc.* **48**, 237.

BURGEFF, H. (1932). *Saprophytismus und Symbiose*. Jena: Fischer.

BURGEFF, H. (1936). *Samekeimung der Orchideen*. Jena: Fischer.

COCKAYNE, L. (1967). *New Zealand Plants and Their Story*. Wellington, New Zealand: Owen.

DAFT, M. J. & NICOLSON, T. H. (1966). Effect of *Endogone* on plant growth. *New Phytol.* **65**, 343.

DOWDING, E. S. (1959). Ecology of *Endogone*. *Trans. Br. Mycol. Soc.* **42**, 449.

GARRETT, S. D. (1939). Soil conditions and take-all disease of wheat. IV. Factors limiting infection by ascospores of *Ophiobolus graminis*. *Ann. appl. Biol.* **26**, 47.

GARRETT, S. D. (1956). *Biology of Root-infecting Fungi*. Cambridge University Press.

GARRETT, S. D. (1970). *Pathogenic Root-infecting Fungi*. Cambridge University Press.

GERDEMANN, J. W. (1955). Relation of a large soil-borne spore to phycomycetous mycorrhizal infections. *Mycologie* **47**, 619.

GERRETSEN, F. C. (1937). Manganese deficiency of oats and its relation to soil bacteria. *Ann. Bot.* N.S. **1**, 208.

GERRETSEN, F. C. (1948). The influence of microorganisms on phosphate intake by the plant. *Pl. Soil* **1**, 51.

GIBBS, I. N. (1967). A study of the epiphytic growth habit of *Fomes annosus*. *Ann. Bot.* **32**, 755.

GODWIN, H. (1956). *The History of the British Flora*. Cambridge University Press.

HALLSWORTH, E. G. (1958). Nutritional factors affecting nodulation. In *Nutrition of Legumes*. Ed. E. G. Hallsworth. London: Blackwell.

HARLEY, J. L. (1969). *The Biology of Mycorrhiza*, p. 334. London: Leonard Hill.

HARLEY, J. L. & LEWIS, D. H. (1969). The physiology of ectotrophic mycorrhizas. *Adv. Microbial Physiol.* **3**, 53.

HARLEY, J. L. & McCREADY, C. C. (1952). The uptake of phosphate by excised mycorrhizal roots of the beech. III. *New Phytol.* **51**, 342.

HENRY, A. W. (1931). Occurrence and sporulation of *Helminthosporium sativum* in the soil. *Can. J. Res.* **5**, 407.

HILTNER, L. (1904). Über neuere Erfahrungen und Probleme auf dem Gebiet der Bodenbakteriologie und unter besonderer Berücksichtigung der Gründüngung und Brache. *Arb. dt. LandwGes.* **98**, 59.

KATZNELSON, H. (1965). The nature and importance of the rhizosphere. In *Ecology of Soil-borne Plant Pathogens*. Ed. K. F. Baker and W. C. Snyder. University of California Press.

KOCH, H. (1961). Untersuchungen über die Mykorrhiza der Kulturplanzen unter besonderer Berücksichtigung von *Althaea officinalis* L., *Atropa belladonna* L., *Helianthus annuus* L. und *Solanum lycopersicum* L. *Gastenbaumiss* **26**, 5.

LEACH, W., BALMAN, R. & KROCKER, J. (1954). Studies in plant mineral nutrition. 1. An investigation into the cause of grey speck disease of oaks. *Can. J. Bot.* **32**, 358.

LEWIS, D. H. & HARLEY, J. L. (1965a). Carbohydrate physiology of mycorrhizal roots of beech. I. Identity of endogenous sugars and utilization of exogenous sugars. *New Phytol.* **64**, 224.

LEWIS, D. H. & HARLEY, J. L. (1965b). Carbohydrate physiology of mycorrhizal roots of beech. II. Utilization of exogenous sugars by uninfected and mycorrhizal roots. *New Phytol.* **64**, 238.

LEWIS, D. H. & HARLEY, J. L. (1965c). Carbohydrate physiology of mycorrhizal roots of beech. III. Movement of sugars between host and fungus. *New Phytol.* **64**, 256.

LISTER, G. R., SLANKIS, V., KROTKOV, G. & NELSON, C. D. (1968). The growth and physiology of *Pinus strobus* seedlings as affected by various nutritional levels of nitrogen and phosphorus. *Ann. Bot.* N.S. **32**, 33.

MAGROU, J. (1936). Culture et maculation du champignon symbiotique de l'*Arum maculatum C. r. hebd. Séanc. Acad. Sci., Paris* **203**, 887.

MARX, D. H. (1966). *The role of ectotrophic mycorrhizal fungi in the resistance of pine roots to infection by Phytophthora cinnamonii Rands.* Ph.D. thesis, North Carolina State University.

MEJSTRIK, V. & BENECKE, U. (1969). The ectotrophic mycorrhizae of *Alnus viridis* and their significance in respect of phosphorus uptake. *New Phytol.* **68**, 14.

MELIN, E. (1963). Some effects of forest tree roots on mycorrhizal basidiomycetes. In *Symbiotic Associations*, p. 125. Ed. P. Nutman and B. Mosse. Cambridge University Press.

MELIN, E. & NILSSON, H. (1957). Transport of C^{14} labelled photosynthate to the fungal associate of pine mycorrhiza. *Svensk. bot. Tidskr.* **51**, 166.

MIKOLA, P. (1965). Studies on the ectendotrophic mycorrhiza of pine. *Acta for. fenn.* **79**, 2.

MOSSE, B. (1963). Vesicular–arbuscular mycorrhiza: an extreme form of fungal adaptation. In *Symbiotic Associations*, p. 146. Ed. P. S. Nutman and B. Mosse. Cambridge University Press.

MOSSE, B. & BOWEN, G. D. (1968). The distribution of *Endogone* spores in some Australian and New Zealand soils and in an experimental field soil at Rothamsted. *Trans. Br. Mycol. Soc.*, **51**, 485.

MUGHOGHO, L. K. (1968). The fungus flora of fumigated soils. *Trans. Br. Mycol. Soc.* **51**, 441.

NELSON, C. D. (1964). The production and translocation of photosynthate C^{14} in conifers. In *Formation of Wood in Forest Trees*. Ed. H. H. Zimmerman. New York: Maria Moors Cabot Foundation.

NICOLSON, T. H. (1967). Vesicular–arbuscular mycorrhiza – universal plant symbiosis. *Sci. Prog., Lond.* **55**, 561.

NUTMAN, P. S. (1963). Factors influencing the balance of mutual advantage in legume symbiosis. In *Symbiotic Associations*, p. 51. Ed. P. S. Nutman and B. Mosse. Cambridge University Press.

NUTMAN, P. S. (1965). The relation between nodule bacteria and legume host in the rhizosphere and in the process of infection. In *Ecology of Soil-borne Pathogens*, p. 231. Ed. K. F. Baker and W. C. Snyder. University of California Press.

PEUSS, H. (1958). Untersuchungen zur Ökologie und Bedeutung der Tabakmycorrhiza. *Arch. Mikrobiol.* **29**, 112.

REID, C. P. P. & WOODS, F. W. (1969). Translocation of C^{14} labelled compounds in mycorrhizae and its implications in interplant nutrient cycling. *Ecology* **50**, 179.

REISSEK, S. (1847). Endophyten de Pflanzenzelle. *Naturw. Abh., Berl.* **1**, 31.

REYNOLDS, E. S. (1931). Studies on the physiology of plant disease. *Ann. Mol. Bot. Gdn.* **18**, 57.

RIFAI, M. A. (1969). A revision of the genus *Trichoderma*. *Mycol. Pap.* 116.

RISHBETH, J. (1950). Observations on the biology of *Formes annosus* with particular reference to East Anglian pine plantations. *Ann. Bot.* **14**, 365.

ROVIRA, A. D. (1965*a*). Plant root exudates and their influence on soil microorganisms. In *Ecology of Soil-borne Plant Pathogens*, p. 170. Ed. K. F. Baker and W. C. Snyder. University of California Press.

ROVIRA, A. D. (1965*b*). Interaction between plant roots and soil microorganisms. *Ann. Rev. Microbiol.* **19**, 241.

ROVIRA, A. D. & BOWEN, G. D. (1966). Phosphate incorporation by sterile and non-sterile plant roots. *Aust. J. Biol. Sci.* **19**, 1167.

VAN SCHREVEN, D. A. (1958). Some factors affecting the uptake of nitrogen by legumes. In *Nutrition of Legumes*, p. 328. Ed. E. G. Hallsworth. London: Butterworth.

SCHRADER, R. (1958). Untersuchungen zur Biologie der Erbsmycorrhiza. *Arkiv Mikrobiol.* **32**, 81.

SHIROYA, T., LISTER, G. R., SLANKIS, V., KROTKOV, G. & NELSON, C. D. (1962). Translocation of products of photosynthesis in roots of pine seedlings. *Can. J. Biol.* **40,** 1125.

SIMMONDS, P. M. & LEDINGHAM, R. J. (1937). A study of the fungus flora of wheat roots. *Sci. Agric.* **18,** 49.

SMITH, D. C., MUSCATINE, L. & LEWIS, D. C. (1969). Carbohydrate movement from autotrophs to hetrotrophs in parasitic and mutualistic symbiosis. *Biol. Rev.* **44,** 17.

SMITH, S. E. (1966). Physiology and ecology of *Orchis* mycorrhizal fungi with reference to seedling nutrition. *New Phytol.* **65,** 488.

SMITH, S. E. (1967). Carbohydrate translocation in *Orchid* mycorrhiza. *New Phytol.* **66,** 371.

STAHL, M. (1949). Die Mykorrhiza der Lebermoose mit besonderer Berücksichtigung der thallosen formen. *Planta* **37,** 103.

TIMONIN, M. I. (1941). The interactions of higher plants and soil microorganisms. III. Effect of biproducts of plant growth on activity of fungi and actinomycetes. *Soil Sci.* **52,** 395.

TIMONIN, M. I. (1947). Microflora in the rhizosphere in relation to the manganese-deficiency disease of oats. *Soil Sci. Soc. Am. Proc.* **11,** 284.

WRIGHT, J. M. (1956). Biological control of a soil-borne *Pythium* infection by seed inoculation. *Pl. Soil* **8,** 132.

ZAK, B. (1964). The role of mycorrhizas in root disease. *A. Rev. Phytopath.* **2,** 377.

MICROBIAL PRODUCTIVITY IN POLAR REGIONS

JOHN BUNT

Institute of Marine Sciences, University of Miami, U.S.A.

INTRODUCTION

Conventionally, the term 'productivity' is taken to mean the rate of synthesis of organic materials by processes such as photosynthesis. Very often, it is understood to imply some sort of tangible yield although this is not necessarily the case. For many purposes, this view of productivity is too restrictive. Large numbers of microorganisms display productivity in catabolic processes which provide raw materials needed in synthesis by other species. All forms of life, the simple and the complex, perform a wide variety of controlling functions which act to influence the fabric of ecosystems. Although many of these influences may be subtle and difficult to measure, they are none the less productive. Reasoning in this fashion leads one to the position that mechanisms of synthesis represent only one form of productivity. In this account, productivity will be interpreted to cover any contribution made by microorganisms with recognizable, if not readily measurable, consequences. Among microorganisms, I have chosen to admit for discussion any virus, bacterium, fungus, alga or protozoan, alone or in association, without regard to size.

Minimally, a satisfactory treatment of regional microbial productivity should provide the reader with an account of the magnitude and efficiency of primary production attributable to photosynthetic and chemosynthetic autotrophs, the relative quantitative importance of microorganisms in overturn of organic materials and the proportion of the total biomass which is microbial. It should offer, also, incisive information on contributions made by microorganisms to the overall organization, character and stability of the ecosystem and should identify, in predictable terms, at least the major factors which limit microbial activity in the region under consideration. Our knowledge of the polar regions is too limited to meet these requirements. The foundations, however, are well-enough established to aim in the directions indicated, to identify outstanding problems, and to speculate on unresolved issues.

THE POLAR ENVIRONMENT

The environmental boundaries of the polar regions are complex. For convenience, I will include as polar those seas whose surface temperatures never exceed 0° by more than a degree or two, the lands comprising Antarctica in the south and those occupied by tundra in the north, including the highlands beyond the treeline within the Arctic Circle.

Together, these regions represent almost 14 % of the surface of the globe, or a total of roughly 27×10^6 square miles. Of this area, the arctic lands cover 5×10^6 square miles, the Arctic Ocean $5 \cdot 3 \times 10^6$ square miles, Antarctica $5 \cdot 2 \times 10^6$ square miles, and the Antarctic Ocean $11 \cdot 6 \times 10^6$ square miles. Only an estimated 7,700 square miles of Antarctica, largely within the Antarctic Peninsula, is exposed, the remainder being mantled permanently by a thick ice sheet. In the Arctic, however, less than 40 % of the land surface is covered with ice.

Because many excellent popular and technical accounts deal with polar geography, only salient features will be reiterated. Much of the Arctic land is low-lying, treeless tundra, extensively covered with small ponds and shallow lakes, carrying the unmistakable signs of past glaciations and subject to the action of frost, leading to a variety of remarkable land forms with inherent instability. Although precipitation is severely limited, a good deal of the tundra is marshy, a condition brought about partly by low air temperatures and partly by the underlying permafrost which impedes drainage. Where drainage is adequate, and particularly at higher latitudes and elevation, the boulder and gravel-covered surfaces support only hardy lichens. Within the region as a whole are to be found some 900 species of vacular plants, 500 species of mosses and upwards of 2,000 lichen species. There exists a diversity of plant communities highly adapted to growth during the brief summer, and a populous if not highly diverse fauna, much of it migratory. Polunin (1959) may be consulted for a description of the arctic flora, and Britton (1957) for an excellent account of the vegetation of arctic Alaska. According to Johnson (1969), who lists the major pedological literature, soils in the Arctic include lithosols, podzol-like, upland tundra, arctic brown, meadow tundra, half-bog and bog. With their acid, shallow, surface organic horizon, the pedologically immature tundra soils are most wide-spread.

Air temperatures are a poor indication of thermal conditions immediately above the ground surface and within the soil. Bliss (1962) has recorded summer temperatures as high as 38° at the surface, a stratum subject also to rapid and pronounced thermal fluctuation. Below the surface, on the other hand, abundant moisture commonly keeps temperatures lower than conventional meteorological data indicate.

Within the tundra, shallow lakes and ponds are a prominent feature of the landscape. These bodies, which frequently remain frozen throughout the year, at least at the surface, and whose temperatures during the warmest months commonly do not exceed 4°, are generally oligotrophic (Dunbar, 1968). Because of their thermal characteristics, vertical mixing is active in summer, with little indication of a thermocline.

In one sense, the ice-free areas of Antarctica may be considered as a harsher extension of the terrestrial Arctic. Thermally, the environment is certainly more extreme, even though air temperatures above zero are recorded in favoured situations (e.g. see references in Bunt, 1967). The soils, however, are distinctive and cannot be related, basically, to those of the north, although MacNamara (1969) has noted Arctic affinities in the soils of Enderby Land. Tedrow & Ugolini (1966) recognize several types of so-called cold desert soils which are virtually abiological; the pockets of soil (protoranker) found beneath isolated areas of moss and lichen; ornithogenic soils so named because their existence is restricted to penguin rookeries; occasional regosols and, most commonly, lithosols in which soil-forming processes are principally physical in nature. Such formations are almost entirely restricted to coastal areas. Those receiving moisture during the brief summer depend on local snow melt or glacial thaw. Precipitation is extremely light. Few chemical analyses have been made on Antarctic soils. Data on stored samples have been published by Piper (1938), who noted quite narrow carbon:nitrogen ratios. More recent information is available from Blakemore & Swindale (1958), Boyd, Staley & Boyd (1966), Tedrow & Ugolini (1966) and Rudolph (1966), who found it possible to grow *Poa pratensis* (Blue grass) successfully in artificially warmed soil from Cape Hallett; a direct qualitative indication of inherent local fertility.

Ponds and small lakes are not uncommon in the ice-free areas of Antarctica. Generalized statements concerning their character are not feasible because few have been studied. Furthermore, variations in size, depth and general appearance are obvious to the casual observer. All are partially or completely frozen for the greater part of the year and it is probable that, in some, complete thawing is very rare. The lakes of the Taylor, Wright and Victoria dry valleys at the base of the Ross Sea have received particular attention because of their accessibility from the U.S. Base in McMurdo South (see, for example, Armitage

& House, 1962; Goldman, 1964; Dillon, Bierle & Schroeder, 1968; Dillon, Walsh & Heth, 1969).

Just as Antarctica surpasses the Arctic lands in environmental severity, so, broadly, may the Arctic Ocean be viewed with respect to the Antarctic Ocean. The Arctic Ocean shares to some extent, the isolation of Antarctica and occupies the equivalent geographical location. Each of the two ocean systems has distinctive features. For detailed accounts of the physics and chemistry of the Southern Ocean the reader is referred to Deacon (1937, 1963), Wyrtki (1960), Kort (1962) and Mosby (1968). Short descriptive statements may be found by Deacon (1964) and Holdgate (1967) among others.

Because the continental shelf of Antarctica is very narrow and, over large areas, obscured by barrier and shelf ice, the water masses south of the Antarctic Convergence constitute a deep oceanic system. Defined by a surface feature, namely the Convergence, the system is not a separate entity, but continuous and in active circulation with the global ocean. Three layers are recognizable: the shallow, eastward- and northward-moving fertile surface water with its seasonally variable ice cover; the southward-moving, nutrient-enriched intermediate water with origins traceable far into the northern hemisphere; and the northward-moving bottom water. The surface and bottom layers stem from the rising intermediate water in the region of the Antarctic Divergence south of which lie the coastal waters of what may be termed the east-wind region. In the surface waters immediately south of the convergence, summer temperatures may reach 5°. Close to the coast, however, temperatures are consistently low and seldom exceed $-1 \cdot 5°$.

Some years ago, Mackintosh & Herdman (1940) estimated the maximum extent of southern pack ice at around 26×10^6 km.^{2} (10×10^6 square miles) and its summer minimum at $2 \cdot 6 - 5 \cdot 2 \times 10^6$ km.^{2}. A more recent estimate based on satellite data (Predoehl, 1966) puts the maximum within the limits $16 \cdot 8 - 19 \cdot 8 \times 10^6$ km.^{2}. The coverage is not complete, even during winter, and the state of the ice, including its thickness, age, and snow cover, quite variable.

For incisive biological studies, our knowledge of this vast region is inadequate. Specifically, reliable estimates of rates of water movement, particularly in the surface layers, are not available; hydrological studies have never been made across the pack-ice zone during the winter; and observations on the sea ice, an important habitat for microalgae, seriously limited.

Accounts of the water masses and water and ice movements of the

Arctic Ocean have been offered by Coachman (1963, 1969) and Mosby (1963). Bilello (1961) and Untersteiner (1963, 1969) have discussed the pack ice, while data on the chemistry may be found in Dunbar (1953) and English (1961).

Briefly, the Arctic Ocean is divided by submarine ridges into several deep basins as well as the Barents, Kara, Laptev, East Siberian and Chukchi Seas, which occupy rather shallow water including a broad continental shelf along the Eurasian coast. Also included are the waters of the Canadian Arctic archipelago. Three major water masses include the cold Arctic surface layer some 200 m. thick, the warmer Atlantic layer at depths from 200–900 m. and, below that, the cold, but not freezing bottom water. The surface waters accept run-off mainly from the Eurasian land mass, and low-salinity water entering through the Bering Strait. Outflow takes place at the surface between Greenland and Spitzbergen. Subsurface water of Atlantic origin enters the system in the same sector, the only major contact with the global ocean. Generally, major nutrients throughout the water column, and particularly in the surface layers, are less abundant than in Antarctic waters. Ice cover at the surface is variable but mostly heavy, permanent and extensive, notwithstanding partial melt in summer.

In concluding these introductory remarks about the environment of the polar regions, it is scarcely necessary to remind the reader of two outstanding features – the continuous exposure to solar radiation over the greater part of the summer, and the darkness of winter. This seasonality exerts deep influence on the character of the physical environment and on the dependent biota. Exposure is complicated by atmospheric conditions, surface relief and, particularly in the sea, by the existence and distribution of ice and snow at the surface.

COMPONENTS OF MICROBIAL PRODUCTIVITY

Photosynthesis

In the marine environment, with few exceptions, primary productivity can be attributed entirely to microbial activities or to organisms whose affinities with the microbial world are close. I will not labour the needless schisms created by biological distinctions based on size. Frequently in lakes and generally on land, it is impractical to distinguish the productivity of algae and lichens from that of all other plant life. This is not the case in many parts of the Arctic and Antarctic where only the simplest of plants can survive, although very few measurements have been attempted in such extreme habitats.

Reviewing the ecology of arctic and alpine plants, Billings & Mooney (1968) quote figures for shoot productivity by vascular species in the tundra between 3 and 128 g. dry matter/m.^{2}/year. Those values, as they point out, are misleading because production is limited to a quite short growing period. None the less, whether associated lichens and algae might match these rates can only be conjectured. No comparative measurements are known to me. Although growth in lichens exposed to extreme conditions is so rare that thalli up to 4,500 years old have been reported (see Billings & Mooney, 1968), there can be no doubt that, under more favourable circumstances, their production, if not their productivity, is important. Certain species serve as grazing for caribou, musk oxen and other animals (see Llano, 1944; Kursanov & Diachkov, 1945; Scotter, 1963). A preliminary description of studies with antarctic lichens has been given by Gannutz (1970), but no rate data are listed. Mikhaylov (1958) has described lichen growth under the protective cover of snow in the Arctic.

There has been a long-standing interest in the floristics of algae which grow on snow (e.g. see Lagerheim, 1883). To my knowledge, only Fogg (1967) has attempted measurements of metabolic activity, estimating productivity at around 10 mg. carbon/m.^{2} snow surface/day on Signy Island in the South Orkneys. While such contributions may appear small, their relative importance may be considerable in marginal environments.

Although still limited, rather more information is available on microbial primary productivity in Arctic and Antarctic lakes. From *in situ* studies near Point Barrow, Alaska, where up to 90 % of the Tundra is aquatic, Comita & Edmondson (1953) reported rates of net oxygen exchange under ice-free conditions ranging between 120 and 1,750 mg./m.^{2}/day and a mean of 750 mg./m.^{2}/day. In a later study in the same lake, but using ^{14}C-labelled substrates Kalff (1967a) obtained values for daily carbon fixation of 44–120 mg./m.^{2}. In the same area, also with *in situ* experiments, Kalff (1967b) reported maximum rates of no more than 30 mg. carbon/m.^{2}/day and average daily rates, based on two seasons, of only 2·6–6·1 mg./carbon/m.^{2} (less than Fogg's (1967) values with snow algae). Annual production of 380–850 mg. carbon/m.^{2} also was less than Hobbie's (1964) reported 900 mg./m.^{2}/year for Lake Peters in the Brooks Range of Alaska.

Working on the meromictic Lake Bonney, Antarctica, Goldman (1964) found a 4 m. surface ice cover in late November (the austral summer) and an inverse thermal stratification with 7·5° at 10 m. below the ice. Under these conditions, *in situ* radioactive carbon measurements indicated a daily production rate of 30·7 mg./carbon/m.^{2} with a pronounced maximum at 5 m. Rates up to 17·7 mg. carbon/m.^{3}/day were obtained close to the shoreline. In Lake Vanda, which is also meromictic and ice-covered, carbon fixation rates varied with time and depth, yielding values up to 80 mg. carbon/m.^{3}/day late in December. At that time, the photic zone extended down to 25 m., with temperatures throughout the water column above 5° except for the gradient at the ice interface. Nitrate additions were distinctly stimulatory, while a trace-metal mix decreased carbon fixation, and phosphate supplement had no effect. Before further discussing conditions in the lakes, I should like to consider the polar seas.

Early studies of polar marine phytoplankton were devoted largely to questions of taxonomy and species distribution. Investigations of productivity have been relatively recent, although observations of standing stocks are of long standing, e.g. Gran (1904). As late as 1963, however, Ryther was forced to report a possibly true but very tentative 1 g. carbon/m.^{2}/year for the Arctic Sea and to use Hart's (1942) chlorophyll estimations to speculate on the possible annual primary production in the Antarctic Ocean. The figure of 100 g. carbon/m.^{2} then proposed remains open to debate. On the basis of observations on both sides of the Antarctic Peninsula, El-Sayed (1968a) has reached the view that Ryther's (1963) estimate may be conservative, although data from other sectors are much less promising.

As things stand, we are scarcely justified in making predictions for such a large region. Uptake of ^{14}C and related measurements have been made mainly in the area of the Antarctic Peninsula, parts of the Weddell Sea, the northern part of the Pacific Sector and in the Ross Sea, with a scattering in the northern Indian and Atlantic sectors. This leaves a truly vast area unexplored. Furthermore, practically all observation has been restricted to summer cruises working transects largely in open water. Consequently, very little information is available on seasonal developments, on year-to-year differences which may be considerable, or on events within and beneath extensive sea ice and pack ice.

Of much greater immediate interest is the remarkable variability in reported standing stocks of chlorophyll and rates of primary production. From various authors, Bunt (1968a) listed values for chlorophyll a in summer ranging from 0·01 to 55 mg./m.^{3}. Similarly, productivities have been recorded in open water as low as a mean 0·34 mg. carbon/m.^{3}/h. in saturating light (Saijo & Kawashima, 1964) to more than 20 mg. carbon/m.^{3}/h. *in situ* by Horne, Fogg & Eagle (1969). Some of the factors likely to exert influences on the phytoplankton have been discussed by El-Sayed (1968b), Horne *et al.* (1969), and Bunt (1968a), who stressed the likely importance of algae in sea ice as a seeding population for the water column. More recent data have strengthened this view.

There are at least two habitants in sea ice suitable for the growth of microalgae, namely the layer of loosely aggregated ice crystals associated with the undersurface (Bunt, 1963) and the snow-ice interface described by Fukushima (1961), Meguro (1962) and Burkholder & Mandelli (1965). Similar ice communities in the Arctic have been studied by Gran (1904), English (1961), Apollonio (1961), Meguro, Ito & Fukushima (1967) and Allen (1970). Partly by direct inspection, Bunt & Lee (1970) found that the thickness of the ice-crystal layer in McMurdo Sound varies with location and with time, so that the volume available for algal growth is not constant. Furthermore, the amount of growth achieved is closely dependent on the presence or absence of snow cover at the surface. A sampling site with surface snow yielded a standing stock of 0·5 g. carbon/m.³ (essentially algal) compared with 10·4 g. carbon/m.³ from another site free of snow. Part of this difference may have been caused by thicker ice at the snow-covered site, although other observations indicate that this factor is relatively unimportant.

Its stability probably affected by currents, the ice crystal layer in McMurdo Sound is dispersed suddenly in December or January down into the water column, many weeks before the break-up of the main ice layer. Under favourable conditions, the phenomenon releases large numbers of physiologically active algal cells into the water column. It is possibly these populations which give rise to some of the large standing stocks recovered in open water and whose variable density partly may reflect unevenness in distribution at sources in the ice. Only systematic observations with time across the pack-ice zone will enable evaluation of this issue.

Since estimates of annual primary production are derived from short-term measurements, some comment on the treatment and interpretation of this information seems justified. In seeking a possible value for regional production, El-Sayed (1968a) based his calculations on the mean of his inshore and offshore daily production rates (g. carbon/m.²/day) and assumed that the rates were maintained. The result, however, is misleading since the inshore zone makes up but a small proportion of the total area and has, on average, a quite short season free of ice at the surface. It is likely, also, that events inshore are subject to variation since El-Sayed's (1968a) values for daily production/m.^{2} are appreciably higher than those of Mandelli & Burkholder (1966) but somewhat lower than those of Horne *et al.* (1969). By making *in situ* measurements with time at a single station, Horne *et al.* (1969), furthermore, were able to show that both daily rates of production and assimilation quotients changed considerably over the experimental period. Bunt & Lee (1970) obtained similar results with the ice communities in McMurdo Sound.

Horne *et al.* (1969) estimated conservatively that production at their Station 2 off Signy Island amounted to 130 g. carbon/m.^{2} during 125 days when the water was clear of ice. Assuming a growing season of similar length, a comparable figure would emerge from El-Sayed's (1968*a*) inshore rates, while his mean offshore value of 0·42 g. carbon/ m.^{2}/day would, if sustained, produce 84 g. carbon during a 200-day summer season. Data from Crooks (1960) and Saijo & Kawashima (1964), however, reflect very much lower production rates in other sectors. These values should be considered in relation to Ryther's (1969) suggested world production levels of 50, 100 and 300 g. carbon/ m.^{2}/year for the open ocean, coastal zones and upwelling areas respectively.

Another, and perhaps more satisfactory, basis for evaluating the relative productivity of the antarctic phytoplankton depends on comparing available solar radiation, and the possible photosynthetic yield during a given period, with observed yield of cell material. Assuming an ideal environment and taking into account the likely quantum requirement for photosynthesis, variations in energy per quantum with wavelength and several other factors, Ryther (1959) arrived at a graphical relationship between gross as well as net daily organic production as g./m.^{2} and total incident visible solar radiation in g. cal./ cm.^{2}/day. The relationship assumed a 5 % back-scattering and reflexion loss at the air/sea interface.

In the southern Weddell Sea, El-Sayed & Mandelli (1965) recorded an incidence of total solar radiation of 470 g. cal./cm.^{2}/day. Allowing a high 50 % loss at the surface by reflexion (El-Sayed, 1968*b*), this represents an actual radiation incidence, in Ryther's (1959) terms, close to 235 g. cal./cm.^{2}/day. Under these conditions, one would predict a theoretical daily production of 5 (net) to 12·5 (gross) g. carbon/m.^{2}. El-Sayed & Mandelli (1965), however, measured a daily production of 0·68 g. carbon/m.^{2} based on *in situ* exposures made over a 24 h. period. With generally comparable total exposures, Ryther (1959) listed annual average gains of 0·20 (net) and 0·44 (gross) g. carbon/m.^{2}/day in the Sargasso Sea, where maximum values up to

2·0 (gross) g. carbon/m.²/day were obtained. The maximum of approximately 2·7 g. carbon/m.²/day *in situ* recorded by Horne *et al.* (1969) occurred with an incident solar radiation level close to 177 g. cal./cm.²/ day and is approximately 77 % of the theoretical net production. Data collected by Bunt & Lee (1970) on levels of submarine radiation and standing stock in a highly shaded ice habitat indicate that 50 % of the theoretical production was achieved under these conditions. It must be noted, however, that mean daily radiation levels reaching the ice layer at that site did not exceed 2 g. cal./cm.². Clearly, Ryther's (1959) respiratory correction is not applicable under these circumstances, the percentage efficiency having been referred to theroetical gross, not net production. A mechanism to explain respiratory inhibition observed in some algae at low light intensities has been suggested by Hoch, Owens & Kok (1963).

It is worth noting that the value of 1 g. carbon/m.² for annual primary production in the Arctic Ocean put forward by Apollonio and English and reported by Ryther (1963) is exactly the value obtained by Bunt & Lee (1970) at their highly shaded ice station in McMurdo Sound. If conditions in the Arctic Ocean are generally comparable with those at the McMurdo Sound site, as seems likely, it is doubtful whether the current estimate can be much in error, although the efficiency of utilization of available light energy may be quite high.

As might be expected, the remarkable efficiency of the ice algae at environmental temperatures continuously at or very close to $-2\cdot0°$ is maintained only when light is limiting. At saturating intensities, performance can be maintained only by raising temperatures to the optimal range in the vicinity of 5·0°. This fact has been demonstrated in the laboratory by Bunt (1968 *b*) in measuring autotrophic growth down to 0·001 g. cal./cm.²/min., and may help to explain the excellent rates of production reported by Horne *et al.* (1969) and some of those from El-Sayed (1968 *a*).

Returning to a brief consideration of lake productivities, Comita & Edmondson (1953) obtained daily rates of production in Lake Imikpuk as high as 1·7 g. carbon/m.² at visible radiation levels up to 300 g. cal./cm.²/day, while Kalff (1967 *b*), working in tundra ponds in the same region, found rates no higher than 30 mg. carbon/m.²/day and usually very much less. In comparison with lakes elsewhere, these rates are low and, in the case of the ponds, reflect especially poor utilization of solar energy. As Kalff (1967 *b*) remarks, nutrient limitations appear mainly responsible. Figures for annual production are also extremely low simply because the period of thaw is so short. Without information on the penetration of light through the thick ice of the lakes of the dry valleys near McMurdo Sound, comment on daily production rates obtained by Goldman (1964) is difficult, although values close to 30 mg. carbon/m.^{2}/day suggest that the utilization of light may be comparable to that in the sea ice.

One further comment is needed concerning potential productivity. Calculations of the type made by Ryther (1959) give no information on standing stocks minimally necessary for maximal utilization of solar energy. According to Eppley & Strickland (1968), growth constants achieved by micro-algae under laboratory conditions can be as high as $k = 0.1$ derived from the expression

$$k = \frac{1}{t - t_0} (\ln n_t - \ln n_{t_0}),$$

where n_{t_0} and n_t refer to the concentration of any cell constituent, or cell numbers, between time t_0 and t in hours. Values for k as low as 0.002 have been obtained by Bunt (1968 b). On this basis may be listed the standing stock of chlorophyll a/m.^{2}, assuming a carbon:chlorophyll a ratio of 50, that would be required, for several values of k, to produce 12 g. carbon/m.^{2}/day (according to Ryther, 1959), the theoretically gross production possible for a total daily incident radiation of 200 g. cal./cm.^{2}:

Value of k	Standing stock chlorophyll a/m.^{2} (mg.)
0·1	94
0·01	940
0·001	9,400

Note, for example, that the mean standing stock of chlorophyll a recorded by El-Sayed (1968 a) in inshore waters was only 19·05 mg./m.^{2}, and far too low to produce the expected yield even allowing very rapid growth. For a standing stock of this dimension to produce the observed 1·23 g. carbon/m.^{2}/day, again assuming a carbon:chlorophyll a ratio of 50, would require a mean value of $k = 0.052$, a remarkably high value for the conditions.

Nitrogen fixation

To the best of my knowledge, nitrogen-fixing microorganisms have not been recovered from polar seas, although members of this group, particularly the heterocystous members of the Cyanophyta, the free-living bacterial genera *Azotobacter* and *Clostridium* (certain species), and symbiotic rhizobia all are present in the terrestrial environment. Symbiotic associations involving rhizobia do not occur in Antarctica.

Knowledge of symbiotic nitrogen-fixing plant associations in the Arctic has been reviewed by Allen, Allen & Klebesadel (1964), who expressed the view that their activities are 'the hub of an efficient nitrogen source for the entire ecosystem'. At least ten species among four genera of indigenous legumes have been found with nodulated roots in Arctic Alaska. Some non-leguminous species also are known to carry nodules. Among these, although not strictly in the full Arctic environment, stands of Sitka alder have been estimated to add 65 kg. of nitrogen/acre in leaf fall. Observations well within the tundra suggest that symbiotic nitrogen fixation is a factor of significance. Quantitative data are, however, lacking.

Examining soils from Northern Greenland, Jensen (1951) was unable to isolate *Azotobacter* although *Clostridium butyricum* was common. Another report by Isachenko & Simakova (1934) discusses the distribution of free-living nitrogen-fixing bacteria in the Russian Arctic but the details are unknown to me. *Azotobacter* seems to be rare in the soils of Antarctica, although Boyd *et al.* (1966) were able to make isolations in penguin rookeries and small ponds on Ross Island in McMurdo Sound. One suspects that this group of bacteria is unlikely to be an important contributor of nitrogen in the polar environment.

As Holdgate (1967) has pointed out, the sea is probably the most important source of nutrients, including nitrogenous nutrients, under Antarctic conditions. None the less, the microbial contribution cannot be overlooked, especially in so far as blue-green algae are concerned. These organisms are prominent in the scant vegetation of the ice-free areas of Antarctica, either free-living or as lichen symbionts, and appear widespread in the Arctic tundra. Recent studies by Fogg & Stewart (1968), using ^{15}N, demonstrated *in situ* nitrogen fixation in *Nostoc commune* and the lichens *Collema pulposum* and *Stereocaulon* spp. Active fixation occurred in *Nostoc commune* at temperatures as low as $1 \cdot 0°$. *Collema pulposum* was also effective in fixation at low temperatures although it had a high temperature coefficient, an undoubted advantage at times when insolation results in local warming at the ground surface. *Nostoc commune* also occurs on the soil surface and in ponds in the McMurdo Sound area. In fact, this was the only nitrogen-fixing species isolated from a large number of samples by Holm-Hansen (1963). I can testify to the luxuriant growth of this organism in favourable locations, not only in this sector, but also around the coastline of Wilkes Land, MacRobertson Land and Enderby Land.

Heterotrophy and chemoautotrophy: the cycling of organic materials
The Free-Living Microflora and Fauna

As the reader will discover, our comprehension of the importance of heterotrophic microorganisms in polar ecosystems, in a quantitative sense, is extremely limited. None the less, surveys have revealed a good deal about the general nature of the populations, including their size, diversity and varying complexities, stemming from a comparatively long history of interest. Among pioneering studies may be mentioned those of Ehrenberg (1869) (see Sandon, 1924) at Spitzbergen, and Gazert (1912), Tsiklinsky (1908) and Ekelöf (1908) in Antarctica. Apart from a general description of isolates, Ekelöf (1908) made counts of the soil flora and noted a tenfold increase in numbers during the summer.

Since that time, an assortment of information has accumulated on the microbiology of soils and fresh waters. Information on the protozoa may be found in Sandon (1924), Sandon & Cutler (1924), Dillon et al. (1968) and Dillon et al. (1969). A review by Rybalkina (1957), not accessible to me, deals with Russian research on the microbial ecology of Tundra soils and the role of microorganisms in soil processes. Jensen (1951) recorded denitrification, cellulose decomposition and nitrifying activity in soils from Northern Greenland raised to $25\cdot0°$ in the laboratory. Plate counts revealed $2\cdot2–14\cdot1 \times 10^6$ bacteria/g. moist soil. Perotti & Verona (1933) made counts of bacteria, actinomycetes and fungi in soils from Spitzbergen and apparently found nitrifiers to be scarce.

Tedrow & Douglas (1959) found that in Arctic soils at $3°$ (the *in situ* temperature in mid-June), rates of decomposition of organic matter were very low. Apart from limitations imposed by temperature, moisture content was critical and inhibitory under conditions of saturation. Carbon dating established the age of the humic fraction in an arctic brown soil as $2,000 \pm 150$ years. Seal carcasses have been found in the dry valley systems of Antarctica which are about the same age (Llano, 1962). Under those conditions, resistance to decomposition may be ascribed as much to the extreme dryness of the environment as to low temperature. In this regard, it must be pointed out that psychrophilic microorganisms are much more common in the polar marine environment than on land. For example, Boyd et al. (1966) in one soil from Ross Island (McMurdo Sound) found that, numerically, psychrophiles were only $0\cdot14\%$ of the mesophilic population. However, abundances appear to vary since, in another soil examined by this group, the mesophile count was 57·5 % of the psychrophile count. Nevertheless, it is doubtful whether facultative or obligate psychrophily offers any real advantage under terrestrial conditions in Antarctica where temperatures are either well below freezing or, over brief periods, within the temperature range of mesophiles through local insolation.

It is not feasible to draw conclusions on the rates of individual soil microbial processes and total metabolic activity from the survey conducted by Boyd et al. (1966). This study, however, discloses a number of features which deserve mention. Physiological groups detected included heterotrophic nitrogen-fixers, nitrifiers and sulphate-reducers apart from an assortment of saprophytes with unspecified functions. Methanogenic forms and chemolithotrophic sulphur bacteria were not found. Although they should be listed elsewhere in this account, one notes that green and purple sulphur and non-sulphur bacteria were found in an organically enriched pond. I have seen evidence of these forms elsewhere on the coastline of Antarctica, especially in polluted melt water.

The numbers of heterotrophic bacteria, determined by plate count on several media at several temperatures, sometimes revealed the presence of thermophilic forms, although only in areas of human activity, and demonstrated generally small populations. Many soils of the dry valleys appeared sterile. Filamentous fungi were uncommon in comparison with soils of warmer latitudes, although Cameron, King & David (1968) found ascomycetous forms in some samples as well as actinomycetes and yeasts, and Gressitt (1964) considers microfungi important as food for tardigrades and springtails. I found pigmented yeasts dominating plate counts in a pocket of highly saline soil at Mawson in MacRobertson Land. Microfungi are apparently abundant in the soils of Signy Island (see Holdgate, 1967).

Cameron et al. (1968) recognize a number of environmental factors which they consider interact to control microbial population density and diversity in the dry valleys of McMurdo Sound. Under the most favourable conditions where the growth of mosses (and blue-green algae) is allowed, populations do not exceed 10^5/g. soil although the food web appears quite complex. Under extreme conditions they, like Boyd et al. (1966), encountered either sterility or communities made up of non-pigmented, aerobic heterotrophs. Strangely, anaerobes were not found although they certainly occur in other situations, notably penguin rookeries. In the dry valleys, protozoa were restricted to flagellates and amoebae.

Holdgate (1967) has listed estimates of terrestrial standing stocks of microfauna for Signy Island. These include: protozoa, 4 g./m.^{2}; nematodes, 0·5 g./m.^{2}; Collembola, 0·4 g./m.^{2}; acarina, 0·2 g./m.^{2}; and tardigrades and rotifers, perhaps 0·1 g./m.^{2}. This, the full extent of the permanent Antarctic fauna, is considered largely to rely, directly or indirectly, on the microflora for sustenance. Using the figure 10^6 microbial cells/g. soil, assuming such a population to exist in the surface 5 cm., and taking bacterial cells as cylinders measuring 1×3 μm, it may be calculated that, in reasonably favourable areas, the microfloral standing stock may amount to 100 g. fresh weight/m.^{2}.

In parts of Signy Island (Heal, Bailey & Latter, 1967) plate and direct counts indicate clearly that the microbial biomass may be very considerably higher. This sector of the Antarctic, it must be emphasized, is not typical of the small and largely desolate ice-free south polar lands which, together, must be considered to play an insignificant role microbiologically in the total southern ecosystem. The same cannot be said for the comparatively fertile and extensive tundra although I have been unable to locate data for a substantive discussion of saprophytic and other soil microbial processes in that region.

As part of his treatment of the Antarctic ecosystem, Holdgate (1967) has suggested likely values for standing stocks of the macroscopic components of the Antarctic marine food web and also the phytoplankton. It is revealing, however, that an estimate of the non-photosynthetic microbial populations has not been attempted although their presence is acknowledged. We are, in effect, most uninformed on the proportion of the annual primary production which is channelled through microorganisms, in the Arctic as well as the Antarctic Oceans.

Treating this topic, it is logical to begin with the widely known and controversial work of Kriss and his colleagues (Kriss, 1963; Kriss *et al.* 1967) recounting Russian investigations which have extended from the central Arctic Ocean to the shores of Antarctica. One of their major conclusions was that, in high latitudes, microbial populations tend to be low. One finds this puzzling, even though the media used in counting were selective. At Mawson, on the coast of MacRobertson Land, Bunt (1960) examined with phase microscopy high-speed centrifugates of sea water throughout 1956 and almost always observed bacteria, especially in samples from the upper 5 m. Counts were not made, although I recall noting large numbers associated with colourless microflagellates which, incidentally, reached densities as high as $22·5 \times 10^6$/l. at the height of summer. I have also encountered bacteria, and sometimes yeasts, growing prolifically in enrichment cultures of McMurdo Sound ice algae in mineral media at $-2·0°$. Algal extracellular products appear to be the source of nutrients. Should this be so, these bacteria may be regarded as coproducers rather than saprophytes since they are establishing rather than reconverting particulate biomass.

It is obvious that, as is the case in soils, plate counts give very low estimates of total populations. For example, in the central Arctic Ocean, where plate counts indicated 3,000 microbial cells/l., Kriss (1963) reports that direct counting revealed from 35 cells/ml. at 3,400 m. to more than 39,000 cells/ml. at the surface during the summer. As biomass, it was estimated that these populations represent a total standing stock for the entire water column of approximately 10 mg./m.^{2}. One may consider this figure in relation to primary productivity under similar conditions averaging 10 mg. carbon/m.^{2}/day in the months June to August (see Ryther, 1963). Standing stocks of phytoplankton beneath the ice in July typically were 7 mg./m.^{2}, estimated as chlorophyll a in the surface 250 m. Based on the examination of submerged glass slides, it was estimated (Kriss, 1963) that the mean daily increase in microbial biomass throughout the water column was close to 40 %, i.e. 4 mg./m.^{2}, approximately 20 % of the daily primary production of organic matter. In view of the techniques employed, this estimate must be regarded as speculative. It is encouraging to learn that an alternative experimental approach has been used by Pomeroy et al. (1969) in the Antarctic Ocean although the results of their study are not yet available. Among other things, it will be interesting to discover whether these workers have been able to confirm earlier suggestions (Kriss, 1963; Walsh, 1969) that heterotrophic activities are highest in Antarctic intermediate waters. We do not know the relative contributions made to regenerative processes by the various biological components of the polar oceans, and we have minimal understanding of the constraints to which they are exposed.

No mention has so far been made in this review of microbial activities in the benthos. For information on this topic, the reader is referred to a paper by Zatsepin (1970). Similarly, the importance of psychrophily has not been considered. Many of the yeasts isolated by Fell (Fell & van Uden, 1968) in Antarctic waters have low temperature requirements. All of the protozoa found in sea ice appear to be obligately psychrophilic (J. S. Bunt, unpublished data). Information on marine psychrophilic bacteria has been reviewed by Morita (1966).

Host-related non-photosynthetic microorganisms

The potential influence exerted on polar ecosystem dynamics by host-related non-photosynthetic microorganisms usually is not considered, their function being assessed in much narrower contexts. It seems to me important that they be recognized.

Microorganisms pathogenic to homeothermic animals including man are well established in the polar regions and have received a great deal of study. As generally representative, may be mentioned accounts of an influenza epidemic (Anonymous, 1953*a*); poliomyelitis (Anonymous, 1953*b*); entozoa in eskimos (Laird & Meerovitch, 1961) transmitted, incidentally, from poikilotherms; tuberculosis (Porsild, 1930); venereal infections (Ronnenberg, 1957); and gastro-enteritis (Williams, 1950). Laird (1961) has discussed the lack of avian and mammalian haematozoa in the Antarctic and the Canadian Arctic. Margni & Castrelos (1964) have described earlier work and their own studies on the microflora of the upper respiratory tracts of various Antarctic birds.

Easily overlooked are the symbiotic protozoa and bacteria within the digestive tracts of the large numbers of herbivorous animals so prominent in the Arctic tundra during the summer. Ecologically, these symbionts are important and interesting because the association in which they are involved literally removes much of the turnover of organic material to a highly desirable thermal environment. Presumably their requirements and activities are close to those which are associated with domestic and wild animals in less severe climates. I do not know of any investigations in this area, although Lubinsky (1958, 1963) has described some of the rumen ciliates from reindeer and musk-oxen.

Savile (1963) has reviewed the status of mycology in the Arctic, including discussions of the biology of forms which exist as parasites of higher plants.

Other interbiotic relationships

As I have attempted to stress at the beginning of this review, productivity may be influenced through a variety of controls other than the direct production or conversion of biomass. Often, it is difficult to assess, in quantitative terms, the consequences of control activities. However, I should like to mention one phenomenon of interest.

Following observations by Bunt (1955) and others, that some common antarctic birds lack an aerobic microflora in their gastro-intestinal tracts, Sieburth (1960) traced the cause to acrylic acid, incorporated in the avian diet, and originating in the mucilage produced by *Phaeocystis*, an organism which is at times prominent in the phytoplankton. The significance of the effect for the bird populations is not known, although it has been suggested that algal species successions and productivity may be affected materially wherever *Phaeocystis* is abundant.

CONCLUSION

It should be clear that the indigenous productivity of the Arctic tundra and the Antarctic Ocean far exceeds the capacities of the highly stressed Antarctic terrestrial system and Arctic Ocean. Micro-algae are responsible almost exclusively for primary production in the Antarctic Ocean but appear of little significance in the tundra except as lichen symbionts. The factors which limit biological activity are, in many respects, distinct for each environment. Far more attention needs to be given to the activities of the heterotrophic populations and to microbial control processes. Until this is done, our comprehension of the dynamics, organization and global significance of polar ecosystems will remain superficial. Projections for the Arctic have been made by Johnson (1969). A current overview of Antarctic ecology (see El-Sayed, 1969) is expected.

REFERENCES

ALLEN, E. K., ALLEN, O. N. & KLEBESADEL, L. J. (1964). An insight into symbiotic nitrogen-fixing plant associations in Alaska. *Science in Alaska. Proc. 14th Alaskan Science Conference, Anchorage, Alaska*, p. 54.

ALLEN, M. B. (1970). Metabolic activities of phytoplankton associated with Arctic sea ice. (In the Press).

ANONYMOUS (1953a). Influenza virus epidemic at Victoria Island, N.W. Territories, 1949. *Polar Rec.* **6,** 680.

ANONYMOUS (1953b). Poliomyelitis epidemics in the Canadian Eastern Arctic. *Polar Rec.* **6,** 679.

APOLLONIO, S. (1961). The chlorophyll content of Arctic sea ice. *Arctic* **14,** 197.

ARMITAGE, K. B. & HOUSE, H. B. (1962). A limnological reconnaisance in the area of McMurdo Sound, Antarctica. *Limnol. Oceanogr.* **7,** 36.

BILELLO, M. A. (1961). Formation, growth and decay of sea ice. *Arctic* **14,** 3.

BILLINGS, W. D. & MOONEY, H. A. (1968). The ecology of Arctic and alpine plants. *Biol. Rev.* **43,** 481.

BLAKEMORE, L. C. & SWINDALE, L. D. (1958). Chemistry and clay mineralogy of a soil sample from Antarctica. *Nature, Lond.* **182,** 47.

BLISS, L. C. (1962). Adaptations of Arctic and alpine plants to environmental conditions. *Arctic* **15,** 117.

BOYD, W. L., STALEY, J. T. & BOYD, J. W. (1966). Ecology of soil microorganisms of Antarctica. *Antarctic Res. Ser.* **8,** 125.

BRITTON, M. E. (1957). Vegetation of the Arctic tundra. *Oregon State Univ. 18th Biol. Colloquim, 'Arctic Biology'*, p. 26.

BUNT, J. S. (1955). A note on the faecal flora of some Antarctic birds and mammals at Macquarie Island. *Proc. Linn. Soc. N.S.W.* **80,** 44.

BUNT, J. S. (1960). Introductory studies: Hydrology and plankton, Mawson, June 1956 – February 1957. *A.N.A.R.E. Rep.* B 3, pp. 1–135.

BUNT, J. S. (1963). Diatoms of Antarctic sea ice as agents of primary production. *Nature, Lond.* **199,** 1255.

BUNT, J. S. (1967). Thermal energy as a factor in the biology of the polar regions. In *Thermobiology*, p. 555. Ed. A. H. Rose. New York: Academic Press.

BUNT, J. S. (1968a). Microalgae of the Antarctic pack ice zone. In *Symposium on Antarctic Oceanography*, Santiago, Chile, p. 198. Ed. R. I. Currie. Cambridge: Scott Polar Research Institute.

BUNT, J. S. (1968b). Some characteristics of microalgae isolated from Antarctic sea ice. *Antarctic Res. Ser.* **11**, 1.

BUNT, J. S. & LEE, C. C. (1970). Seasonal primary production within Antarctic sea ice. *J. Mar. Res.* (In the Press.)

BURKHOLDER, P. R. & MANDELLI, E. F. (1965). Productivity of microalgae in Antarctic sea ice. *Science, N.Y.* **149**, 872.

CAMERON, R., KING, J. & DAVID, C. (1968). Soil microbial and ecological studies in Southern Victoria Land. *Antarctic J.* **3**, 121.

COACHMAN, L. K. (1963). Water masses of the Arctic. *Proc. of the Arctic Ocean Symposium*, Oct. 1962, 143.

COACHMAN, L. K. (1969). Physical oceanography in the Arctic Ocean, 1968. *Arctic* **22**, 214.

COMITA, G. W. & EDMONDSON, W. T. (1953). Some aspects of the limnology of an arctic lake. In *Current Biol. Res. in the Al. Arctic*, p. 7. Ed. Ira L. Wiggins. Stanford University Press.

CROOKS, A. D. (1960). *Oceanic Observations in Antarctic Waters*, M. V. Magga Dan, 1959. Oceanographical Station List, no. 44. C.S.I.R.O. Australia.

DEACON, G. E. R. (1937). The hydrology of the Southern Ocean. *'Discovery' Rep.* **15**, 1.

DEACON, G. E. R. (1963). The southern ocean. In *The Sea 2*, 281. Ed. M. N. Hall. New York: Interscience Publishers.

DEACON, G. E. R. (1964). Antarctic Oceanography: the physical environment. In *Antarctic Biology*, p. 81. Ed. R. Carrick. Paris: Hermann.

DILLON, R. D., BIERLE, D. & SCHROEDER, L. (1968). Ecology of antarctic protozoa. *Antarctic J.* **3**, 123.

DILLON, R. D., WALSH, G. L. & HETH, S. R. (1969). The ecology of antarctic protozoan ponds. *Antarctic J.* **4**, 114.

DUNBAR, M. J. (1953). Arctic and subarctic marine ecology: immediate problems. *Arctic* **6**, 75.

DUNBAR, M. J. (1968). *Ecological Development in Polar Regions*. Englewood Cliffs, N.J.: Prentice-Hall Inc.

EHRENBERG, C. G. (1869). Die mikroscopische Lebensverhältnisse auf der Oberfläche der Insel Spitsbergen. *Monatsb. Akad. Berlin.*

EKELÖF, E. Bakteriologische Studien wahrend der schwedischen Südpolar Expedition 1901–03. *Wiss. Ergebn. schwed. Südpolarexped.*, **4**, Stockholm.

EL-SAYED, S. Z. (1968a). On the productivity of the Southern Atlantic Ocean and the waters west of the Antarctic Peninsula. *Antarctic Res. Ser.* **11**, 15.

EL-SAYED, S. Z. (1968b). Prospects of primary productivity studies in Antarctic waters. In *Symposium on Antarctic Oceanography*, Santiago, Chile, p. 227. Cambridge: Scott Polar Research Institute.

EL-SAYED, S. Z. (1969). Antarctic ecology. *Bioscience* **19**, 1031.

EL-SAYED, S. Z. & MANDELLI, E. F. (1965). Primary production and standing crop of phytoplankton in the Weddell Sea and Drake Passage. *Antarctic Res. Ser.* **5**, 87.

ENGLISH, T. S. (1961). Some biological oceanographic observations in the Central North Polar Sea, Drift Station Alpha, 1957–8. Arctic Institute of North America. *Scientist Rep.* no. 15.

EPPLEY, R. W. & STRICKLAND, J. D. H. (1968). Kinetics of marine phytoplankton growth. *Adv. Microbiol. Sea* **1**, 23.

FELL, J. W. & VAN UDEN, N. (1968). Marine yeasts. *Adv. Microbiol. Sea* **1**, 167.

FOGG, G. E. (1967). Observations on the snow algae of the South Orkney Islands. *Phil. Trans. Roy. Soc. Lond.* B **252**, 279.

FOGG, G. E. & STEWART, W. D. P. (1968). *In situ* determinations of biological nitrogen fixation in Antarctica. *Br. Antarctic Surv. Bull.* **15**, 39.

FUKUSHIMA, H. (1961). *Antarctic Record*, no. 11, p. 164.

GANNUTZ, T. P. (1970). Photosynthesis and respiration of plants in the Antarctic Peninsula area. *Antarctic J.* **5**, 49.

GAZERT, H. (1912). Untersuchungen öber Meeresbakterien und ihren Einfluss auf der Stoffwechsel in Meere. *Deutsche Sudpolar-Expedition* 1091-03, vol. 1, p. 1. Berlin: Reimer.

GOLDMAN, C. R. (1964). Primary productivity studies in Antarctic lakes. In *Antarctic Biology*, p. 291. Ed. R. Carrick. Paris: Hermann.

GRAN, E. H. (1904). Diatomaceae from the ice-floes and plankton of the Arctic Ocean. In F. Nansen, *Scient. Results. Norw. N. polar Exped.* **4**. (11).

GRESSITT, J. L. (1964). Ecology and biogeography of land arthropods in Antarctica. In *Antarctic Biology*, p. 211. Ed. R. Carrick. Paris: Hermann.

HART, T. J. (1942). Phytoplankton periodicity in Antarctic surface waters. ' *Discovery*' *Rep.* **21**, 261.

HEAL, O. W., BAILEY, A. D. & LATTER, P. M. (1967). Bacteria fungi and protozoa in Signy Island soils compared with those from a temperate moorland. *Phil. Trans. Roy. Soc. Lond.* B **252**, 191.

HOBBIE, J. E. (1964). Carbon-14 measurements of primary production in two Arctic Alaskan lakes. *Verh. int. Verein. theor. angew. Limnol.* **15**, 360.

HOCH, G., OWENS, O. VAN H. & KOK, B. (1963). Photosynthesis and respiration. *Archs Biochem. Biophys.* **101**, 171.

HOLDGATE, M. W. (1967). The Antarctic ecosystem. *Phil. Trans. Roy. Soc. Lond.* B **252** 363.

HOLM-HANSEN, O. (1963). Algae: Nitrogen fixation by Antarctic species. *Science, N.Y.* **139**, 1059.

HORNE, A. J., FOGG, G. E. & EAGLE, D. J. (1969). Studies *in situ* of the primary production of an area of inshore antarctic sea. *J. mar. biol. Ass. U.K.* **49**, 393.

ISACHENKO, B. L. & SIMAKOVA, T. L. (1934). *Bakteriologicheskie issledovaniia pochv Arktiki*. Leningrad. Vsesoiuznyi arkticheskii institut. Trudy, T.9, 107.

JENSEN, H. L. (1951). Notes on the microbiology of soil from Northern Greenland. *Meddr. Grønland* **142**, 23.

JOHNSON, P. C. (1969). Arctic plants, ecosystems and strategies. *Arctic* **22**, 341.

KALFF, J. (1967a). Phytoplankton dynamics in an Arctic lake. *J. Fish. Res. Bd Can.* **24**, 1861.

KALFF, J. (1967b). Phytoplankton abundance and primary production rates in two Arctic ponds. *Ecology* **48**, 558.

KORT, V. G. (1962). The Antarctic Ocean. *Scient. Am.* **207**, 113.

KRISS, A. E. (1963). *Marine Microbiology*. Edinburgh and London: Oliver and Boyd.

KRISS, A. E., MISHUSTINA, I. E., MITSKEVICH, N. & ZEMTSOVA, E. V. (1967). *Microbial Population of Oceans and Seas*. New York: St Martin's Press.

KURSANOV, A. L. & DIACHKOV, N. N. (1945). Lichens and their utilization. *Akademiia nauk SSSR* **54**, 2.

LAGERHEIM, N. G. (1883). A contribution to knowledge of the snow flora of Lulea Lappmark. *Bot. notiser*, p. 230.

LAIRD, M. (1961). A lack of avian and mammalian haematozoa in the Antarctic and Canadian Arctic. *Can. J. Zool.* **39**, 209.

LAIRD, M. & MEEROVITCH, E. (1961). Parasites from northern Canada. I. Entozoa of Fort Chino eskimos. *Can. J. Zool.* **39**, 63.

LLANO, G. A. (1944). Lichens, their biological and economic significance. *Bot. Rev.* **10**, 1.

LLANO, G. A. (1962). The terrestrial life of the Antarctic. *Scient. Am.* **207**, 213.

LUBINSKY, G. (1958). Ophryoscolecidae (ciliate) of the reindeer from the Canadian Arctic. *Can. J. Zool.* **36**, 937.

LUBINSKY, G. (1963). *Metradinium caudatum* sp.n. – a rumen ciliate of muskox from northern Canada. *Can. J. Zool.* **41**, 28.

MACKINTOSH, N. A. & HERDMAN, H. F. P. (1940). Distribution of the pack ice in the Southern Ocean. '*Discovery*' *Rep.* **19**, 287.

MACNAMARA, E. E. (1969). Pedology of Enderby Land, Antarctica. *Antarctica J.* **4**, 208.

MANDELLI, E. F. & BURKHOLDER, P. R. (1966). Primary productivity in the Gerlache and Bransfield Straits of Antarctica. *J. mar. Res.* **24**, 15.

MARGNI, R. A. & CASTRELOS, O. D. (1964). Quelques aspects de la bactériologie Antarctique. In *Antarctic Biology*, p. 121. Ed. R. Carrick. Paris: Hermann.

MEGURO, H. (1962). Plankton ice in the Antarctic Ocean. *Antarctic Rec.* **14**, 1192.

MEGURO, H., ITO, K. & FUKUSHIMA, H. (1967). Bottom type 'plankton ice' in the Arctic Ocean. *Antarctic Rec.* **28**, 33.

MIKHAYLOV, I. S. (1958). Natural glasshouses in Severnaya Zemlya. *Priroda* (Nature), no. 2, p. 118.

MORITA, R. Y. (1966). Marine psychrophilic bacteria. In *Oceanogr. Mar. Biol. Ann. Rev.* **4**, 105.

MOSBY, H. (1963). Water, salt and heat balance in the North Polar Sea. *Proc. of the Arctic Basin Symp.*, Oct. 1962. Washington, D.C.: Arctic Institute of North America, p. 69.

MOSBY, H. (1968). Bottom water formation. In *Symp. on Antarctic Oceanography*, Santiago, Chile, p. 47. Ed. R. I. Currie. Cambridge: Scott Polar Research Institute.

PEROTTI, R. & VERONA, O. (1933). Bacteriological report on some soils of West Spitsbergen. *Annali Fac. Agr. Univ. Pisa* **9**, 23.

PIPER, C. S. (1938). Soils from subantarctic islands. *Rep. B.A.N.Z. Antarct. Res. Exped.* series A, pt. 2.

POLUNIN, N. (1959). *Circumpolar Arctic Flora*. Oxford University Press.

POMEROY, L. R., WIEBE, W. J., FRANKENBERG, D. & HENDRICKS, C. (1969). Metabolism of total water columns. *Antarctic J.* **4**, 188.

PORSILD, M. P. (1930). Are there any tubercule bacteria in Disko, Greenland? *Dansk bot. ark.* **6**, 7.

PREDOEHL, M. C. (1966). Antarctic pack ice: Boundaries established by Nimbus I pictures. *Science, N.Y.* **153**, 861.

RONNENBERG, J. (1957). *Problems i bekaempelsen af kønssydomne* (sic) *i Grønland*. Atuagagdliutit Grønlandsposten, nr. 24, 12.

RUDOLPH, E. D. (1966). Terrestrial vegetation of Antarctica. *Antarctic Res. Ser.* **8**, 109.

RYBALKINA, A. V. (1957). Microflora of tundra, podsol and chernozem soils. *Akad. nauk SSSR. Pochvennyi institut V. V. Dokuchaeva*, p. 5.

RYTHER, J. H. (1959). Potential productivity of the sea. *Science, N.Y.* **130**, 602.

RYTHER, J. H. (1963). Geographic variations in productivity. In *The Sea*, vol. 2, p. 347. Ed. M. N. Hill. New York: Interscience.

RYTHER, J. H. (1969). Photosynthesis and fish production in the sea. *Science, N.Y.* **166**, 72.

SAIJO, Y. & KAWASHIMA, T. (1964). Primary production in the Antarctic Ocean. *J. Oceanogr. Soc. Japan* **19**, 22.

SANDON, H. (1924). Some protozoa from the soils and mosses of Spitsbergen.

Results of the Oxford Univ. Expedition to Spitsbergen, no. 27. *J. Linn. Soc.* (*Zool.*) **35**, 449.

SANDON, H. & CUTLER, D. W. (1924). Some protozoa from the soils collected by the 'Quest' Expedition (1921–22). *J. Linn. Soc. J.* (*Zool.*) **36**, 1.

SAVILE, D. B. O. (1963). Mycology in the Canadian Arctic. *Arctic* **16**, 17.

SCOTTER, G. W. (1963). Growth rates of *Cladonia alpestris*, *Cladonia mitis* and *Cladonia rangiferina* in the Taltson River region, N.W.T. *Can. J. Bot.* **41**, 1199.

SIEBURTH, J. MCN. (1960). Acrylic acid, an 'antibiotic' principle in *Phaeocystis* blooms in Antarctic waters. *Science, N.Y.* **132**, 676.

TEDROW, J. C. F. & DOUGLAS, L. A. (1959). Organic matter decomposition rates in Arctic soils. *Soil Sci.* **88**, 305.

TEDROW, J. C. F. & UGOLINI, F. C. (1966). Antarctic soils. *Antarctic Res. Ser.* **8**, 161.

TSIKLINSKY, Mlle. (1908). La flore microbienne dans les régions du Pôle Sud. *Expédition antarctique française*, 1903–05, vol. 3. Paris: Masson et Cie.

UNTERSTEINER, N. (1963). Ice budget of the Arctic Ocean. *Proc. Arctic Basin Symp.* Oct. 1962, p. 219.

UNTERSTEINER, N. (1969), Sea ice and heat budget. *Arctic* **22**, 195.

WALSH, J. J. (1969). A statistical analysis of the phytoplankton community within the Antarctic Convergence. Ph.D. dissertation, University of Miami.

WILLIAMS, R. B. (1950). Summary of *Salmonella* and *Shigella* of Alaska. *N.W. Med.* **49**, 340.

WYRTKI, K. (1960). The Antarctic circumpolar current and the antarctic polar front. *Dt. hydrogr. Z.* **13**, 153.

ZATSEPIN, V. I. (1970). On the significance of various ecological groups of animals in the bottom communities of the Greenland, Norwegian and the Barents Seas. In *Marine Food Chains*, p. 207. Ed. J. H. Steele. Berkeley and Los Angeles: University of California Press.

LIMITS OF MICROBIAL
PRODUCTIVITY IN THE OCEAN

W. VISHNIAC

Department of Biology, University of Rochester,
Rochester, N.Y., U.S.A.

The preceding contributions to this symposium have made it abundantly clear that oceanic productivity is a critical part of future food production. I shall not comment here on the unequal distribution of food, but rather consider the statistical observation that for the past decade or two, for the first time since the beginning of the industrial revolution, the food productivity per capita has not increased. The year 1969 saw for the first time a net decrease in the food output of the world. Regardless of whether this decrease is temporary or dictated by other than technological factors, it reminds us how uncomfortably close man may have come to a maximal exploitation of the earth for food production. It is no wonder that human hope has been placed on the use of the ocean as a food source, either by the expansion of existing fishing techniques, or by the development of systematic 'ocean farming'. The prospects of progressing from fishing to ocean farming has been compared to the step that man took when he abandoned hunting and took up agriculture. The optimistic proponents of ocean exploitation point to the limited area of the oceans of the world which have been fished in a systematic fashion, and compare it to the vast areas in which commercial fisheries are not carried out. I propose to examine the ability of the ocean to produce organic matter, and to compare it to our present rate of exploitation.

My own interest in the subject was stimulated, I might as well confess it here, by a science fiction novel (Clarke, 1970) entitled *The Deep Range*. I wish to take this opportunity to express my thanks to friends and colleagues with whom I have discussed this subject and from whom I have received valuable criticism. I am especially grateful to Dr W. T. Russell-Hunter, Syracuse University, and Dr E. C. Raney, Cornell University, in whose company some of these topics were first discussed, and to Dr C. Istock, University of Rochester. By singling out these men I do not mean to slight the many friends and colleagues who have contributed their criticisms and suggestions. At the same time I must accept full responsibility for conclusions drawn and opinions expressed in this contribution.

Table 1. *Important primary producers*

Marine algae which have been shown to produce large amounts
of carbohydrates, much of which is excreted (Provasoli, 1963).

Chlorophytes
  Dunaliella euchlora
  Chlorella sp. (no. 580, Indiana University)
  Chlamydomonas sp. (strain Y, Ralph Lewin)
  Chlorococcum sp.
  Pyramimonas inconstans

Diatoms
  Cyclotella sp.
  Nitzschia brevirostris
  Melosira sp.

Chrysomonads and Cryptomonads
  Isochrysis galbana
  Monochrysis lutheri
  Prymnesium parvum
  Rhodomonas sp.

Dinoflagellates
  Amphidinium carteri
  Prorocentrum sp.
  Katodinium dorsalisulcum

The low concentration of organic compounds found in the ocean at any one time allows us to consider the ocean as a vast hydroponic culture in which carbon dioxide is the sole source of carbon. The microorganisms to which I will be referring are the microscopic algae which are the chief primary producers in the sea. Table 1 lists some of the identified primary producers, which are roughly of bacterial dimensions, without being an exhaustive list. The extent to which these organisms are able to convert mineral matter into organic compounds is frequently limited by growth factors. Microscopic algae which have been grown in pure culture have frequently demonstrated a requirement for a variety of growth factors, among which vitamin B_{12} and thiamin are prominent (Provasoli, 1963). Their existence is therefore dependent on the rate of production of these growth factors by other marine organisms, or by the contribution of vitamins made by the fresh waters discharged into the ocean. Of more general significance is the limited availability of nitrogen and phosphorus. The nitrogen content of the sea, mostly ammonium salts and nitrate, is 60 mg. $N/m.^3$ according to Harvey (1957), but varies greatly (0–500 mg./m.^{3}) with depth, season and latitude. The sources and distribution of nitrogen and phosphorus have been reviewed by Redfield, Ketchum & Richards (1963). While ammonium salts and nitrate are generally quite soluble, the limitation in the amount of phosphorus derives in part from the insolubility of many phosphates at the alkaline pH of sea water. Since carbon dioxide, either as dissolved CO_2 or as bicarbonate, is present in amounts that are far in excess of what can be utilized even if all nitrogen and phosphorus were converted to organic matter, it has been suggested that we 'fertilize' selected portions of the ocean with these deficient minerals. A limited number of experiments have been carried out (Gross, 1947), but not with the success that would instil great optimism in us.

Assuming that limiting growth factors, limiting minerals and other requirements could be met, what is the ultimate level of organic matter which the ocean could produce? Clearly, the limit which cannot be exceeded is the one dictated by incident sunlight upon which all ocean productivity ultimately depends. The radioactive output of the sun is 4×10^{26} W, or $3 \cdot 50 \times 10^{27}$ kWh. per year. The earth intercepts only a small fraction of this energy; in the course of one year it is estimated that $1 \cdot 25 \times 10^{24}$ cal. reach the upper atmosphere (Rabinowitch, 1948). Scattering and absorption by the clouds and by the atmosphere result in a 60 % loss, so that 5×10^{23} cal. per year reach the earth's surface. Fifty per cent of this incident radiation is in the infrared region, and hence inaccessible to photosynthesis. Conservatively, a 20 % loss can be attributed to absorption by rocks and other non-arable surfaces and to reflexion by ice and snow, so that only 2×10^{23} cal. per year fall on arable land and the oceans. About $1 \cdot 6 \times 10^{23}$ calories can therefore by assigned to the ocean, where an additional 10 % is probably lost by reflexion from the water surface and absorption by water and dissolved ions. Probably not more than $1 \cdot 2 \times 10^{23}$ cal. per year are available for absorption by photosynthetic pigments in the ocean. This amount must next be corrected by the efficiency with which photosynthesis takes place. Under field conditions photosynthetic efficiency is about 2 % (Rabinowitch, 1948), so that the conversion of radiant into chemical energy corresponds to about $2 \cdot 5 \times 10^{21}$ cal. per year. This then is the theoretical maximum of energy available for primary productivity in the ocean.

To estimate the amount of organic matter that could be produced at the expense of this energy requires a knowledge of the energy which is needed to convert carbon dioxide into living matter. Rather than use the heat of combustion for such an estimate a better approximation can be reached if we consider the major biochemical steps which take place when a photosynthetic organism fixes carbon dioxide and converts it into the various constituents of the living cell. For the sake of convenience photosynthetic growth may be divided into three stages, as summarized in Table 2. The first stage consists in the reduction of carbon dioxide to the carbohydrate level, according to the Calvin–Benson pathway (Calvin & Bassham, 1962). The sequence of events which leads to the formation of carbohydrate can be summarized by the equation

$$3CO_2 + 9ATP + 2NADPH \rightarrow \text{glyceraldehyde-3-phosphate} + 8PO_4^{3-} + 9ADP + 2NADP^+.$$

From this formula it appears that for each gram-atomic weight of

Table 2. *Minimum energy requirement for photoautotrophic growth*

		Required per g. atomic wt. of C	
Stage in biosynthesis	On basis of:	Cofactors	Energy (kcal.)
$CO_2 \rightarrow$ carbohydrates	Calvin–Benson pathway*	3·0 ATP	24
		2·0 NADPH	108
Carbohydrates $\rightarrow$ amino acids	Survey of known pathways	~0·5 ATP	4
Amino acids $\rightarrow$ whole cells	Molar growth yield†	2·4 ATP	19
			155

The requirement of 155 kcal. per gram-atomic weight of carbon in the conversion of CO_2 to living cells is equal to 13 kcal./g. of carbon.

* Calvin & Bassham (1962).
† Bauchop & Elsden (1960).

carbon which is converted from CO_2 to carbohydrate, 3 moles ATP and 2 moles NADPH are required. For the purposes of this calculation the value of a phosphate anhydride in ATP has been put at 8 kcal. and the reductions of $NADP^+$ by the reaction

$$NADP^+ + H_2O \rightarrow NADPH + H^+ + \tfrac{1}{2}O_2$$

at 54 kcal. The second stage consists in the conversion of carbohydrate to amino acids. A survey of the known biosynthetic pathways by which amino acids are synthesized suggests that on the average a fraction of 1 molecule ATP per carbon atom is required; as an approximation 0·5 ATP molecules per carbon atom has been used in this calculation. The final stage represents the synthesis of living cells from amino acids. The energy requirement for this process can be calculated from the results of Bauchop & Elsden (1960), who measured the molar growth yield of bacteria under conditions of a non-limiting supply of amino acids; that is, conditions which allowed the limiting substrate – for example, glucose – to be used solely as an energy source. Taking their yield of about 10 g. dry weight of cell material per ATP, and assuming that half of the dry weight is carbon, we arrive at an energy requirement of 2·4 molecules ATP per atom of carbon. From Table 2 it then appears that the conversion of 1 mole of CO_2 into living matter requires 155 kcal., or $1·3 \times 10^4$ cal./g. C. While this calculation ignores the formation of purine and pyrimidine bases, certain cell-wall constituents, and other substances, a more detailed calculation is not likely to alter this figure significantly.

The two sets of calculations above – namely that of the energy available for photosynthesis, and that of the energy required to convert CO_2 to living matter – allow us to set an upper limit to the amount of living matter which may be produced in the ocean by the microscopic algae responsible for primary productivity:

$$\frac{2 \cdot 5 \times 10^{21} \text{ cal./year available for photosynthesis}}{1 \cdot 3 \times 10^4 \text{ cal./g. C}} = 1 \cdot 9 \times 10^{11} \text{ metric tons carbon/year.}$$

This maximum primary productivity could be obtained if nitrogen and phosphorus were not limited.

Table 3. *Ocean productivity*

Average carbon assimilation per year (12)	150 g. C/m.^{2}
Area of oceans*	$3 \cdot 5 \times 10^{14}$ m.^{2}
Annual productivity†	$5 \cdot 3 \times 10^{10}$ tons C

* Not counting permanently ice-covered waters.
† To sustain this rate of productivity the limiting amounts of nitrogen and phosphorus are assumed to turn over 2–10 times each year.

It is instructive to compare these figures of potential productivity with direct measurements, using a variety of techniques and methods of evaluating them (Steeman Nielson & Jensen, 1957; Ryther & Yentsch, 1957; Riley, 1956). The productivity of the oceans in terms of carbon converted from carbon dioxide to organic matter has been estimated to be 20–200 g./m.^{2}/year. A critical review of these figures (Russell-Hunter, 1970) suggests that the low figures are very likely much too low, and that a productivity of 100–200 g./m.^{2}/year may well be an appropriate average for the oceans of the world. For the summary in Table 3, a productivity of 150 g. C/m.^{2}/year has been assumed, and from this it appears that the observed ocean productivity is within a factor of 3·5 of the estimated theoretical maximum. This observed productivity must support the exploitation of the ocean; that is, it forms the ecological basis for the environment which sustains the world fisheries. It must be understood that the figures on annual productivity do not mean that such an amount of organic matter is actually present in the ocean at any one time. The limited supply of phosphorus and nitrogen indicates that there is in the course of the year, depending on the geographic location, a turnover of organic matter of 2–10 times a year. The figures of annual productivity are therefore a representation of the synthesis of new organic matter in the course of one year which proceeds concurrently with fishing, predation, death and decay.

The exploitation of the oceans has increased sharply since World

Table 4. *Ocean harvest and required productivity*

	Catch		
Calendar year	10^6 metric tons wet weight	10^6 tons C	Required productivity (10^{10} tons C)
1948	19·6	2·35	1·30
1952	25·0	3·00	1·65
1955	28·9	3·46	1·90
1958	33·2	3·98	2·19
1961	43·4	5·21	2·87
1964	52·7	6·32	3·48
1967	60·5	7·26	3·99

Fish catch is given in million tons of wet weight and the equivalent amount of carbon, assuming a 12 % carbon content. It is assumed that half the fish catch occurs at the third trophic level and half at the fourth trophic level. The efficiency of transfer at each trophic level is assumed to be 10 %. The last column lists the primary productivity required to sustain the fish catch in any given calendar year (Schaefer, 1965; F.A.O. 1968).

War II. Table 4 summarizes the increase between 1948 and 1967 (Schaefer, 1965; F.A.O. 1968). It is assumed for the purposes of the present calculation that 12 % of the wet weight of fish is carbon. To relate this exploitation to the primary productivity requires the assumption of several uncertain figures. One assumption has to be made concerning the distance of the fish, in terms of trophic levels, from the primary producers. In other words, how many links are there in the food chain between the initial conversion of CO_2 to microscopic algae and the growth of commercially important fish? Clearly, this number is not the same for all commercial fish; and furthermore, many fish cannot be assigned to a single trophic level. A fish which feeds simultaneously on planktonic algae, planktonic shrimp, and certain small fish operates simultaneously at several trophic levels. It seems to be fair to assume that about half of the fish catch is at the third trophic level and about half at the fourth (Russell-Hunter, 1970; Schaefer, 1965).

The other controversial figure is the efficiency with which organic matter is transferred from one link in the food chain to another. Figures as low as 5 % and as high as 35 % have been cited, but the best documented values seem to lie between 8 and 10 % (Slobodkin, 1961). For simplicity's sake a 10 % efficiency is assumed here, and it must be understood that the documentation of a lower efficiency would paint a gloomier picture of the world's food supply. Thus the difference between a 5 % and a 10 % efficiency, computed over three links in the food chain, is almost one order of magnitude. The last column in Table 4 represents the primary productivity which was required to sustain the fish yield in the tabulated years, making the assumption of a 10 % transfer efficiency and dividing the fish catch evenly between the third and the fourth trophic level. On this basis, the required annual productivity in 1967 began to approach the observed annual productivity, and is only a factor of 4 below the theoretical maximum productivity (Fig. 1). It is quite clear that these figures are uncertain in many respects, but it is equally clear that even in the event of an error by one order of magnitude we are rapidly approaching a rate of ocean exploitation which is limited

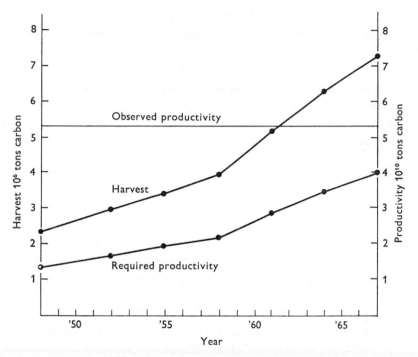

Fig. 1. Ocean harvest 1948–67 in relation to ocean productivity. Harvest, in 10^6 tons of carbon, is plotted against the left ordinate for each calendar year. Productivity is plotted against the right ordinate in units of 10^{10} metric tons of carbon. 'Observed productivity' is based on 150 g. $C/m.^2/year$. 'Required productivity' is based on the amount necessary to sustain the harvest in each calendar year. The theoretical maximal productivity, larger than the observed one by a factor of 3·5, is discussed in the text.

by the amount of sunlight falling on the ocean. It must furthermore be remembered that much of the fish catch is not directly used for human consumption, but goes into the production of fish meal, which in turn is used to fatten chickens and other domestic animals. Thus man may be removed by yet another trophic level from the catch. To cite one example, the United States figures as a major fishing country only because of the menhaden catch in the Atlantic Ocean, and yet the menhaden is entirely reduced to fish meal.

On the basis of the figures cited here, the maximum fish catch which could be realized if the annual productivity were to approach the computed maximal productivity would be about 3×10^8 tons per year. Schaefer (1965), on the basis of different calculations, also cites this figure as a probable maximum in annual fish catch. Should the estimate of the trophic levels at which fish and other seafood are caught be too high, then the potential yield of the ocean is considerably larger than estimated here. Thus, should the fish catch lie between the second and third trophic level, instead of between the third and fourth, then the potential fish catch is larger by one order of magnitude. A deliberate increase in the efficiency of ocean exploitation could consequently be achieved by utilizing organisms at lower trophic levels. One example of such exploitation is the harvest of planktonic crustaceae which is practised in some parts of Indonesia. Fine-meshed nets are placed in rapid tidal currents, and the resultant mass of planktonic shrimp and other crustaceae is dried to an edible, pink, leathery crust; the equivalent of marine jerky (sun-dried meat). Techniques for increasing the primary productivity of the ocean, or at least of selected areas, have been proposed. The preliminary experiments with artificially added nitrates and phosphates have already been cited. In addition, it has been proposed to create artificial upwellings by placing heat sources in suitable locations at some appropriate depth. Such upwellings would carry to the surface organic detritus and mineral particles, introduce them into the illuminated zones, and hence increase the rate of photosynthesis. Several such proposals have been recently reviewed by Russell-Hunter (1970).

The emphasis on factors which limit ocean productivity have omitted any mention of the favourable circumstances under which mankind exists. A biologist with a cosmic, rather than merely global, view might point out that mankind is extraordinarily fortunate to live on a planet which is so richly endowed with water. Life as we know it requires an immense amount of water, when compared to the requirement for any other molecule. It is on earth that creatures such as blue whales may exist; each individual requires about $5 \mathrm{km.}^3$ for its support. The examination of lunar rock samples revealed a complete absence of water (Keil, Prinz & Bunch, 1970). Not only was there no water of crystallization, but no hydroxyl radicals could be found in those minerals that might conceivably have contained them. Admittedly these samples were taken from the very surface of the moon, but the present attempts to reconstruct the geological history of the moon suggest that the moon at no time possessed significant amounts of water (Anders, 1970). This view, if it is substantiated, would rule out earlier hypotheses that an atmosphere and liquid water were present on the moon at one time and might conceivably have led to an early evolution of life (Sagan, 1961). The mere absence of water may thus have obviated previously proposed plans for lunar palaeontology.

Turning from the moon to Mars, a planet which conceivably may support life, since it possesses CO_2, an illuminated surface, and at least some water in the atmosphere, we are dealing at present with much more limited information. For instance, there is at present no evidence for nitrogen or any nitrogen compound in the Martian atmosphere. Such water as has been observed on Mars is limited to the atmosphere, and represents an amount which, if precipitated uniformly on the surface of the entire planet, would make a layer 20 μm. deep (Schorn et al. 1967). However, at the time of this writing, it is not known whether the water vapour is uniformly distributed over the atmosphere, or whether it is confined to particular strata – possibly the boundary layer just above the surface. Average conditions on Mars correspond to the region below the triple point for water, thus the existence of liquid water on the surface of Mars seems impossible. However, the 12–15 km. differences in altitude of the Martian surface (Shapiro, 1968) would correspond to great differences in pressure: ranging from 3 to 4 mb. at elevations to perhaps 20 mb. in the deepest depressions. Thus conditions under which liquid water may exist for at least a short time may be found on Mars. It is suggested that the diurnal temperature fluctuations freeze out water from the atmosphere every evening, and that at sunrise the ice would not instantaneously sublime but, because mixing with the atmosphere takes a finite length of time, would permit the formation of liquid water for a short period. Some water may therefore soak into the soil and be available for a short time every day.

Nothing is known at present about the composition of Martian soil. Let us assume that nitrogen and phosphorus are present in the Martian minerals. Should the soil contain soluble electrolytes, films of brine might exist near the surface and support populations of halophilic organisms. Other mechanisms exist for the retention of water besides high osmotic pressure. For example, it has been suggested on the basis of the polarization of reflected light from the Martian surface that limonite constitutes at least part of the Martian surface (Dollfuss, 1951).

This controversial hypothesis may of course be wrong, but it serves to illustrate one mechanism by which water storage may occur in a biologically accessible form. The composition of limonite is given as $Fe_2O^3 \cdot H_2O \cdot nH_2O$, where n is usually near 2 (Deer, Howie & Zussman, 1962).

In other words limonite can be considered as a ferric oxide with one molecule of water of crystallization (goethite), plus an undefined additional amount of adsorbed water. The energy with which this additional water is bound to limonite is small enough to be susceptible to biological attack. Particularly, should there be a respiration in which oxidized iron serves as an electron acceptor, the resultant reduction of ferric oxide to ferrous oxide would release water which is biologically available (Vishniac, 1965). Such a hypothetical iron cycle may be illustrated by Fig. 2. At our present state of ignorance it may turn out that no limonite covers the surface of Mars, but conceivably other minerals may play an equivalent role.

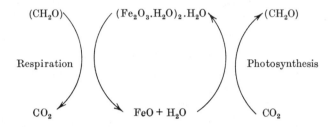

Fig. 2. Model of an iron-based ecology. Iron acts as the electron buffer between photosynthesis and respiration. Without balancing the equation, the left-hand side shows the respiratory oxidation of organic matter to CO_2 with iron acting as the ultimate electron acceptor. On the right side iron acts as the electron donor in photosynthetic biosynthesis. (Vishniac, 1965).

Such speculation suggests terrestrial laboratory experiments. We have been able to maintain, as a mixed culture, a bacterial population under anaerobic conditions with ferric iron as the only electron acceptor, using ethanol as the electron donor. The compensating process, a photosynthetic oxidation of iron, is likewise a process that could be studied on earth by traditional enrichment culture techniques. An examination of Mars is instructive from the point of view of global productivity, because Mars represents a world in which the required mineral factors for primary productivity are conceivably present in excess over one identifiable limiting component – water. Therefore the study of the chemical complexity of the Martian surface might teach us more about prebiology if no living organisms were detected there, than if life were to exist and transform its surface in a way in which the terrestrial surface has been transformed by living organisms. Thus the calculation of the primary productivity on earth leads us to questions of the origin and evolution of life. My personal dream is the construction of a curve against an ordinate which represents chemical complexity, the highest of which would be the existence of living organisms and the lowest the existence of purely inorganic compounds, and an abscissa representing limiting physical parameters. We do not know what such limitations are; they would presumably include abundance of water, temperature regime, and relative abundance of essential elements. We are acquainted with only two points on such a curve, two worlds: Earth, with its highly complex chemistry; and the moon with its low carbon content, its lack of an atmosphere, and its lack of water. But where do Mars and Venus fit on to this curve? And, more importantly, what are the limiting physical parameters which determine whether or not chemical evolution reaches a complexity that we recognize as living organisms? To me these comparisons represent the major excitement of the space age. The energy flow through earth on which life feeds inevitably directs our attention to other members of the solar system. It directs our thoughts to the unity of biology and cosmology.

REFERENCES

ANDERS, E. (1970). Water on the Moon? *Science, N.Y.* **169**, 1309.
BAUCHOP, T. & ELSDEN, S. R. (1960). The growth of microorganisms in relation to their energy supply. *J. gen. Microbiol.* **23**, 457.
CALVIN, M. & BASSHAM, J. A. (1962). *The Photosynthesis of Carbon Compounds.* New York: Benjamin.
CLARKE, A. C. (1970 reissue). *The Deep Range.* New York and London: Harcourt Brace.
DEER, W. A., HOWIE, R. A. & ZUSSMAN, J. (1962). *Rock-forming Minerals*, vol. 5, p. 124. London: Longmans.
DOLLFUS, A. (1951). La polarisation de la lumière renvoyée par les différentes régions de la surface de la planète Mars et son interprétation. *C. r. hebd. Séanc. Acad. Sci., Paris* **233**, 467.
FOOD AND AGRICULTURE ORGANIZATION (1968). *Yearbook of Fishery Statistics.* Rome: Food and Agriculture Organization of the United Nations.
GROSS, F. (1947). An experiment in marine fish cultivation. *Proc. Roy. Soc. Edin.* **63B**, 1.
HARVEY, H. W. (1957). *The Chemistry and Fertility of Sea Waters.* Cambridge University Press.
KEIL, K., PRINZ, M. & BUNCH, T. E. (1970). Mineral chemistry of lunar samples. *Science, N.Y.* **167**, 597.
PROVASOLI, L. (1963). Organic regulation of phytoplankton fertility. In *The Seas*, vol. 2, pp. 165–219. Ed. M. N. Hill. New York: Wiley.
RABINOWITCH, E. I. (1948). *Photosynthesis and Related Processes*, vol. 1, pp. 8–9. New York: Interscience.
REDFIELD, A. C., KETCHUM, B. H. & RICHARDS, F. A. (1963). The influence of organisms on the composition of sea water. In *The Seas*, vol. 2, pp. 26–27. Ed. M. N. Hill. New York: Wiley.

RILEY, G. A. (1956). Oceanography of Long Island Sound, 1952–1954. IX. Production and utilization of organic matter. *Bull. Bingham oceanogr. Coll.* **15**, 324.

RUSSELL-HUNTER, W. N. (1970). *Aquatic productivity.* New York, London and Toronto: Macmillan.

RYTHER, J. H. & YENTSCH, C. S. (1957). The estimation of phytoplankton production in the ocean from chlorophyll and light data. *Limnol. Oceanog.* **2**, 281.

SAGAN, C. (1961). *Organic Matter and the Moon.* Washington: Nat. Acad. Sci. publ. no. 757.

SCHAEFER, M. B. (1965). The potential harvest of the sea. *Trans. Am. Fish. Soc.* **94**, 123.

SCHORN, R. A., SPINAD, H., MOORE, R. C., SMITH, H. J. & GIVER, L. P. (1967). High dispersion spectroscopic observations of Mars. II. The water vapor variations. *Astrophys. J.* **126**, 231.

SHAPIRO, I. I. (1968). Radar observations of the planets. *Sci. Am.* **219**, 28.

STEEMAN NIELSEN, E. & JENSEN, E. A. (1957). Primary oceanic production, the autotrophic production of organic matter in the oceans. '*Galathea*' *Rep.* **1**, 49.

SLOBODKIN, L. B. (1961). *Growth and Regulation of Animal Populations.* New York: Holt.

VISHNIAC, W. (1965). Bacterial ecologies in limonite. In *Life Sciences and Space Research*, vol. 3, pp. 139–141. Ed. M. Florkin. New York: Wiley.

INDEX

abortion, enzootic, of sheep, 129
acetate: glutamate from, 82; lysine from, 83; in ruminants, 150, 156, 158–9
Acetobacter, food spoilage by, 179
acetylene, reduced by nitrogenase, 290, 291, 294, 307
Achromobacter aquamarinus, 248
Achromobacter nematophilus, 211
Acinetobacter spp., food spoilage by, 179, 182
acrylic acid, product of *Phaeocystis*, 349
acrylonitrile, reduced by nitrogenase, 291, 301
Actinomycetes: in polar regions, 345, 346; in soil, 265, 272
adenosine monophosphate (AMP), cyclic, 97
adenylosuccinate synthetase, in nucleotide synthesis, 89, 90
Aerobacter spp., insect pathogens, 202
Aerobacter aerogenes, 256, 270; nucleotide synthesis in, 86, 87
Aeromonadaceae, medium for selection of, 191
Aeromonas, food spoilage by, 182
Agaricaceae, in mycorrhizas, 316
Agathis, mycorrhiza and phosphate uptake by, 319
Alcaligenes group, food spoilage by, 179, 182
aldehydes, methylobacteria oxidize alkanes to, 42
algae: on alkanes, protein content of, 22; as food source, 47–75; filamentous, 69, 72, 73; on snow, 338, 340; symbiotic, 324; *see also* phytoplankton
n-alkanes, microbial growth on, 16–28
allene, reduced by nitrogenase, 291, 301
allomucic acid, in toxin of *B. thuringiensis*, 207
allulose pathway of metabolism, in methylobacteria, 35, 36, 39
Alnus: in colonization, 326, 327; mycorrhiza and phosphate uptake by, 319, 328; root nodules and nitrogen fixation in, 315, 326, 344
Alternaria, food spoilage by, 182
amino acids: essential, in microbes grown on hydrocarbons, 23, 41; microbial production of, 77, 78–87; microbial uptake of, from sea water, 248; needs of ruminants for, 165–70
D-amino acids, from mucoproteins, 22
amino polysaccharides, 22
ammonia: metabolism of, in rumen, 160, 161, 162; in microbial growth on hydro-carbons, 20, 33, 37, 39; represses forma-tion of nitrogenase, 290, 294, 298; in sea water, and growth rate of phytoplankton, 242–3; in soil, 262, 272
Amphidinium, biotin requirement of, 243, 244
Anabena: nitrogen fixation by, 296; nitro-genase of, 288
Anabenopsis, nitrogen fixation by, 296
Anacystis nidulans, yield of, 65
animal feeds: algae for, 63–6; antibiotics in, 4, 127, 143–5; costs of different pro-teins in, 66–7
antagonism: between food spoilage mic-robes, 184; between fungi, 312
anthrax, 134, 139
antibiotics: in animal feeds, 4, 127, 143–5; for bovine mastitis, 142; microbial pro-duction of, 77; from mycorrhiza fungi, 321; in soil, 262, 269, 273
antioxidants, microbial production of, 77
apple: mycorrhiza and growth of, 318; *Venturia* infection of, 110
aquatic environments, microbial activity in, 231–53, 355–66
arabitol, fungal metabolite, 117, 324
arbovirus diseases of livestock, 136
'area wasted', as measure of crop loss, 107
Armillaria mellea, in soil, 274
Arthrobacter: food spoilage by, 182; medium for selection of, 191
Arthrobotrys spp., in soil, 277
arthropods: estimate of biomass of, in soil, 265, 347
Ascomycetes: breaking of dormancy of spores of, 261; in polar soils, 346
asepsis, in fermentation of alkanes, 25, 27
Ashbya gossypii, riboflavin producer, 92, 94, 95, 96
aspartokinase, in *E. coli* and in lysine pro-ducers, 85, 86
Aspergillus spp.: food spoilage by, 182; in soil, 257, 277
Athiorhodaceae, nitrogen fixation by, 296, 297
ATP: estimates of biomass in water by determination of, 233–4; estimates of respiratory activity of plankton by deter-mination of, 237–8; in oxidation of hydro-carbons, 18, 34; partially reverses inhibi-tion of RNA polymerase by toxin of *B. thuringiensis*, 208; required for nitro-genase, 290, 292, 293, 297; requirements for, in conversion of carbon dioxide to living matter, 358

Aulosira, nitrogen fixation by, 296
autolysis, of methylobacteria, 38
azide, reduced by nitrogenase, 290
azoferredoxin, in nitrogenase, 289
Azotobacter: cysts of, in soil, 256, 270; nitrogen fixation by, 267, 293, 294, 296, 298; in polar regions, 343, 344; travels along root surfaces, 273
Azotobacter chroococcum, A. vinelandii, nitrogenase of, 287, 288
Azotobacteriaceae, nitrogen fixation by, 294, 296, 297
Azotococcus, nitrogen fixation by, 294, 296
Azotomonas, nitrogen fixation by, 294, 296
Azotomonas fluorescens, A. insolita, not nitrogen-fixers, 294

Bacillaceae, medium for selection of, 191
Bacillus spp., spores of: in food, 181, 182; in soil, 256
Bacillus amyloliquefaciens, 8
Bacillus cereus, insect pathogen, 202
Bacillus fribourgensis, insect pathogen, 202, 203
Bacillus lentimorbus, control of scarabaeid beetles by, 202, 203, 216, 223
Bacillus macerans, nitrogen-fixer? 296
Bacillus megaterium, digestibility of, 24
Bacillus polymyxa: nitrogen fixation by, 295, 296; nitrogenase of, 288
Bacillus popilliae, control of scarabaeid beetles by, 193, 202, 203–4, 216, 218, 223
Bacillus stearo-thermophilus, alkane-utilizing strain of, 22
Bacillus subtilis: nucleotide synthesis by, 88, 89, 90; in soil, 256, 277
Bacillus thuringiensis: control of lepidopterous insects by, 198, 202, 204–5, 222, 223–4; field application of, 208–10, 216; housefly strain resistant to, 221; safety of, 218; toxin of, 202, 205–8
bacteria: genetic manipulation of, 25; on hydrocarbons, 16, 21, 22, 23; in polar regions, 345, 347; in sea and lakes, 231, 232, 245; in soil, 256–7, 259–60, 261, 265, 268, 271, 272, 277
barley: assessment of losses of, from disease, 104–5, 105–6, 108–9; infection and photosynthesis in, 114, 115; soil microbes and uptake of phosphorus by, 314
Basidiomycetes: in mycorrhizas, 316, 317, 325; in soil, 277
Bdellovibrio, 25
beans, effects of infections on, 114–15, 116, 119
Beauveria bassiana, fungus infecting insects, 197, 210, 216, 219
Beijerinckia, nitrogen fixation by, 296
benzoic acid, for food preservation, 186

biomass of microbes, assessment of: in sea and lakes, 232–4; in soil, 258, 265, 377
biotin: glutamate producers and, 79, 81, 83, 96; and growth rate of phytoplankton, 243, 244
bluetongue virus, 136, 137, 139
Boletaceae, in mycorrhizas, 316
Boletus elegans, in *Larix*, 325
Boletus variegatus, in *Pinus*, 322
Bombyx mori: resistant strains of, 200; toxin of *B. thuringiensis* in, 205, 207, virus disease of, 215
Botrytis spp.: on broad beans, 111; food spoilage by, 182
Brevibacterium, medium for selection of, 191
Brevibacterium ammoniagenes, 'salvage syntheses' by, 91
Brevibacterium flavum, lysine producer, 81
Brevibacterium lactofermentum, glutamate producer, 81
Brevibacterium liquefaceiens, produces cyclic AMP, 97
Brevibacterium thiogentalis, oleate-requiring mutant of, 82
brucellosis of cattle, 129, 134
butyric acid, in caecum of fowl and pig, and rumen of sheep, 150

caecum, site of microbial activity in pig and horse, 149, 150
Calothrix, nitrogen fixation by, 296
calves, neonatal mortality of, 128, 129
Candida spp.: on alkanes, 16; food spoilage by, 182; production of riboflavin by, 92
Candida lipolytica, 18
Candida tropicalis, 18
carbon dioxide: algae *v.* higher plants for fixation of, 49–54; in alkaline growth media (for algae), 72; evolved from soil, 258–9; and food spoilage, 183; in seawater, 356; in soil, and microbial activity, 261
carbon monoxide, inhibits reduction of nitrogen by nitrogenase, 291
carbohydrates: accumulation of, in leaves of infected plants, 116, 117–19; conversion of, to fatty acids in ruminants, 152, 154–7; of host plants, supplied to mycorrhizas, 322–4; in microbes on alkanes, 24; produced by marine algae, 356
β-carotene: in *Chlorella*, 55; as colouring agent, 77
Casuarina: in colonization, 327; root nodules and nitrogen fixation in, 315
catalase-negative cocci, food spoilage by, 179
catalase-positive cocci: food spoilage by, 179; medium for selection of, 191
cattle: losses from disease among, 128, 129; tuberculosis of, 130, 134

Ceanothus: in colonization, 327; root nodules and nitrogen fixation in, 315

cell membranes: in aerobic nitrogen-fixers, 299; in lithotrophic bacteria, 32; in methylobacteria, 19, 31, 36

cells walls: of microbes on alkanes, 19; of *Nocardia* on different media, 24

cells: disintegration of, 7–8, 24; efficiency of conversion of hydrocarbons to, 21, 40

cellulase, *Scenedesmus* cell walls not digested by, 59

cellulose, digestion of: by *Rhizoctonia*, 317; in sheep, pig, and rat, 149–52

centrifuging, cost of, 51, 68

Cephalosporium, in soil, 257, 277

cereals, lysine supplementation of, 83

chemical preservatives for foods, 185–6

chemostat method of continuous culture, 20, 24, 245

chemotherapy, of animal diseases, 130

Chlamydomonas mundana, yield of, 65

chlamydospores, of fungi, in soil, 257, 258, 274, 275

Chlorella spp.: drying of, 63; light-limited growth of, 242; night temperature and yield of, 71

Chlorella ellipsoidea, protein from, 59, 67

Chlorella pyrenoidosa: decolorization of, 60; heterotrophic growth of, 48; indigestible, 55, 56, 59; milling of, 55, 58; protein from, 64, 65, 67; protein efficiency ratio of, 63; yields of, *in vitro* and from lagoon, 66

Chlorella vulgaris, protein efficiency ratio of, 63

Chlorobium thiosulfatophilum, nitrogen-fixing strain of, 295, 296

Chlorogloea, nitrogen fixation by, 296

chlorophyll *a*, estimates of biomass of phytoplankton by determination of, 233, 242; in polar waters, 339, 343, 348

Chlorophytes, carbohydrate production by, 356

Chloropseudomonas ethylicum, nitrogen-fixing strains of, 296

Chromatium, nitrogenase of, 288

Chromatium minutissimum, *C. vinosum*, nitrogen-fixing strains of, 297

Chroococcales, nitrogen fixation by, 296

Chrysomonads, carbohydrate production by, 356

ciliates, from rumen of reindeer and musk ox, 349

citrate, glutamate from, 82–3

citrate permease, 83

Citrobacter, food spoilage by, 182

Citrus, zinc deficiency in, 313

Cladosporium, food spoilage by, 182

Cloaca spp., insect pathogens, 202

Clostridium spp.: food spoilage by, 180, 182; nitrogen-fixing, in polar regions, 343; production of riboflavin by, 92; spores of, in foods, 181, and in soil, 256

Clostridium brevifaciens, insect pathogen, 202

Clostridium butyricum, 296; in polar regions, 344

Clostridium malacosomae, insect pathogen, 202

Clostridium pasteurianum: nitrogen fixation by, 296; nitrogenase of, 287, 288

clover: effect of infection on carbohydrates in, 117; mycorrhiza and uptake of phosphate by, 319; soil microbes and uptake of phosphate by, 315.

coccidiosis, antibiotics active against, 4, 144

coelenterates, symbiotic algae in, 324

Collema pulposum (lichen), nitrogen fixation by, 344

compensation: by extra growth of non-infected plants for losses in infected ones, 110, 115; financial, for losses from animal disease or preventive slaughter, 130–1, 134

Conococcum graniforme, mycorrhizal fungus 325

continuous processes, 3, 4, 5, 20, 25

cooling and circulating of algal cultures, 70–1

copper sulphate, as growth promoter for pigs, 144

Coriaria: in colonization, 327; root nodules and nitrogen fixation in, 315

Corticium solani, cause of sharp eyespot, 105

Corynebacterium spp.: on alkanes, 16; food spoilage by, 181, 182; medium for selection of, 191

Corynebacterium glutamicum, lysine production by homoserine-requiring mutants of, 83–5

Corynebacterium hydrocarboclastus, glutamate production by, 83

costs: of centrifuging, 57, 68; of cultivation of microbes, 47; and culture methods, 66–73; of fractionation, 48; of solvent extraction, 60; of texturization of protein, 61

crustacea, harvest of planktonic, 362

Cryptomonads, carbohydrate production by, 356

cutin (from surface of plants), in soil, 276, 277

cyanide: reduced by nitrogenase, 290; released by roots of strain of flax resistant to pathogens, 311

cyanocobalamin, *see* vitamin B_{12}

cyanogen, reduced by nitrogenase, 291

Cyanophyceae (blue-green algae): and colonization, 326, 327; nitrogen fixation by, 294, 296, 297; nitrogenase of, 288, 292; in polar regions, 343, 346; site of nitrogen fixation in, 299

Cyclotella nana, vitamin B$_{12}$ requirement of, 243, 244

Cylindrocarpon, in soil, 274

Cylindrospermum, nitrogen fixation by, 296

cysts: of *Azotobacter*, 256, 270; of methylobacteria, 32–3

cytochromes: in nitrogen fixation? 293; in oxidation of alkanes, 17; protect nitrogenase proteins, 288

cytokinin, in infected plants, 119

Debaromyces: on alkanes, 16; food spoilage by, 182

dehydrogenation: of alkanes, 17–18; in metabolism of methylobacteria, 33

depth in ocean: and biomass, 234; and microbial activity, 249–50; and rate of respiration, 238

Derxia, nitrogen fixation by, 296

Derxia gummosa, 298

Desulfovibrio africanus, not nitrogen fixer, 295, 296

Desulfovibrio desulfuricans, nitrogen fixation by, 295, 296; nitrogenase of, 288

Desulfovibrio gigas, nitrogen-fixing strains of, 295, 296

Desulfovibrio salexigens, not nitrogen fixer, 295, 296

Desulfovibrio vulgaris, nitrogen-fixing strains of, 295, 296

Desulfotomaculum nigrificans, not nitrogen fixer, 295, 296

Desulfotomaculum orientis, *D. ruminis*, nitrogen fixation by, 295, 296

diacetates, for food preservation, 186

diatoms, carbohydrate production by, 356; vitamin B$_{12}$ requirement of, 243

dicarboxylic acid pathway, plants employing, 51, 53

dihydropicolinate synthetase, inhibited by lysine in *E. coli* but not in lysine producers, 85

O-dimethylase, 35

dimethylether, growth of methylobacteria on, 29, 34

Dinoflagellates, carbohydrate production by, 356

diphenyl, for food preservation, 186

Discaria: in colonization, 327; root nodules and nitrogen fixation in, 315

dithionite, as reductant for nitrogenase, 292

DNA, incorporation of, into microbes, 25

dormancy of microbes: constitutive, 260–1; exogenous, 261–2

Dryas, root nodules and nitrogen fixation in, 316

Dryas octopetala, distribution of, 326

drying of algae, 63

earthworms, estimate of biomass of, in soil, 265

East Coast fever of livestock (Africa), 134, 136

efficiency: of food utilization, in pig, fowl, and ruminant, 153–4; photosynthetic, 49, 50–1, 357; of transfers in food chains, 360

Eleagnus spp.: in colonization, 327; root nodules and nitrogen fixation in, 315, 326

electron microscopy: of chloroplasts from diseased plants, 114; of crystalline protein toxin of *B. thuringiensis*, 206; of diseased plant cells, 111–12; of soil microbes, 270, 271, 273; of yeasts grown on hydrocarbons, 19

electron transport system: for nitrogenase, 293; in plankton, 237; in *Pseudomonas oleovorans*, 17

Endogone, fungus of vesicular–arbuscular mycorrhiza, 316, 318, 324–5

energy: maximum available to phytoplankton, 357; minimum required for photoautotrophic growth, 357–8

Enterobacter, food spoilage by, 182

Enterobacteriaceae: as index organisms in control of food processing, 188; medium for selection of, 191

enzymes: for breakdown of resistant soil substances, 278–9; large-scale production of, 8, 77; lysis of microbial cells by, 59–60

Eremothecium ashbyii, riboflavin producer, 92, 94

Erisiphales, supply of carbohydrate to, 323

Erisiphe graminis, cereal mildew, 105, 113, 114, 115, 117

Erwinia carotovora, soft-rot bacterium, 111

Escherichia coli: alteration of permeability of, 81; large-scale production of, 7, 8; syntheses in, (lysine) 82–3, (threonine) 86–7

ethane, microbes oxidizing, 34

ethanol: extraction of *Chlorella* by, 60; glutamate production from, 83

ethirimol, fungicide, 106

Eucalyptus, mycorrhiza and growth of, 318

Fagus: mycorrhiza and phosphate uptake by, 319, 320, 321; translocation of food to mycorrhiza from, 322

fatty acids: in blood of germ-free and normal fowls, 151; derivatives of, alter permeability in glutamate producers, 79, 81, 96–7; digestibility of, in different animals, 170–1; production of, in digestive tracts of different animals, 150, 151–2, 154–9; unsaturated, diminish methane production in rumen, 157

fermentation: of algal cells, 60; foods produced by, 3
ferredoxin, electron transport factor for nitrogenase, 292, 293
fibre: digestion of, in different animals, 149; interaction of fat and, in rumen, 170–2
Fischerella, nitrogen fixation by, 296
fisheries, harvest of, in relation to ocean productivity, 359–32
flagellates, in polar waters, 347
Flavobacter, food spoilage by, 182
Flavobacterium spp., in soil, 277
flax, cyanide-releasing strain of, 311
fleas, transmission of myxomatosis by, 199, 200
flocculation, harvesting of small algae by, 68
Fomes annosus, on pine roots, 311
food: biological losses of, 177; microbial production of additives for, 77–101; preservation of, 177–95; world production of, 354, 355
Food and Agriculture Organization (FAO), 127, 128, 132
foot and mouth disease, 128, 130, 132, 133, 136, 137; vaccination for, 138, 139
forests, mycorrhizas and, 327, 328
formaldehyde, in metabolism of methylobacteria, 33, 35, 36
formic acid, in metabolism of methylobacteria, 33
fowl plague, 136
fowls: digestibility of fats and fatty acids in, 170–1; efficiency of food utilization in, 153–4; microbial activity in caecum of, 150, 151
fractionation of microbes, 48; of algae, 54, 55, 57–61
fulvic acid, in humus, 276, 277
fungi: on alkanes, 17, 22; food spoilage by, 179, 180, 182; medium for selection of, 191; in polar regions, 345, 346; in soil, 257, 258, 265, 271, 272; 'sugar' and 'lignin' types of, 272, 273, 277; testing dried foods for presence of, 188–9
Fusarium spp.: food spoilage by, 182; in soil, 257, 274, 277; strain of flax resistant to, 311
Fusarium oxysporum, in soil, 262, 275

Galleria mellonella (waxmoth): susceptible to virus disease of *Bombyx*, 215; toxin of *B. thuringiensis* in, 207
gallic acid esters, for food preservation, 186
gas–oil process for growth of microbes, 27–6
genetic manipulation of microbes, 25
gibberellic acid, microbial production of, 77
gibberellins, in infected plants, 119
Gleocapsa, nitrogen fixation by, 299

Glomerella cingulata, inhibitors of germination of, 262
glucose: effects of addition of, to soil, 270, 278, 279; production of fatty acids from, in ruminants, 158–9; production of glutamate from, 80; rates of entry and oxidation of, in sheep, 156; synthesized from propionate in ruminants, 155, 158; translocation of, from mycorrhiza to host, 317; uptake of, by aquatic microbes, 246–9
glutamic acid: microbial production of, 78–83, 96; nitrogen assimilation of *Corynebacterium* via, 84
glycine: hydroxymethylation of, to serine, 35, 36; and riboflavin production, 94
glycogen, fungal metabolite, 323
Gram-negative rod-shaped bacteria: food spoilage by, 179, 180, 181; medium for selection of, 191
granuloses (insect viruses), 212
growth patterns, of soil microbes, 272–4
growth rates: of aquatic microbes, 239–40; of different microbes, and type of food spoilage, 183; of plants with and without mycorrhiza, 318; of soil microbes, calculated by various methods, 266–8
guanosine monophosphate (guanylic acid, GMP), flavour enhancer, 87, 89
guanosine monophosphate reductase, 87, 89
Gunnara, in colonization, 327
gypsy moth, virus disease of, 212

Hafnia, food spoilage by, 182
halophilic yeasts, food spoilage by, 182
Hapolosiphon, nitrogen fixation by, 296
harvesting: of filamentous algae, 69; of microbes grown on alkanes, 22; of small algae, 68
heat: loss of, during ruminant digestion, 157–8; processing of foods by, 181
Heliothis virescens (tobacco budworm), control of, by *B. thuringiensis*, 209, 224
Heliothis zea (cotton bollworm), specificity of virus disease of, 214, 215
Helminthosporium sativum: strain of flax resistant to, 311; on wheat, 111, 310–11
hemicellulose, digestion of, in sheep, pig, and rat, 149–50
heterocysts, site of nitrogen fixation in Cyanophyceae, 299
hexose monophosphate pathway, in respiration of plant and pathogen, 116
Hippophae: in colonization, 327; root nodules and nitrogen fixation in, 315, 326
Hippophae rhamnoides, distribution of, 326
histidine, limiting amino acid for lactating cows? 167

homoserine dehydrogenase: deficient in lysine producers, 83, 84; not inhibited by threonine in threonine producers, 87

Hormidium, filamentous alga, 65, 69

hormones, in infected plants, 119

horse, microbial activity in caecum of, 149

horsesickness, African, 130, 133; insect vector of, 136; vaccination for, 139

housefly: resistant strain of, 221; toxin of *B. thuringiensis* in, 207

human cells, diploid, for production of viral vaccines, 5

humic acid, in humus, 276, 277, 278

Humicola spp.: in soil, 277

humus, radiocarbon dating of, 276, 345

hydrocarbons: production of glutamate from, 83; as source of microbial protein, 15–46

hydrogenase, association of nitrogenase with, 290, 293

hydroperoxidation, of alkanes, 17

p-hydroxybenzoic acid, for food preservation, 186

β-hydroxybutyrate polymers, in microbes, 24, 32, 41, 271

hydroxylation, of alkanes, 17

β-indolyl acetic acid (IAA), in infected plants, 119

inhibitors of microbial growth, in soil, 262–3, 269

inosine monophosphate (inosinic acid, IMP), microbial production of, 78

inosine monophosphate dehydrogenase, in nucleotide synthesis, 89, 90, 91

insects: biological control of, 202–29; development of resistance in, 220–1

International Biological Programme, 265

International Organization of Biotechnology and Bioengineering (IOBB), 10

intestinal flora, and nutrition of host, 149–75

iron: model of ecology based on, 363–4; in nitrogenase, 289, 301–2; in riboflavin production, 94, 96; in rubredoxin, 17

isoascorbic acid, antioxidant, 77

isocyanide, reduced by nitrogenase, 270, 272, 300–1

isoenzymes: of aspartokinase, 85, 86; of homoserine dehydrogenase, 86

isoleucine: auxotrophy for, in threonine producers, 87; in microbes grown on hydrocarbons, 23

α-ketoglutarate-aspartate transaminase, in glutamate producers, 81

α-ketoglutarate dehydrogenase, deficient in glutamate producers, 79, and in many methylobacteria, 39

Klebsiella: food spoilage by, 182; nitrogen fixation by, 295

Klebsiella pneumoniae: nitrogen fixation by, 296; nitrogenase of, 288

Klebsiella rubiacearum, in leaf-nodule symbiosis, 296, 297

Lactobacillaceae: food spoilage by, 179, 180, 182; medium for selection of, 191

Lactobacillus plantarum, food spoilage by, 183

Larix, mycorrhizal fungus of, 325

leaf-nodule symbiosis, 296, 297

legumes: in colonization, 326, 327; effect of infections on carbohydrates in, 118; haemoglobin of, 299; mycorrhizas as well as nodules on roots of, 319, 328

leucine, in microbes grown on hydrocarbons, 23

Leucopaxillus cerealis, mycorrhiza fungus, 321

Leucothrix, growth rate of, 268

lichens: nutrition in, 324; on tundra, 334, 338

light: and growth of phytoplankton, 242; percentage absorption of, by algae, 69–70

lignin: breakdown products of, inhibit germination of fungal spores, 262; in soil, 276, 277, 278

'lignin fungi', 272, 277

linolenic acid, diminishes production of methane in rumen, 157

lipids: of algae, 55, 57–8, 60; biotin deficiency and synthesis of, 82; interaction of fibre and, in rumen, 170–2; manganese deficiency and synthesis of, 92; of microbes grown on alkanes, 24

Liriodendron, mycorrhiza on: and growth rate, 318; and phosphate uptake, 319

Listeria, food spoilage by, 182

lithotrophic bacteria, cell membranes in, 32, 38

livestock: control of epidemics among, 131–9; intensive rearing of, 140–5; microbial disease and productivity of, 125–31; intestinal flora in nutrition of, 149–75

lumpyskin disease, 137

lysine: in microbes grown on hydrocarbons, 23; microbial production of, 78, 83–5, 96; not limiting amino acid for lambs, 167, 169, 170; and threonine production, 87

lysosomes, in diseased plant cells, 111

magnesium: required for nitrogenase, 290; in 'salvage synthesis' of nucleotides, 91

maize, mycorrhiza and growth of, 318

Manduca sexta (tobacco hornworm), control of, by *B. thuringiensis*, 209, 224 manganese: disease of oats due to deficiency of, 313; in 'salvage synthesis' of nucleotides, 91, 92, 97

mannitol, fungal metabolite, 117, 317, 323, 324

Margarinomyces, food spoilage by, 182

Mars, conditions on, 363–4

Mastigocladus, nitrogen fixation by, 296

mastitis, bovine, production losses from, 129, 140–1

media, for grouping of food spoilage microbes, 190, 191

Metarrhizium anisopliae, fungus infecting insects, 210

methane: growth of microbes on, 28–43; microbes producing, 30; produced in rumen, 157, 159, 172

methane oxidase, 37, 39

methanol, methylobacteria and, 29, 31, 33, 34, 37, 40

methanol dehydrogenase, 33

Methanomonas methanooxidans, 29, 40

methionine: limiting amino acid for growth of lambs? 167, 168, 169; and lysine production, 85; in microbes, 23, 24, 63; response of wool growth to dosage with, 166, 176; in rumen organisms, 164, 165; as supplement to *Chlorella* for rats, 56; and threonine production, 87

methionine synthetase, deficient in threonine producers, 87

methylene tetrahydrofolic acid, 35, 36

Methylobacter, 31, 32

methylobacteria, 30–3

Methylococcus, 31, 32

Methylococcus capsulatus, 29, 42

Methylococcus minimus, 31

Methylocystis, 31

Methylocystis parvus, 31, 32

Methylomonas, 31, 32

Methylosinus, 31, 40

Methylosinus sporium, 32

Methylosinus trichosporium, 32, 37

mevalonic acid, 97

Microbacterium, food spoilage by, 179, 182

Micrococcus spp.: on alkanes, 16; food spoilage by, 179, 181, 182

milling, of algal cells, 55, 58

molasses, growth of glutamate producers on, 82–3

molluscs, estimate of biomass of, in soil, 265

molybdoferredoxin, in nitrogenase, 288

monensin, antibiotic active against coccidiosis, 4

Monilia, food spoilage by, 182

Monochrysis lutheri, vitamin requirements of, 243, 244

mono-oxygenase, in growth on alkanes, 17

monosodium glutamate, synergistic flavour effect of nucleotides and, 3

moon, conditions on, 362–3

morphogenic factors, produced by microbes associated with roots, 317

mortality rates, from animal diseases, 128

Mortierella marburgensis, in soil, 277

mosquitoes, transmission of myxomatosis by, 199, 200

mosses, in polar regions, 334, 346

mucopeptides, of no nutritional value? 22

Mucor spp.: food spoilage by, 182; in soil, 257

Mucor ramannianus, in soil, 257, 273–4

Mycobacterium spp., on alkanes, 16

Mycobacterium flavum: nitrogen fixation by, 294, 296; nitrogenase of, 288

mycoplasmal diseases of poultry, antibiotics active against, 144

mycorrhizas, 309, 316–17; co-existent with nodules, 319, 328; and disease, 321–4; ectotrophic and vesicular–arbuscular, 317–20; specificity of fungi in, 324–6

'mycostasis' in soil, 262, 269, 271

mycotoxins, in foods, 189

Myrica: colonization by, 327; root nodules and nitrogen fixation in, 315–16, 326

Myrsine, mycorrhiza and growth of, 318

myxomatosis in rabbits, 199–201, 221

Myxomycetes, in soil, 273

NAD, in growth on hydrocarbons, 17, 18, 33

NADH dehydrogenase, and nitrogenase, 288, 292–3

NADH oxidase, in lithotrophs and methylobacteria, 39

NADH and NADPH reductases, in plankton, 237; in *Pseudomonas*, 17

natural gas, possibilities of microbial growth on, 41–3

Nectria galligena, apple canker, 110

nematodes: control of insect pests by, 210–12, 216; estimates of biomass of, in soil, 265, 347; toxicity tests on, 219

Neoaplectania carpocapsa, nematode infecting insects, 211

nitrate, in sea water, and rate of growth of phytoplankton, 242–3

nitrite: for food preservation, 186; formation of, in soil, 272; oxidized to ammonia by methylobacteria, 37

Nitrobacter, growth rate of, 268

nitrogen: complexes of transition metals with di- form of, 300–2; fixation of, 267, 287–307, 343–4; metabolism of, in rumen, 159–65; in sea water, 356, 359

nitrogenase, 287–93, 303; in methylobacteria 37

Nitrosomonas, growth rate of, 268

nitrous oxide, reduced by nitrogenase, 290

Nocardia spp.: on alkanes, 16, 18; nitrogenous constituents of, 24; not nitrogen fixers, 294, 296

Nostoc, nitrogen fixation by, 296

Nostoc commune, in polar regions, 344

Nostocales, nitrogen fixation by, 296

nucleic acids: in algae, 57; limit to human intake of, 2, 20, 47; in rumen microbes, 163

nucleotidase, bacterial mutant deficient in, 89

nucleotides (purine): enhance flavour, 3, 87; extracellular 'salvage synthesis' of, 91–2; microbial production of, 77, 78, 87–92

nutrients: for plants, effects of rhizosphere microbes on absorption of, 312–15; for soil microbes, 263–5, 270–9

oats: crown rust of, 112; manganese-deficiency disease of, 313

Office International des Epizooties (OIE), 127, 132

oleate, mutant of *Brevibacterium thiogentalis* requiring, 82

Ophiobolus graminis, cereal 'take-all', 105, 311

orchids, endotrophic mycorrhiza of, 216, 217

Orchis purpurella, 317

Orychtes rhinoceros (coconut rhinoceros beetle), control of, by virus disease, 224

osmophilic yeasts, food spoilage by, 182

over-production, by microbes, 77

oxidation: of alkanes, 17–19; of methane, 33–5

oxygen: in deep water, 249; and microbial activity in soil, 261; molecular, incorporated in growth on alkanes, 18, and on methane, 34, but not in growth on methanol or dimethyl ether, 35; in nitrogen fixation, 298; and types of food spoilage, 182–3

pantothenate, in 'salvage synthesis' of nucleotides, 91

parasporal body, in *B. popilliae* and *B. lentimorbus*, 203

parvovirus, of pigs, 129

pathogen-free animals: fatty acids in blood of (fowls), 151; repopulation of herds with, 143

pathogens: handling of, on large scale, 4–5, 6; for higher animals, in polar regions, 349; for livestock, productivity effects of, 125–47; for plants, control of, by mycorrhizas, 321–4, and by rhizosphere populations, 310–12; for plants, productivity effects of, 103–23

penicillin; alters permeability of glutamate producers, 79, 96; in animal feeds, 144; and nucleotide producers, 92

Penicillium spp.: antagonistic to growth of *Pythium*, 312; food spoilage by, 184; in soil, 257, 273, 277, 279

Penicillium chrysogenum, nutrients for, 258, 270

permeability: of glutamate producers, 79; of nucleotide producers, 92, 97

pesticides: possible toxicity of, to plants, 105, 106; regulation of, in U.S.A., 216–20; survival of, in soil, 276–7

pH: of algal cultures, and carbon-dioxide reserve, 72; and microbial activity in soil, 262; and types of food spoilage, 180, 183, 184

Phaeocystis, acrylic-acid producer in phytoplankton, 349

phages, 7, 25

Phaseolus vulgaris, effects of infections on, 116, 119

phenyl alanine, in microbes grown on hydrocarbons, 23

phosphates: mycorrhizas and plant uptake of, 319, 327; rhizosphere microbes and plant uptake of, 312–15; in sea water, 356, 359

phosphatidyl choline, in membranes of some lithotrophs and methylobacteria, 39

phosphoribosyl pyrophosphate amidotransferase, in nucleotide synthesis, 89

phosphoribosyl pyrophosphate kinase, in 'salvage synthesis' of nucleotides, 92

phosphorylation, in infected plants, 114, 116

photochemical equivalence, Einstein law of, 51, 52

photosynthesis: effects of pathogens on, 114–15; efficiency of, 49, 50–1, 357; by phytoplankton, 235; in polar regions, 337–43

Phytophthora, mycorrhiza antagonistic to, 312

Phytophthora cactorum, dormancy period of spores of, 260–1

Phytophthora cinnamonii, mycorrhiza antagonistic to, 321

Phytophthora infestans, potato blight, 105, 107–8

phytoplankton: estimates of mass of, 233; growth rate of, 242–3; photosynthesis by, 235; in polar regions, 339–43, 347–8; water-soluble vitamins as growth factors for, 243–4

Pichia, on alkanes, 16

Pieris brassicae: resistant strains of, 220–1; virus disease of, 212–13

pigs: digestibility of fats and fatty acids in, 170–1; diseases of, 128, 129, 142;

pigs:
efficiency of food utilization by, 153–4; medicated feeds for, 144; microbial activity in caecum of, 149, 150, 151–2; pathogen-free, 143

pilot plants for production of microbes, 2, 4, 5, 6–9, 10, 11

Pinus: effect of mycorrhiza on growth of, 318, and on phosphate uptake by, 319; translocation of food to mycorrhiza in, 322, 323

Pittosporium, mycorrhiza and growth of, 318

plankton: method of concentrating, 236; respiration of, 236–8; *see also* phytoplankton, zooplankton

plants (higher): *v.* algae for carbon dioxide fixation, 49–54; growth of, in sterile and infected soil, 311, 313; growth of, with and without mycorrhiza, 318; microbial disease and productivity of, 103–23; primary production of, in different communities, 264–5; of tundra, 334; *see also* phytoplankton, roots

pleuropneumonia, contagious bovine, 132; eradication of, 133, 137

Podocarpus, mycorrhiza and growth of, 318

polar regions, microbial productivity in, 333–54

polyhedroses (insect viruses), 212, 213

polysaccharides, in microbes, 24

Polystictus spp.: in soil, 277

Polystictus versicolor, in soil, 278

Popillia japonica (Japanese beetle), bacterial diseases of, 203, 216

Poria subacida, in soil, 277

Porphyra, filamentous alga, 69

Porphyra tenera, marine alga eaten in Japan, 55

poultry: medicated feeds for, 144; pathogen-free, 143; *see also* fowls

price of crops, effect of losses from disease on, 106–7

Prodenia spp. (army worm), virus disease of, 215

productivity: distinction between increase in production and in, 127; of livestock, microbial disease and, 125–47; of microbes, 1–13; of microbes in the ocean, limits of, 355–66; of microbes in polar regions, 333–54; of microbes in soil, 255–86; of plants, microbial disease and, 103–23

propionic acid; in caecum of fowl and pig, and rumen of sheep, 150; for food preservation, 186; synthesis of glucose from, in ruminants, 155, 158

protein (1): in different crops, 64, 67; metabolism of, in rumen, 159–65

protein (2), microbial: amino-acid composition of, 23, 163–5; biological value of,

28, 56, 63; cost of, 66–7; digestibility of, 24, 28, 55, 56, 63, 163, 164; limitations of, 47–8; percentage content of, 22, 24, 41, 54; yield of, per unit area of different crops, 64, 65, 67

protein efficiency ratios, for algae and ordinary foods, 63

Proteus, food spoilage by, 182

Proteus mirabilis, P. rettgeri, P. vulgaris, insect pathogens, 202

proteolysis, of algae, 61

protozoa: on algae, 69; estimate of biomass of, in soil, 265; in polar regions, 345, 346, 347, 348

Pseudomonas spp.: on alkanes, 16; food spoilage by, 179, 180, 182; in soil, 277

Pseudomonas aeruginosa: insect pathogen, 202; medium for selection of, 191

Pseudomonas azotocolligans, P. azotogenesis, not nitrogen fixers, 294, 296

Pseudomonas chlororaphis, insect pathogen, 202

Pseudomonas coronafaciens, on oats, 111

Pseudomonas fluorescens: food spoilage by, 132; insect pathogen, 202

Pseudomonas (Methanomonas) methanica, 29, 30, 38, 40

'*Pseudomonas methanitrificans*', nitrogen fixation by, 294, 296

Pseudomonas oleovorans, electron transport in, 17

Pseudomonas phaseolicola, and photosynthesis in host, 114

Pseudomonas putida, P. reptilivora, P. septica, insect pathogens, 202

psychrophilic microbes: in food spoilage, 182; in polar regions, 345–6, 348

Puccinia carthami, on safflower, 116, 117

Puccinia graminis, and photosynthesis in host, 114

Puccinia helianthi, on sunflower, 112

Puccinia poarum, and carbohydrates of host, 117

Puccinia punctiformis, and gibberellin content of host, 119

Puccinia recondita, and photosynthesis in host, 114

Puccinia striiformis (yellow rust), 105; effects of, on host, 112–13, 114, 118

Pullularia, not nitrogen fixer, 294

pyruvate, natural reductant for nitrogenase, 292

Pythium, fungi antagonistic to, 311–12

quarantine measures, against animal diseases, 130, 137

Quercus: mycorrhiza and growth of, 318; mycorrhiza and phosphate uptake by, 319

rabbits, myxomatosis in, 199–201, 221
radiocarbon dating: of humus, 276, 345; of lichens, 338
rats, feeding of: with algal protein, 56; with cellulose, 150; with fats and fatty acids, 170–1; with rumen microbes, 164, 165–6
redox potential: of foods, and type of spoilage, 180, 184; in nitrogen fixation, 300
research, use of microbiological materials in, 5–9
resistance: of insects to microbial control agents, 220–1; of rabbits to myxomatosis, 201
respiration: of aquatic microbes, 235–8; of mycorrhiza, 323; of plants infected with pathogens, 115–16; of soil microbes, 258–60
Rhizobium, 273, 300, 325; specificity of, 326
Rhizobium japonicum + soya bean: nitrogen fixation by, 300; nitrogenase from, 288
Rhizoctonia spp., in soil, 257
Rhizoctonia solani, digestion of cellulose by, 317
rhizomorphs, of fungi in soil, 274
Rhizopogon roseolus, mycorrhizal fungus, 322
Rhizopus spp., food spoilage by, 182
Rhizopus nigrificans, fermentation of masses of, 60
rhizosphere, 309–10; population of, and host, 310–15
Rhodomicrobium, nitrogen-fixing strains of, 296
Rhodopseudomonas spheroides, nitrogen-fixing strains of, 296
Rhodospirillum rubrum: nitrogen-fixing strains of, 296; nitrogenase of, 288
Rhodotorula: on alkanes, 16; in soil, 277
riboflavin, microbial production of, 77, 78, 92–6
riboflavin synthetase, 94
ribosomes, protein of, 20
rinderpest, 128, 130, 133; eradication of, 133, 137; vaccination for, 139
RNA polymerase (DNA-dependent), inhibited by toxin of *B. thuringiensis*, 208
root nodules, 299; co-existent with mycorrhiza, 328; in non-leguminous plants, 315–16, 326, 344
roots: associations of microbes and, 309–22; material exuded by, 264, 309–10; and microbial activity, 271; respiration of, 258–9; in rusted cereals, 113, 118
rubredoxin, in *Pseudomonas oleovorans*, 17
rumen microbes: effect of lipids on, 172; nitrogen-fixers among, 297; protein of, 163–5
ruminants: algae as feed for, 56, 66, 73; amino-acid needs of, 165–60; convert non-protein nitrogen into high quality protein, 126; efficiency of food utilization in, 153–4; metabolism in rumen of, 149, 154–65

Saccharomyces, food spoilage by, 182
Salmonella typhimurium: in foods, 184, 185; nucleotide synthesis in, 88
Salmonellae, in foods, 184–5
sampling, in control of food processing, 187–8
sanitation, of food processing plants, 186–7
sawflies, control of, by virus diseases, 213, 221, 223
Scenedesmus: as animal feed, 56; installation to grow, 69;
Scenedesmus obliquus: action of cellulase on, 59; drying of, 63; milling of, 58
sclerotia of fungi, in soil, 257
Sclerotium spp., in soil, 257
Scytonema, nitrogen fixation by, 296
semen of livestock, transmission of infection in, 136
serine, and riboflavin production, 94
serine pathway of metabolism, in methylobacteria, 35, 36
Serratia: food spoilage by, 182; growth rate of, 268
Serratia marcescens, insect pathogen, 202
sheep: amino-acid needs of, 165–70; digestibility of fats and fatty acids in, 170–1; microbial activity in rumen of, 150
Shepherdia: distribution of, 326; root nodules and nitrogen fixation in, 315, 326
smuggling of livestock, introduction of disease by, 132
soil: microbial productivity in, 255–86; in polar regions, 334, 344, 345–7
solvent extraction, of algal cells, 58, 59, 60
sorbic acid, for food preservation, 186
soya beans: protein from, 61, 67; *Rhizobium* in, 288, 300
specificity, of symbiotic microbes, 324–6
Spirillum, growth rate of, 268
Spirogyra, filamentous alga, 69
Spirulina spp., 56, 69
Spirulina maxima, yield of protein from, 65
Spirulina platensis, yield of protein from, 64, 67
spoilage associations, of different foods, 179–84
spores: of food spoilage microbes, 181; of insect pathogens, 204, 205, 206; of *Methylosinus*, 32; of soil bacteria, 256–7; of soil fungi, 257, 273–4
Sporotrichum, food spoilage by, 182
Staphylococcus aureus, medium for selection of, 191
starch, in rumen, 155

starvation, symptoms of, in diseased plants, 113, 118
Stereocaulon spp. (lichens), nitrogen fixation by, 344
Stigeoclonium, filamentous alga, 65, 69
Stigonema, nitrogen fixation by, 296
strawberry, mycorrhiza and growth of, 318
Streptococci: food spoilage by, 182; medium for selection of, 191
Streptomycetes, in soil, 257, 273, 274; inhibitors for germination of, 262, 269, 270; and resistant substances, 277
succinate dehydrogenase, in plankton, 237
sugar-beet yellows, 165
'sugar fungi', 272, 273, 277
sulphites, for food preservation, 186
sulphur, soils deficient in, 302
sulphur bacteria, in polar regions, 346
swine fever, African, 133
swine fever, eradication of, 137
Synechococcus lividus (alga), yield of, 65
synergism: between food spoilage microbes, 183; between monosodium glutamate and nucleotides, in imparting flavour, 3

tannins: in foods, 180–1; in soil, 277, 278
temperature: and cytoplasmic breakdown in diseased plant cells, 111–12, for growth on alkanes, 22; and microbial activity in soil, 261–2, 263; at night, and yield of *Chlorella*, 71; optima of, for different algae, 65, 66, 71; of polar soils, 334, and waters, 335, 336; and respiration rate of plankton, 238; and riboflavin production, 94–6; and types of food spoilage, 182
tetracycline, in animal feeds, 144
texturization, of algal protein, 61–3
Thamnidium, food spoilage by, 182
Thermoactinomyces vulgaris, spores of, in soil, 257
thermophilic microbes: alkane-oxidizing, 22; no nitrogen fixers among? 295
thiamine: and glutamate production from hydrocarbons, 83; and growth rate of phytoplankton, 243, 244, 356; in 'salvage synthesis' of nucleotides, 91
threonine: and lysine production, 85; microbial production of, 85–7; in microbes grown on hydrocarbons, 23; and riboflavin production, 94; second limiting amino acid for sheep? 169–70
threonine deaminase, deficient in threonine producers, 87
ticks, infections carried by, 136
tissue cultures, for virus production, 5, 223
tobacco: control of insect pests of, 209, 224; mycorrhiza and growth of, 318
Tolypothrix, nitrogen fixation by, 296

tomato: control of insect pests of, 209; soil microbes and phosphate uptake of, 314, 315
Torulopsis: on alkanes, 16; food spoilage by, 182
toxicological tests, on microbes as food, 24–5; on microbial control agents for pests, 216–20
toxins, crystalline protein, of *B. thuringiensis*, 202, 205–6
transport in plants, effects of pathogens on, 116–20
trehalase, fungal metabolite, 117, 317, 323, 324
Trichoderma, in soil, 257
Trichoderma viride: antagonistic to growth of other fungi, 311–12; antibiotic production by, 273; survival of conidia of, 275
trichomoniasis of cattle, 129
Trichoplusia ni (cabbage looper): control of, by *B. thuringiensis*, 209; virus disease of, 215
Trifolium subterraneum, see clover
tropical crops, estimate of disease losses from, 109, 110
trypanosomiasis, 135
tsetse flies, 135–6
tuberculosis of cattle, 130, 134
Tussilago farfara, effect of infection on carbohydrates in, 117

Ulothrix, filamentous alga, 65
urea, metabolism of, in ruminants, 160, 161
Urediales, supply of carbohydrate to, 323
Uromyces fabae, and cytokinin content of host, 119
Uromyces phaseoli: and carbohydrate content of leaves, 116; and cytokinin content of host, 119; and photosynthesis, 114
Uromyces trifolii, and carbohydrates in host, 117
Uronema, filamentous alga, 65, 69
Uronema, marine protozoan, 241
Ustilago nuda, and carbohydrates in host, 117

vaccination, in control of animal diseases, 4, 130, 137–9
vaccines, production of, 4–5; extraneous infections in, 136
valine, in microbes grown on hydrocarbons, 23
Vaucheria, filamentous alga, 69
vector control, and animal diseases, 130
Venturia inaequalis, on apple leaves, 110
vesicular–arbuscular mycorrhiza, 316, 318, 324–5
vesicular exanthema, eradication of, 137
Vibrio spp., food spoilage by, 180

Vibrio parahaemolyticus, medium for selection of, 191
vibriosis of cattle and sheep, 129
Vicia faba, effect of infection on, 119
viruses: for control of insect pests, 212–16, 224; of fungi, 9; toxicity tests on, 219–20
vitamin B$_{12}$: in caecum bacteria of pig, 152; and growth rate of phytoplankton, 243, 244, 356; microbial production of, 77, 78
vitamins: synthesized by flora of gut in fowl, 157, and in ruminants, 152; water-soluble, as growth factors for phytoplankton, 243–4.

wastes, from factory farming, 73, 145
water: cosmic view of, 362–5; in foods, and types of spoilage, 179–80, 183; of polar seas, movements of, 336–7; in soil, and microbial activity, 261
Westiellopsis, nitrogen fixation by, 296
wheat: effects of infection on, 112–13, 114, 117, 118; soil microbes and, 310–11, 314, 315

wild animals, as reservoirs of infection, 135
World Health Organization (WHO), 132

xanthosine monophosphate (xanthylic acid, XMP), flavour enhancer, 87
xanthosine monophosphate aminase, bacterial mutants lacking, 89

yeasts: on alkanes, 16, 21, 22, 26, 27; amino-acid content of, 23, 24; food spoilage by, 179, 180, 181; as index organisms in control of fruit-juice processing, 108; medium for selection of, 191; not nitrogen-fixers, 294, 295, 296; not susceptible to phages, 25; in polar regions, 346, 347, 348; production of, 2, 66; production of riboflavin by, 92; as source of protein, 22, 48, 67

zinc, disease of citrus due to deficiency of, 313
zooplankton, growth rate of, 240–1
Zygorhynchus, in soil, 274